# PILOTS' INFORMATION FILE 1944

## THE AUTHENTIC WORLD WAR II GUIDEBOOK FOR PILOTS AND FLIGHT ENGINEERS

**Schiffer Military/Aviation History**
Atglen, PA

Library of Congress Catalog Number: 94-74327



Printed in China.
ISBN: 0-88740-780-3



Published by Schiffer Publishing Ltd.
77 Lower Valley Road
Atglen, PA 19310
Please write for a free catalog.
This book may be purchased from the publisher.
Please include $2.95 postage.
Try your bookstore first.

# TABLE OF CONTENTS *Pilots' Information File*

**When you place this form No. 24 in your PIF, remove the old form 24, dated May, June, July, 1944, sign it, and give it to your base operations officer.**

In occordance with the provisions of AAF Regulation 62-15, and 62-15a, all AAF pilots, designated flight surgeons, designated aviation medical examiners, and flight engineers in the United States will certify that they have read and understand all instructions and information contained in Pilot's Information File. They will do so by signing in the space provided at the end of the Table of Contents. Remove the old Table of Contents for May, June, and July, 1944 now. Sign it and give it to your Operations Officer for placement in your Form 5 file.

Subjects preceded by an asterisk (*) have been revised or added to PIF since May 1, 1944. Be sure your copy of PIF contains all amendments. Read them carefully before signifying compliance on Form 24A.

Check with operations offices regularly to be sure you have all current amendments to PIF and Table of Contents. The Table will be revised quarterly and distributed on the same basis as the Pilots' Information File.

Blank spaces at the end of sections are for your convenience. Enter new subjects in their proper place as you receive them.

**AUTHORITY FOR PIF—AAF REGULATION 62-15**
**TABLE OF CONTENTS—AAF FORM 24:**

WAR DEPARTMENT
A.A.F. FORM No. 24
APPROVED 4-1-43

## SECTION FOUR—MAN GOES ALOFT

| Subject and Date | PIF No. | Initials Note Compliance |
|---|---|---|
| Effects of High Altitude (Oct. 1, 43) | 4- 1 | |
| *Oxygen Equipment (Aug. 1, 44) | 4- 2 | |
| *G Forces (Aug. 1, 44) | 4- 3 | |
| *Physical Fitness (Aug. 1, 44) | 4- 4 | |
| Vision at Night (Oct. 1, 43) | 4- 5 | |
| *Dangerous Gases (Aug. 1, 44) | 4- 6 | |
| *Heat and Cold (Aug. 1, 44) | 4- 7 | |
| *The Flight Surgeon (Aug. 1, 44) | 4- 8 | |
| *Sense of Position in Flight (Aug. 1, 44) | 4- 9 | |

## SECTION FIVE—POWER PLANT

| Subject and Date | PIF No. | Initials Note Compliance |
|---|---|---|
| Ground Operation of Engines (May 20, 43) | 5- 1 | |
| Flight Operation of Engines (Oct. 1, 43) | 5- 2 | |
| Engine Instruments (Dec. 1, 43) | 5- 3 | |
| Fuel Systems (Oct. 1, 43) | 5- 4 | |
| Oil Systems (May 1, 43) | 5- 5 | |
| Hydraulics (May 1, 43) | 5- 6 | |
| *Propellers (Aug. 1, 44) | 5- 7 | |
| Electrical Power System (Jan. 1, 44) | 5- 8 | |

## SECTION SIX—THE AIRPLANE

| Subject and Date | PIF No. | Initials Note Compliance |
|---|---|---|
| Tricycle Landing Gear (May 1, 43) | 6- 1 | |
| Landing Wheel Brakes (May 1, 43) | 6- 2 | |
| Flaps (Feb. 1, 44) | 6- 3 | |
| Surface Control Locks (May 1, 43) | 6- 4 | |
| How to trim Your Plane (May 1, 43) | 6- 5 | |
| Care of the Airplane (May 20, 43) | 6- 6 | |
| Emergency Exits (May 1, 43) | 6- 7 | |
| Safety Belts (April 1, 44) | 6- 8 | |
| Precautions—Bomb Bay Doors (May 1, 43) | 6- 9 | |
| Precautions Against Fouling Controls (May 1, 43) | 6-10 | |
| Safety Hints (Feb. 1, 44) | 6-11 | |
| Load Adjuster (Oct. 1, 43) | 6-12 | |
| Miscellaneous Accessories (May 1, 43) | 6-13 | |

## SECTION SEVEN—ARMAMENT

| Subject and Date | PIF No. | Initials Note Compliance |
|---|---|---|
| Safety Precautions when Releasing Bombs (Feb. 1, 44) | 7- 1 | |
| Minimum Altitude for Release and Pre-arming of Fuses (May 1, 43) | 7- 2 | |
| Machine Guns and Cannon (May 1, 43) | 7- 3 | |
| Gun Sights (May 1, 43) | 7- 4 | |

## SECTION EIGHT—EMERGENCY

| Subject and Date | PIF No. | Initials Note Compliance |
|---|---|---|
| To Bail or not to Bail (May 1, 43) | 8- 1 | |
| Forced Landings (May 1, 43) | 8- 2 | |
| Ditching (Dec. 1, 43) | 8- 3 | |
| *Parachutes (July 1, 44) | 8- 4 | |
| Pyrotechnic Pistols (May 1, 43) | 8- 5 | |
| Smoke Grenades (May 1, 43) | 8- 6 | |
| Panel Signals (May 1, 43) | 8- 7 | |
| Body Signals (May 1, 43) | 8- 8 | |
| Emergency Kits (Oct. 1, 43) | 8- 9 | |
| *Life Preserver Vest (July 1, 44) | 8-10 | |
| Life Rafts (April 1, 44) | 8-11 | |
| Fire Fighting (May 1, 44) | 8-12 | |
| Kits, First Aid, Aeronautical (April 1, 44) | 8-13 | |
| First Aid in Flight (April 1, 44) | 8-14 | |

*I certify that I have read and understand all subjects in the Pilot's Information File listed in Form 24, dated August, September, October, 1944.*

**When you receive your new Form 24, dated November, 1944, remove this form, sign it, and give it to your operations officer to put in your Form 5 file.**

SIGNED____________________

RANK____________________

ORGANIZATION____________________

DATE____________________

## Directory Chart, Headquarters Army Air Forces

COMMANDING GENERAL

SECRETARY OF AIR STAFF

DEPUTY COMMANDER, AAF & CHIEF OF AIR STAFF

MANAGEMENT CONTROL

DEPUTY CHIEF OF AIR STAFF

DEPUTY CHIEF OF AIR STAFF

DEPUTY CHIEF OF AIR STAFF

DEPUTY CHIEF OF AIR STAFF

ASSISTANT CHIEF OF AIR STAFF
PERSONNEL

ASSISTANT CHIEF OF AIR STAFF
INTELLIGENCE

ASSISTANT CHIEF OF AIR STAFF
TRAINING

ASSISTANT CHIEF OF AIR STAFF
MATERIEL MAINTENANCE & DISTRIBUTION

ASSISTANT CHIEF OF AIR STAFF
OPERATIONS COMMITMENTS & REQUIREMENTS

ASSISTANT CHIEF OF AIR STAFF
PLANS

AIR INSPECTOR

AIR SURGEON

BUDGET AND FISCAL

AIR JUDGE ADVOCATE

OFFICE OF LEGISLATIVE SERVICES

SPECIAL PROJECTS

AIR COMMUNICATIONS OFFICER

FLYING SAFETY

SPECIAL ASSISTANT FOR ANTI-AIRCRAFT

OFFICE OF TECHNICAL INFORMATION

AAF REDISTRIBUTION CENTER

TRAINING COMMAND

FIRST TROOP CARRIER COMMAND

FIRST AIR FORCE

SECOND AIR FORCE

THIRD AIR FORCE

FOURTH AIR FORCE

AIR SERVICE COMMAND

MATERIEL COMMAND

AIR TRANSPORT COMMAND

AAF TACTICAL CENTER
SCHOOL OF APPLIED TACTICS
DEMONSTRATION AIR FORCE

WEATHER WING

PROVING GROUND COMMAND

ARMY AIRWAYS COMMUNICATIONS WING

1 JUNE, 1944

# SECTION 1 GENERAL

# PILOT'S RESPONSIBILITIES

★ The success of any flying mission, whether in training or combat operations, depends primarily upon the pilot. As commander of the airplane you carry heavy responsibilities. The majority of your responsibilities, however, involve only good, sound judgment.

★ All of your duties make up a long and formidable list; but you can discharge each of them by systematic planning, step by step. Don't overlook anything. Establish a routine. Memorize and follow a check list of your responsibilities.

## Before Flight

### PLAN

#### FORM 23

★ Plan your flight on Form 23.
★ Remember that the shortest distance between two points is not always a straight line.
★ Check terrain features on your line of flight: maximum altitude of mountains, facilities for emergency landings.
★ Airways, radio facilities, check points.

#### WEATHER

★ Talk to the forecaster, don't be satisfied with the brief forecast on your Form 23.
★ Know the over-all weather picture. Study the latest weather map.
★ Read the weather sequences yourself.
★ Check winds aloft, and plan your flight to take advantage of the best altitude.
★ Know temperatures, dewpoints, and barometric readings along your line of flight.
★ Altimeter setting at point of departure and destination.

#### CHARTS

★ Secure the latest sectional and regional maps.
★ Check them against the chart in the operations office for changes.
★ Look up the latest information in NOTAMS on danger and caution areas.
★ Draw a pencil line along your intended route, and mark it off in 20-mile intervals.
★ Don't start any flight without proper charts!

# EQUIPMENT

## AIRPLANE

★ Check and know the condition of your plane.
★ Visual check.
★ Freedom of movement of all controls.
★ Check Form 1A and Form 41B.

## FUEL

★ Servicing crews fuel the airplane but the pilot is solely responsible for his fuel supply.
★ Check fuel tanks visually; don't rely solely on fuel gages.
★ If flying a long mission, be sure to study cruise control charts for best power setting at your predetermined altitude.

## WEIGHT AND BALANCE

★ Check the loading of your airplane.
★ Be sure the Form F is properly executed and the final index is within the allowable center of gravity range.
★ Read and understand the provisions of AAF Reg. 55-3.

★ The Weights and Balance officer is charged with the responsibilities of compliance with the regulations; but remember you have to fly the airplane.

## RADIO

★ Test the frequencies you will use.
★ Be sure your modulation is correct.
★ Tune in your nearest radio range.
★ Check command, liaison, interphone, and call.
★ See that your radio compass is working.

## FLIGHT RESTRICTIONS

★ Check and know all flight restrictions and special orders concerned with the type of airplane in which the flight is to be made.

## EMERGENCY EQUIPMENT

★ Check the equipment necessary for the flight:
★ Oxygen.
★ Life rafts.
★ Mae Wests.
Parachutes.
Emergency kits.
First-aid kits.

# CREW

## INSPECT CREW'S CLOTHING AND EQUIPMENT

★ Proper clothing for altitude, temperature, and season.
★ Oxygen masks, properly fitted and in good condition.
★ Parachutes—harness properly fitted. Each crew member instructed in how to inspect and use his parachute.

## DUTIES IN FLIGHT

★ See that each crew member knows his place in the airplane, and his duties in flight.
★ Assure yourself that each crew member understands emergency procedures; that he knows what exit to use in bailout; and is trained in ditching procedure, and forced landing routine.
★ You are responsible that the crew has been properly briefed for the flight.
★ See that each has made the proper preflight inspection of everything he is responsible for.

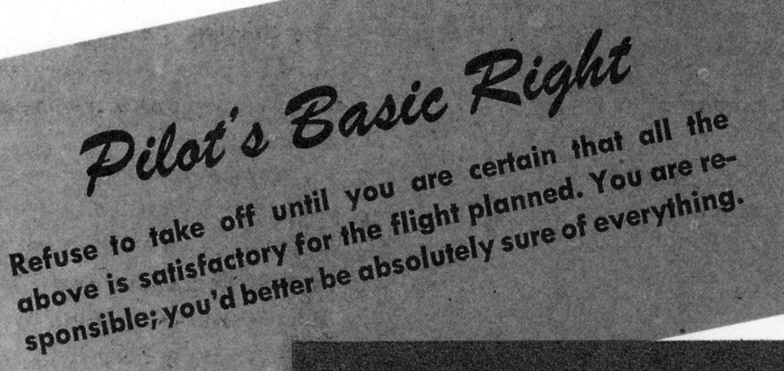


# During Flight

## COMMAND OF THE AIRPLANE

★ During flight you will ordinarily be in command of your airplane, the crew, and any passengers aboard. Army Regulation 95-15 directs that, "The senior member of the operating crew of an aircraft who holds an appropriate military pilot rating will command the aircraft, except when the organization commander responsible for the aircraft specifically designates who shall command."

## CHECK LISTS

★ Use check lists in cockpit for takeoff, cruise, and landing procedure.
★ Use cruise control charts for airplanes in which they are provided.

## FOLLOW YOUR FLIGHT PLAN

★ Adhere strictly to your flight plan as outlined on Form 23, except in an emergency.
★ If it's necessary to change your flight plan, you will get the necessary approval by radio.

## CHECK POINTS

★ Report your position frequently over radio check points, giving:
★ Point of takeoff.
★ First intended landing.
★ ETA (Estimated Time of Arrival).
★ Altitude.
★ Time of report.
★ Any unusual weather encountered.

## NAVIGATION

★ Consult your charts regularly; know your position all the time.
★ Don't navigate solely by radio aids. Check and double check your position frequently.
★ Check your wind drift; don't take it for granted that the winds aloft information you get at the weather office is correct, or that the wind hasn't changed.
★ Check time, speed, and distance, frequently.
★ If flying airways on CFR, maintain the correct odd or even altitude depending on your direction of flight.

## LISTENING WATCH

★Keep your radio tuned to the range and maintain an alert listening watch for "Army Flight Control advises - -". They may be trying to call you.

## WEATHER

★Check weather reports along your line of flight.
★Ask by radio for weather reports whenever in doubt.
★Watch for any change in sky or cloud conditions that may indicate weather change.

## FUEL CONSUMPTION

★Make periodic checks of your fuel supply.
★Check all your fuel gages.
★Check indicated consumption against speed, time, and distance.
★Calculate indicated fuel remaining and distance to go.
★Know most economical power setting.
★Make certain you have fuel reserve.

# *After Landing*

★Complete Forms 1 and 1A before you leave the cockpit.
★Hand the yellow copy of Form 23 to the line chief to turn in at the operations office. It is your responsibility to see that the arrival record is properly filled in.

## REQUEST SERVICE REQUIRED

★Fuel, oil, repairs.
★If fuel and oil must be secured from a commercial contractor, you must sign Form 81. Keep a copy to file at your home station.
★If you have secret or confidential equipment aboard your plane, see that it is properly guarded or removed to a safe.
★Report to the weather office to execute Form 37 (Weather Interview).

## IF OTHER THAN HOME BASE, REPORT:

★Where you will stay.
★Proposed time of departure.
★Proposed next destination.
★Make sure your crew is properly cared for.
★Send, or arrange for operations to send, RON message to your home station. This is the proper form; it may be sent government telegram, collect, TWX, or AACS.

```
Commanding Officer
Attention Operations Officer
Randolph Field, Texas
  Army 43-7436 R.O.N. Pampa AAB
  Pampa, Texas, 13 April 1944
                              DORAN
```

# FLYING SAFETY

PILOT ERROR IS THE CAUSE OF
70 TO 80 PERCENT OF ALL AIRCRAFT ACCIDENTS

In the campaign to defeat these enemies of safety, authorities have prescribed rules, regulations, and standard practices; but they can only point the way—

## PILOT ERROR RESULTS FROM —

★ Ignorance

★ Carelessness

★ Poor physical condition

★ Disobedience

★ Poor judgment

## SAFETY OF FLIGHT DEPENDS UPON YOU

★ Know the rules

★ Abide by the rules

★ Keep constantly on the alert

★ Use considered judgment

★ Keep yourself physically fit

★ Plan in advance for possible emergencies and work out in your own mind procedures you propose to follow for each

General check-out procedure requires a pilot to demonstrate his proficiency on any type of aircraft before being cleared to fly it.

Check-out procedures are prescribed locally by Commanding Officers, setting forth minimum time in cockpit familiarization, taxiing, and minimum number of landings. Careful study of the Handbook of Flight Operating Instructions of the particular model of the airplane to be flown is required.

**Do not be satisfied with minimum requirements for check-out.** No pilot was ever too familiar with his plane.

If you are not certain that you know and understand the airplane you are about to fly, **it is your right and duty to request further instruction and more time.**

Do not be satisfied merely with check-out. After you have been certified as qualified for a type of aircraft, **watch constantly for new information, flight restrictions, and special instructions issued from time to time.** All these will be found in the Transition Flying Training Index.

Finally, you must always check the particular airplane you are about to fly and **make certain that you are familiar with all modifications, special equipment, and the present condition of the airplane.**

There are uniform traffic rules for flying the airways.

**Know the rules.**

**Plan your flight in accordance to the rules.**

**Adhere strictly to your flight plan** (except as provided for in the rules).

**Keep on the lookout for other traffic.**

On local flights, be sure you know the local regulations and restrictions. When in doubt, use common sense and courtesy.

**If possible, ask for traffic information.**

**Conform to local traffic pattern.**

**Keep constantly alert.**

**Keep away from other aircraft in flight.**

**Don't be an Air Hog!**

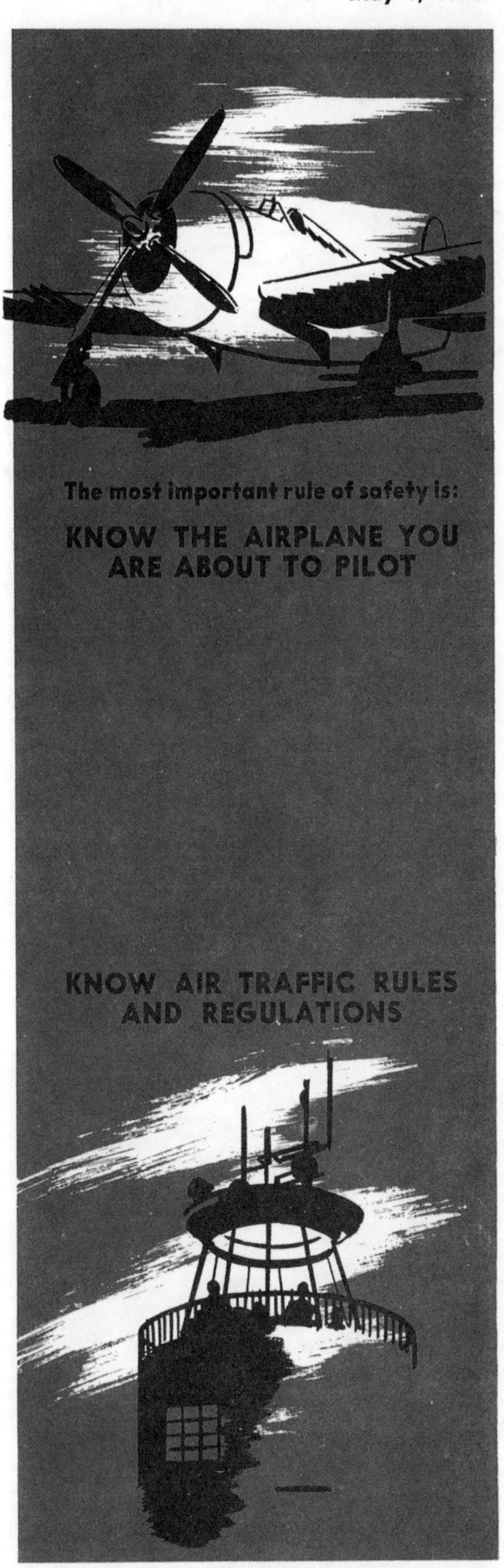

In almost every case where weather is the cause of an aircraft accident, it is chargeable to **Pilot Error:** because the pilot didn't know what kind of weather he was going to encounter.

**Do not take a chance on doubtful weather.**

**Beware of icing conditions.**

**Remember that your carburetor may ice under almost any temperature conditions, summer or winter. All it needs is moisture.**

If you run into bad weather conditions **go around, not through,** or turn back.

**The safest maneuver in doubtful weather is the 180-degree turn!**

Instrument Failure seldom causes aircraft accidents. When it is a contributing cause, it is usually the pilot's fault for not using them correctly.

**Know how to check your instruments.**

**Know instrument procedures.**

**Trust your instruments.**

The most important safety rule in instrument flying is: **have confidence in your instruments, in yourself, and in your ability to fly instruments.**

Such confidence can only be developed by practice; take advantage of every opportunity to use the Link Trainer, to make hooded flights, and to fly on instruments under proper supervision.

Do not be satisfied with the minimum requirements for an instrument rating.

Accidents that are chargeable to lack of oxygen are usually due to **Pilot Error.** There is always more than one source of oxygen. If you know your equipment, you can avoid trouble.

**Be sure your oxygen mask fits properly. Adjust it before take-off.**

**Know how to switch from one source of oxygen to another.**

**Begin the use of oxygen soon enough.**

**Remember that even a short period of oxygen starvation ruins your judgment.**

**KEEP CONSTANTLY ALERT IN TAXIING**

More than half of all aircraft accidents occur on the ground. Practically all taxiing accidents are 100% carelessness.

**Keep your eyes open. Look ahead of you, look behind you, look above you.**

If you are taxiing a "blind" airplane, zigzag enough to know what is in front of you.

Keep your radio tuned to the tower, listen for warnings.

Do not taxi within 100 feet of the line without a crew man in full view in front of you, guiding you by hand signals.

**Taxi slowly.**

Be sure you know how to operate the emergency brake system on your plane.

**USE EXTRA PRECAUTION ON TAKE-OFF**

The chief hazards on take-off arise from engine failure and poor piloting technique.

**Be sure your engines are properly warmed up and functioning perfectly before beginning a take-off.**

**Do not take a chance on an engine that doesn't "rev up" on the ground.**

If an engine fails on take-off **get the nose down** and decide instantly on your procedure:

Is there enough landing field left to stand a chance of getting down?

Can you find a spot approximately straight ahead for a forced landing?

Do **not** try to stretch your glide.

**Above all do not bank.** Remember your plane stalls at higher speed on a bank than in level flight.

If you must crash land, be sure to get the landing wheels up.

**ON LANDINGS, PUT YOUR WHEELS ON THE FIRST THIRD OF THE LANDING STRIP**

Most landing accidents are due to overshooting.

On your approach, **keep your plane enough above stalling speed to have full control.**

If you have misjudged on your approach, **there is no disgrace in going around again.**

***At all Times***

**Do not try to show off.**

**Never try maneuvers with an airplane unless and until you know such maneuvers are permitted in that type of airplane.**

Remember that **altitude above you will never help you get out of trouble.**

***Finally*** **FLYING IS AN EXACTING, SERIOUS BUSINESS. IT DEMANDS EVERYTHING YOU HAVE OF KNOWLEDGE, ATTENTION, EFFORT, JUDGMENT, AND SKILL. IF YOU GIVE IT ANY LESS THAN YOUR BEST, IT EXACTS A HIGH PRICE FOR YOUR MISTAKES.**

# SAFEGUARDING CLASSIFIED MATERIAL

★ Military information is of varying degrees of value to foreign governments, and therefore requires the exercise of varying degrees of caution for its safety. For the purpose of providing necessary safeguards the information is classified as "Restricted," "Confidential," or "Secret." All military information, documents, and material other than that classified as above, are regarded as unclassified. All persons having any contact with classified material should study and become thoroughly acquainted with Army Regulation No. 380-5.

## Secret

| WHO MAY READ IT | HOW TO KEEP IT | HOW TO MAIL IT | MAY BE DISCUSSED WITH |
|---|---|---|---|
| *Only persons directly concerned.* | *Always locked in 3-combination safe when not in use.* | *Use two envelopes.*<br>*—An inner envelope addressed properly, marked or stamped SECRET.*<br>*—Address outer envelope and do not use any markings to indicate classification.*<br>*—Send by Registered Mail.* | *Only persons directly concerned.* |

## Confidential

| WHO MAY READ IT | HOW TO KEEP IT | HOW TO MAIL IT | MAY BE DISCUSSED WITH |
|---|---|---|---|
| *Persons in military establishment and authorized civilians whose duties require that they read it.* | *Same as secret.* | *Same as secret.* | *Persons authorized to read it. Never over telephone.* |

## Restricted

| WHO MAY READ IT | HOW TO KEEP IT | HOW TO MAIL IT | MAY BE DISCUSSED WITH |
|---|---|---|---|
| *Anyone in government service and civilians whose loyalty is unquestioned. But never the press or public.* | *Ordinary precautions, guarded area; locked file or desk, or locked doors, or may be kept on person.* | *First Class Mail, unmarked.* | *Persons authorized to read it.* |

Do not put Secret or Confidential Material into wastebaskets. When it is to be discarded it must be destroyed either by burning or by an approved shredding machine. Until that can be done, it must be torn into small pieces, placed in an envelope, and safeguarded according to classification until destroyed. Restricted material must be torn up before being thrown away.

Safeguarding of Classified Military Information is the responsibility of all military personnel and civilian employees of the War Department.

At every Headquarters an inspection will be made each day immediately before the close of business to insure that all Secret and Confidential documents have been properly put away.

# HOW TO USE PIF

## General

The Pilots' Information File is a manual of instructions and information of a general nature, with which all AAF pilots in the continental United States are required to comply, in accordance with the provisions of AAF Regulation 62-15.

PIF is designed to keep you informed of matters affecting flying safety and operational efficiency. It contains basic material not discussed in detail in Handbooks of Flight Operating Instructions and other publications comprising the Transition Flying Training Index.

Familiarize yourself with both the Pilots' Information File and the Transition Flying Training Index. Check frequently to be sure that you have read and understand the information in each.

## How Subjects Are Numbered

Each subject listing is numbered. **The first number** indicates the section. For example, 2, which is Flight. **The second number** refers you to the subject. For example, 2-20, "Spins and Stalls." **The third number** lists the page. Thus, the third page of "Spins and Stalls" will be marked 2-20-3.

## Revisions

It is particularly important that you keep all of the subjects and pages in your book in proper order, so that you may use PIF as a quick reference guide.

To make the process as simple as possible you receive revisions in an envelope upon which is printed a list of the pages you are to remove from your copy of PIF as you put the new revisions into the book.

Revised sheets bear a new date line in the upper right hand corner thus: REVISED February 1, 1944. If the page is new and does not replace an old one, it will bear a line thus: ADDED February 1, 1944.

Once in a while a whole subject will be revised completely and the revised pages may number more or less than the pages they replace. In such a case all the pages will merely bear the REVISED notation, with the date. So don't be confused if you have more or fewer pages than you have removed.

Revisions and additions from now on will be issued regularly on the first of each month. Get into the habit of asking for revisions at your operations office during the first few days of the month.

## Temporary Revisions

From time to time it is necessary to issue a revised page more rapidly than is possible in the regular process of revision. Such temporary revisions will be issued with a notation at the bottom of the page "This is a temporary page to be inserted facing page (giving number). It will be replaced in an early Revision."

## WANTED: YOUR CORRECTIONS AND CRITICISMS OF PIF

May we call your attention to the provision of Paragraph 2 of AAF Regulation 62-15, which directs all AAF establishments to submit items they desire to have included to the address given below. This also means any criticism of material already in PIF—corrections, questions of interpretation, and mistakes which may have inadvertently crept into the copy or the art.

Sometimes such errors grow out of the lag between publication of T.O.'s and Regulations, and publication of PIF. We are working to reduce that lag. In any case we welcome free criticism and prompt correction of mistakes from Operations Officers and pilots.

Our aim is to keep PIF accurate, current, and fully useful. If you can help us do that, we will appreciate it. Write direct to:

**OFFICE OF FLYING SAFETY**
**PILOTS' INFORMATION FILE**
**BUHL BUILDING**
**DETROIT 26, MICHIGAN**

### To Comply with Revisions

When you receive a set of revision sheets from your operations office, it will be accompanied by a temporary certificate of compliance (Form 24A).

Before you sign it to certify that you have read and understand all the revisions and/or additions, be sure you **do read the revised pages.**

You will find that often only minor changes have been made on some pages. There is no special indication to show what sentences or paragraphs have been revised. It is felt that you should re-read the whole page in order to get the context of the old material in relation to the new.

### Index

An index is provided which lists alphabetically the principal items treated in PIF. It is revised from time to time to keep it as useful for you as possible. You may find occasionally, however, that an index listing is in error; but by using the table of contents you will be able to trace almost any item you are looking for.

When a revised index is printed, it will be distributed along with the regular monthly revisions.

### Distribution of Revisions

Revisions are distributed to individual pilots by the Base Operations Officer. Base Operations receive revisions automatically from the publisher of the File.

If any Operations Officer does not receive the correct number of revisions (plus a 10% overage) he will communicate at once with

Office of Flying Safety
Pilots' Information File
Buhl Building
Detroit 26, Michigan,

stating number of revisions required at his station. He will also send letter request to the above for any copies he may need of the complete File.

Operations Officers will also report promptly on the activation or deactivation of any station.

### The Table of Contents (Form 24)

Every three months you will receive a new Table of Contents in the envelope with the revisions of that month. The new Table of Contents is published on the first of May, August, November, and February.

In order that you may identify it, and be sure that you have the current table in your PIF, the following color key is used.

| | | | | | |
|---|---|---|---|---|---|
| February | : | Red | May | : | Black |
| August | : | Blue | November | : | Yellow |

Check your copy of PIF against the Table of Contents regularly.

Subjects preceded by an asterisk (*) have material revised or new since the last Table of Contents was issued.

You will find that all the pages of any one subject do not bear the same date. But the date following the subject listing in the Table of Contents is the **latest revision date for any of the pages included in that subject.**

### Don't Destroy Table of Contents

When you replace the Table of Contents with a new one, **don't destroy the old one.** Sign it to show that you have read and understand all the subject matter it lists. Then turn it over to your Operations Officer to be placed in your Form 5 File. It is the record of your compliance with PIF in accordance to provisions of AAF Regulation 62-15.

### Operations Officers Responsibilities

1. Operations Officers are responsible that every pilot attached to his base receives a copy of PIF and all revisions.

2. That every pilot on his base signs a compliance form certifying that he has read and understands all material contained in PIF, revisions and additions thereto.

3. That the compliance certificates (Form 24A and Form 24) are placed in the Form 5 Files of the individuals concerned.

When pilots turn in their Forms 24 at the end of the three-months period for which the Forms are the current Tables of Contents for PIF, the Operations Officer will see that previously dated Forms 24 and 24-A which are in the Form 5 Files are removed and destroyed.

### Keep Up to Date

No matter where you are, at your home station or on cross-country, ask the Operations Clerk for any new PIF revisions.

Never fail to ask for new material. It always will be available to you.

# TECHNICAL ORDERS

All pilots must know how to use the Technical Order files. Use of the Pilots' Information File does not obviate the need for using Technical Orders. The Pilots' Information File contains material of a general nature, and does not attempt to supply the specific engineering, maintenance and supply information contained in Technical Orders.

For detailed information on specific pieces of equipment, or how to operate types or models of airplanes, propellers, carburetors, generators, etc., you should read and follow the instructions found in Technical Orders.

### What are Technical Orders?

AAF Technical Orders are directives published by order of the Commanding General, Army Air Forces, for the purpose of issuing specific instructions and information of a technical nature covering the operation, maintenance, storage, inspection, etc., of AAF equipment and materials, and for the establishment of a uniform system of files wherein such data will be readily accessible.

### T. O. Numbering System

Every Technical Order has a series of three numbers separated by dashes, which identify the T. O. This identification number tells you three things about the T. O.:

**The first number** always has two digits. It corresponds to the AAF property classification number. For instance, if the first number in the T. O. designation is 05, the T. O. is concerned with instruments; if it is 02, it is something about engines and engine maintenance parts.

The only number that does not correspond to a property classification is 00, which is used for Technical Orders of a general nature, such as the Technical Order Index (which is 00-1), distribution of Technical Orders, inspection system, kits and such matters.

The property classifications which correspond to the initial number on Tech Orders are:

01- Airplanes and Maintenance Parts.
02- Engines and Maintenance Parts.
03- Aircraft Accessories.
04- Aircraft Hardware and Rubber Materials.
05- Aircraft Instruments.
06- Fuels and Lubricants.
07- Dopes, Paints and Related Materials.
08- Electrical Equipment and Supplies.
09- Aerial Targets and Gliders.
10- Photographic Equipment and Supplies.
11- Aircraft Combat Material.
12- Fuel and Lubricating Equipment and Supplies.
13- Clothing, Parachutes, Equipment and Supplies.
14- Hangars and Demountable Buildings.
16- Balloon Equipment and Supplies.
17- Machinery, Shop and Warehouse Equipment.
18- Special Tools.
19- Flying Field and Hangar Equipment.
21- Cordage, Fabrics and Leathers.
22- Woods.
23- Metal and Composition Material.
24- Chemicals.
25- Office Equipment and Supplies.
26- School Equipment.
27- Excess and Surplus Property.
29- Commercial Hardware and Miscellaneous Supplies.
30- Publications, Processed Motion Picture Films and Film Strips.

Complete breakdown of these classes is listed in T. O. 00-35A-1.

**The second number** of the Tech Order's identification represents the airplane or engine model, or the sub-division of the general property class identified by the first number. It is impractical to list here all the sub-divisions of each property classification; but to show how the system works, here is a breakdown of one classification, 03, Aircraft Accessories:

03-1 General
03-5 Electrical Equipment
03-10 Fuel System
03-15 Oil System
03-20 Propellers and Accessories
03-25 Wheels, Brakes and Struts
03-30 Air and Hydraulic System Accessories
03-35 Ice Eliminating Equipment
03-40 Control Units
03-45 Fire Extinguishers
03-50 Oxygen Equipment
03-55 $CO_2$ Inflation Equipment
03-60 Skiis
03-65 Accessory Power Plant

Such a breakdown for every property classification will be found in the T. O. Index (00-1) as the first item under each property classification number.

Sometimes the second number still does not give enough identification for you to find the equipment you are looking for, and it is necessary to use a letter after the number. For example: 03-10 identifies Fuel Systems. If you are looking for carburetors, however, you will find they are further separated by using B after 10—thus, 03-10B.

**The third number** usually is merely the serial number of the T. O. concerning the general subject identified by the first two numbers. There is one place, however, where the third number is an important addition to the identification, and that is in the series of airplane model manuals of operation, maintenance and repair, and parts catalogs.

These series of basic Tech Orders are identified as follows:

| FIRST NUMBER | SECOND NUMBER | THIRD NUMBER |
|---|---|---|
| 01 (the airplane) | –40A (the airplane model) | –1 (Handbook of Flight Operating Instructions) |
| 01 | –40A | –2 (Erection and Maintenance Manual) |
| 01 | –40A | –3 (Structural Repair Manual) |
| 01 | –40A | –4 (Parts Catalog) |

### Technical Order Index

The Technical Order Index, T. O. 00-1, is the first T. O. in the complete library. In the front of the Index you will find data listing the airplane and its engine, (airplane and engine cross-reference table) with identification of the manufacturer of each, and the Technical Order series covering each. Immediately behind it is the numerical index of Technical Orders.

Hunt first, in numerical sequence, for the basic number (for example, 01), then the second number (for example, 40A), then run through the serial numbers and titles until you find the title of Tech Order you are looking for. Go to the book containing the corresponding number to get your information. All Tech Orders are listed numerically.

Suppose you want to find a list of all the radio compasses in service and all of the Technical Orders pertaining to them. Turn to T. O. 00-2, "Alphabetical Index of AAF Equipment." Under "Compasses, Radio," you will find all types listed, together with numbers of applicable Technical Order Handbooks.

### Where to Find a T. O.

Each Station, Base, or Sub-Depot has four complete files, assigned to the supply office, engineering office, technical inspector's office, and transient aircraft crew.

Complete information about T. O. files in overseas theaters is listed in T. O. 00-25-3.

### Aircraft Files

The following Technical Orders and current amendments thereto must be maintained in each operating airplane:

1. T. O. 00-20A, Visual Inspection System for Airplanes.
2. T. O. 00-20A-2, Airplane Maintenance Instruction Form.
3. Pilot's Handbook of Flight Operating Instructions for the particular airplane model.
4. Erection and Maintenance Manual for the particular airplane.
5. Engine Handbook of Service Instructions for the particular engine.
6. Handbook of Weight and Balance Data.
7. T. O. 08-15-1, Radio Facility Charts.
8. T. O. 08-15-2, Radio Data and Aids to Airways Flying.
9. T. O. 08-15-3, Instrument Approach Procedures.

### Compliance in Combat Areas

Compliance with Technical Orders and Technical Radiograms in combat areas will be subject to local conditions involving tactical employment of the equipment and at the discretion of the respective Air Force or Task Force Commanders concerned.

### Remember

1. In this business, **nobody knows all the answers.**
2. The airplane is as important as you are. **Know your plane. Study everything you can find about it.**
3. Ignorance is dangerous. **Read your Handbook.**
4. One of the best ways to know your plane is to **keep tab frequently on new Technical Orders about the model you are flying.** It might save your neck some day.

REFERENCE: Technical Order 00-5, dated October 7, 1942.

JAP TECH ORDER

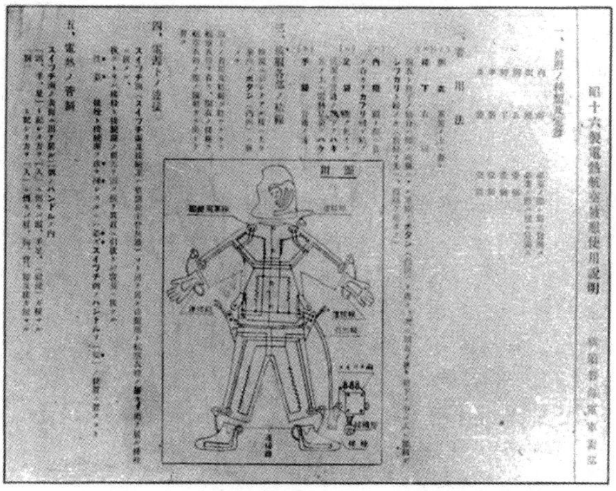

IF YOU THINK OUR T. O.s ARE HARD TO READ—

# UNSATISFACTORY REPORTS

WAR DEPARTMENT
AAF Form No. 54
(Revised 2-18-43)

WAR DEPARTMENT
ARMY AIR FORCES
UNSATISFACTORY REPORT
(See AAF Reg. 15-54 for Information on Proper Use of this Form)

| TO BE FILLED IN BY STATION | |
|---|---|
| STATION SERIAL No. | DATE SUBMITTED |
| 43-115 | 8-10-43 |

| LEAVE BLANK | | |
|---|---|---|
| A. S. C. SERIAL No. | REFER TO | CLASS |

STATION: Sacramento Air Service Command
ORGANIZATION: Maintenance Division
SUBJECT OF REPORT — Property Class—Name: 03-G Fork Assy. Nose Gear Shock Absorber Strut.
Manufacturer: Cleveland Pneumatic
AAF Order or Shipping No.: 26/1
AIRCRAFT—Model & AAF Serial No.: B-26 No. 40-1460
ENGINE—Model & AAF Serial No.:
UNIT OR ACCESSORY—Type, Model and Serial No.: Casting No. 792685
AIRCRAFT REPORTS ONLY — LAST D. I. R.—Depot: None | Date: | Flying Time Since: | Total Flying Time: 229 Hrs.
ENGINE REPORTS ONLY — LAST OVERHAUL—Depot: | Hours Since: | Depots and Hours At Each Previous Overhaul:
PART — Name: Piston - Fork Assy. | Part Drawing, Serial and Specification No.: Part. No. 7926-60
Time in Use: 50 Hrs. | Quantity on Hand: 10 | Quantity Known Defective: 1 | No. Previous Failures: 3 | Manufacturer: Cleve. Pneu. | Inspector's No. or Identification: 10-W

Indicate by "X" Disposition of Exhibit: [X] Photographed and Prints Enclosed; [X] Held for Instructions; [ ] Sent Under Separate Cover; [ ] Sent in Attached Package; [ ] Repaired and Returned to Service; [ ] Disposed of (Explain Below.); [ ] To Overhaul Facility (INITIALS)

GIVE COMPLETE DETAILS, PROBABLE CAUSES AND RECOMMENDATIONS BELOW:
(Use Only Applicable Spaces Above—Avoid Unnecessary Repetition)

EXPEDITE

DESCRIPTION OF TROUBLE:

1. Subject Strut failed by breaking completely off. See Photograph No. 1 and No. 2.

2. Subject failure occurred while the airplane was being taxied at

...indicates the probable originating ... 2.

Information concerning even the smallest failure may be of great value if reported to proper authorities in time. Everyone is encouraged to submit Unsatisfactory Reports whenever he sees an opportunity to contribute to greater efficiency by suggesting correction of faults.

As a pilot, you are in close touch with both methods and machines. A great ground organization is behind the men who fly. But both flying and ground operations always can be improved. **Unsatisfactory Reports** are designed to speed improvements and to permit the individual to present a maintenance problem and his suggested correction, through channels.

Unsatisfactory Reports usually fall into these general classes:

1. Failure of equipment.
2. Unsatisfactory design.
3. Defects due to faulty material, workmanship, or inspection.
4. Unsatisfactory maintenance or supply methods, systems, or forms.

### How to Prepare a U. R.

AAF Form No. 54, obtainable from the Engineering Officer, is used for Unsatisfactory Reports.

Each report must be a complete description of an individual case. It must explain the unsatisfactory condition, including all pertinent information, to enable investigation and correction of the trouble reported without the need for further requests for information. See AAF Regulation 15-54 for details about how to file different types of U.R.'s.

### Coordination

All Unsatisfactory Reports originating at a station are routed through the Engineering Officer, who investigates and enters his endorsement. He sends the U.R. to the Commanding General, Air Service Command at Patterson Field, Fairfield, Ohio.

# FLYING ON A RED DIAGONAL

**"When the 'Status Today' is indicated by a red symbol and an 'exceptional release' has not been granted by an authorized engineering officer, the pilot of the aircraft will sign this release before flight."**

This "exceptional release" appearing on Form 1A is designed to insure that, before the pilot goes aloft, he is aware of any defects in his plane or its equipment.

**Always know "Status Today" before you fly.**

Don't sign an exceptional release as routine. Get full information about the defect noted on Form 1A. Check Form 41B, kept in the airplane, better to determine the condition of your plane. Under certain conditions aloft you might be glad you took the time to check.

When maintenance personnel mark a red symbol on the airplane report, they are fulfilling their responsibility to you and to the safety of flight. You must determine the nature and cause of the trouble and govern your flight accordingly.

### Other Warnings

When an airplane is undergoing work or has parts removed for repair and such status is not readily apparent without careful inspection, appropriate tags will be used by squadron engineering officers in some cases. This applies particularly when work of this nature is performed at other than regularly established air bases.

A red tag secured conspicuously on the control column or ignition switch will bear a notation, "Not to be flown until generator (or radio, carburetor, etc., as the case may be) has been replaced or repaired." This will warn any pilot who might attempt a take-off without knowledge that the airplane is not ready for tactical flight.

Another tag will be secured at or near the location from which parts have been removed. This will permit ready identification for service crew or mechanic.

### A Word to the Wise

**Red means danger. Get the facts before you fly.**

REFERENCE: Technical Order 00-25-23 dated July 23. 1942.

**What Symbols Mean**

|  |  |
|---|---|
| / (diagonal) | A red diagonal shown under "Status Today" on 1A indicates that the airplane is flyable, but is not in perfect condition. |
| + (cross) | A red cross indicates that a major defect exists and the airplane must not be flown. |
| — (dash) | A red dash indicates required inspection has not been made. |
| | Like red traffic lights, these symbols are primarily for your protection. |

# Transition Flying Training

Commanding officers of Army Air Forces stations and organizations will make available to pilots Handbooks, Technical Orders, Operation and Flight Instructions, Special Instructions on Equipment, and all other such publications; and will require that all pilots under their control pursue a prescribed course of transition flying instruction, fitted to the pilot's previous experience, and pass definite tests, when initially assigned to fly aircraft having flying characteristics and controls or equipment with which the pilot is not thoroughly familiar. The scope and duration of this course will be such as to thoroughly qualify the pilot to operate the aircraft safely and efficiently. This course of transition training will cover all maneuvers essential for the safe and efficient operation of the aircraft; in the case of multi-engined aircraft, this course will include flying on the minimum number of engines required to maintain flight during an emergency.

Pilots reporting to an Army Air Forces station other than their home station or to an aircraft factory for ferry duty will demonstrate satisfactory operative knowledge of the type to be flown, to the commanding officer or his representative at that station or to the Army Air Forces representative at the factory. The pilot's statement that entry has been made on his Form 5 to the effect that he is qualified on the type of airplane to be flown, may be accepted provided he has flown the type within the last 30 days.

Pilots operating aircraft in an organization or at a station other than their own will be required to comply with these instructions.

A certificate recording the action taken as directed above, will be prepared under the direction of the commanding officer and filed with the pilot's Form 5 record.

REFERENCE: AAF Regulation 50-16, dated October 26, 1942. This regulation supersedes AC Circular 50-4, dated May 9, 1941

# INSTRUMENT *Flying Training*

It is essential that every AAF pilot be able to fly on instruments. Almost every combat mission, transport, or domestic cross-country flight demands skill in instrument flying.

There is nothing mysterious or supernatural about the ability to fly on instruments. Any pilot can become skilled in flying instruments if he will follow the prescribed course of training, practice instrument procedure, and develop confidence in himself and in his instruments.

## Know Your Instruments

Confidence in your instruments grows out of knowing not only what they can do, but also out of knowing what they cannot do. Learn their limitations, learn how to read them correctly against temperature and altitude, and how to check and cross-check them with each other.

## Creating a Picture

You can develop confidence in yourself by learning to translate what your instruments tell you into a true picture of where you are. In other words, you must learn that the instruments themselves merely give you reference points (like points on a graph). You have to draw the picture in your own mind by connecting the reference points which the instruments establish.

The picture which your mind must learn to draw for you must show you and your plane in a definite relationship to (a) the ground, (b) the range station, (c) the landing field, (d) the points of the compass, and (e) a time relationship to all these.

As you become more skilled in the technique of instrument flying you will create this picture automatically; but at first it is difficult.

Here is a suggestion that may speed your instrument training: with someone else in your plane to watch traffic, fly an orientation problem under contact conditions. Do all your flying by instruments; but from time to time look out and check your position by sight. **Create the picture by instruments first,** then check it by visual reference. Repeat the process with several different orientation problems. You will discover that the picture you are creating by instruments alone is becoming clearer and more accurate.

## Time and Practice

Time and practice are essential for any pilot to gain proficiency on instruments. It is not a skill that can be picked up in a hurry nor retained without constant practice. If you go for any extended period without actually flying instruments, or "brushing up" your technique on the Link or under the hood, you will get rusty.

The regulations provide that an instrument rating is good only for a year. **But do not be satisfied with fulfilling merely the letter of the law.** Take every opportunity to use the Link, to make hooded flights and to keep your hand in at flying on instruments.

## Other Suggestions

1. Never make a steep bank on instruments.
2. Don't overcontrol—be light fingered; use steady, even pressure on controls.
3. When you get off the beam, don't turn too sharply getting back—ease back, a few degrees at a time (try it outside the hood to discover how small a turn you actually need).
4. When bracketing a beam learn to recognize a change of signal early (on the second or third signal). If you catch it at once and **do something about it,** you won't have to make sharp turns. And you won't lose the beam.
5. Remember that your gyro instruments must be watched for occasional correction.
6. Above all, don't work too hard. Give your eyes a rest. Look around the cockpit. Continual close attention is fatiguing. Tired eyes and a tired brain are likely to make mistakes.
7. **Finally use the Link trainer every chance you get. Keep constantly at it.**

# INSTRUMENT FLYING PROFICIENCY

Instrument pilot training is prescribed for all pilots in AAF Regulation 50-3, dated October 15, 1943.

### General

A permanent Board of Officers (competent instrument pilots) will be established at each station, group, squadron or equivalent unit, to examine the instrument proficiency of all rated pilots and service pilots.

Instrument Pilot Certificates, AAF Form 8 (white) and AAF Form 8A (green) are used to indicate the instrument rating of pilots. These, when issued, will be signed by the examining board members and countersigned by the pilot's organization commander.

The certificates will be valid for one (1) year only, from date of issuance. **Pilots and Service Pilots must requalify for their instrument ratings once a year or be removed from instrument status.**

Pilots and Service Pilots as soon as practicable after assignment to duty involving the pilotage of army aircraft will pursue the "Instrument Course" prescribed below, or as much of the course as necessary to complete successfully the "Instrument Flight Test." (Exceptions: Pilots or Service Pilots who hold either a valid CAA Airline Pilot Certificate or a current Air Corps Instrument Card.)

### Instrument Flight Course

This course will be completed within one (1) month of the date it is started, and qualification for a white card will be required within six (6) months after the course is started. Additional instrument training will be given if required for the successful completion of the instrument flight test. The instrument course will consist of the following:

1. At least six (6) hours' practice on basic maneuvers in a suitable airplane using all instruments. Basic maneuvers will include level flight, turns, glides, climbs, and stalls. During this practice, the student will be required to demonstrate his ability by using rate instruments only.
2. At least six (6) hours' practice in a suitable airplane on radio and radio range orientation emphasizing approach and letdown procedures. It will include instruction in the use of all standard radio aids to flying.
3. The completion of a written examination covering air traffic regulations, as prescribed by the Civil Aeronautics Administration.

### Instrument Flight Test

**The white card will be issued to those who have passed successfully the following Instrument Flight Test.** (The test will be given by a designated instrument check pilot and must be given in a suitable airplane and not in a Link trainer or other such device):

### Basic Maneuvers

Basic maneuvers will be accomplished by the use of all instruments, but at various times during these maneuvers the gyro instruments will be caged in order that the student will demonstrate his proficiency in the use of rate instruments. The allowable variations in basic maneuvers will be: air speed within plus or minus 10 miles per hour; altitude within plus or minus 200 feet; straight flight within plus or minus 5°; vertical speed within plus or minus 300 feet per minute; standard rate turns within plus or minus 5° per 90° of turn.

#### Instrument Take-off

The check pilot will align the airplane with the runway. Pilot will set directional gyro either to zero or the nearest 5° of the runway heading, and take off. Pilot should hold heading within 3° either side of initial heading, and attain climbing air speed safely and smoothly.

#### Spiral Climb

Pilot will put the airplane in a standard climbing spiral to the right. After climbing 1000 feet, he will reverse the direction of turn and climb 1000 feet more. Pilot should hold constant rate of turn, proper rate of climb, and air speed.

#### Level Flight

Pilot will fly on a given compass heading for five minutes. Pilot must maintain straight and level flight.

#### 90 and 180 Degree Turns

Pilot will make turns in each direction. Accuracy, maintenance of altitude, and smoothness of control determine proficiency.

#### Steep Banks

Pilot will put the airplane in a bank of 40 to 60 degrees, maintain this bank in a smooth turn, then return to straight and level flight. No specific amount of turn is required. Proficiency will be based on smoothness of turn, maintenance of altitude, and safe air speed. Banks in excess of 40° are not required in aircraft of 20,000 lbs. or over.

#### Stalls

Pilot will place the airplane in a glide without flaps with engine completely throttled, slowly reduce the air speed to a complete stall, then regain normal gliding speed. Proficiency will be based on avoiding a tendency toward a second stall during recovery and on ability to hold the airplane from turning or dropping a wing before the stalling point is reached.

#### Glides

Pilot will place the airplane in a power glide without flaps, with appropriate air speed, safely above stalling speed, and make at least one 90° turn in each direction. Pilot should maintain constant air speed and vertical speed and execute turns smoothly.

#### Recovery from Unusual Maneuvers

The check pilot will place the airplane in an unusual position, then instruct the pilot to take controls, recover, and resume level flight. Proficiency will be based on avoidance of any tendency toward a second stall during recovery. Type of aircraft will govern extent of unusual maneuvers—check pilot will use judgment in execution and allowance for recovery.

### Use of Radio Range and Orientation

#### (Use all instruments)

This portion of the instrument flight test will start from a position unknown to the pilot and within 10 minutes of the radio range station. It will consist of turning the radio to the station, orientation, and bracketing of a beam and following it to the radio range station, recognition of the station, and a let-down using the standard procedure for that range and station.

Proficiency will be based on the closeness and accuracy with which the beam is followed, proper use of volume control, positive recognition of the range station, and rapidity and systematic nature of the pilot's determination of his position over the station. Failure to pick up a beam and follow it to the station is disqualifying.

All the maneuvers during this part of the test must be accomplished within the applicable limits of error as described above. Excessive straying outside or across the radio range will be considered disqualifying. Accidental or unsystematic recognition of the station will not be considered qualifying.

### Green Instrument Card

**The green card will be issued to individuals who have:**

a. Passed successfully the entire "Instrument Flight Test" prescribed above.

b. One hundred (100) hours as pilot in actual instrument weather conditions, airplane flown manually. (Flying on top of an overcast is not considered actual instrument weather condition when used for the purpose of logging actual instrument pilot time.)

c. Passed successfully the **following additional maneuvers** on use of radio range and orientation.

#### Position Plotting by "Intersection"

Take bearings on at least two stations (three if possible) and plot position on D/F chart.

#### Aural Null Orientation and Homing

Using Aural Null, locate station and home. (Link trainers may be used for position plotting by intersection and Aural Null orientation and low approach, provided ADF or loop equipment is not available on aircraft used for test.)

#### Radio Compass Low Approach

This portion of the test is to emphasize the simplicity of executing low approaches using the Radio Compass in "Comp." position. Follow needle to station, turn to reciprocal of station to field course (terrain permitting). Lose ⅔ excess altitude outbound, execute procedure turn, lose remaining excess, cross station, and make final descent to minimum altitude over field. Procedure will closely approximate standard low approach but no reference is made to range legs for lateral corrections of course or headings.

### Practice

All pilots will practice instrument flying under a hood or other suitable enclosure in an airplane or Link trainer. Instrument flying training in excess of the minimum requirements will be encouraged.

### Safety Precautions

Provisions concerning safety precautions prescribed by AAF Regulation No. 60-4 will be observed in connection with instrument flying. Positive communication will be provided at all times between instructor pilot and the pilot.

### Qualification Record

The date a pilot qualifies for an instrument card will be entered on his Form 5 for the month in which he qualifies, and also for the month of June each year.

### Exceptions

The provisions of the above do not apply to student instrument flying training conducted by the AAF Training Command.

## INSTRUMENT AND HOODED FLYING

**An instrument flight** is defined as a flight in an aircraft when natural means of orientation are not visible to the pilot and flight is controlled entirely by reference to instruments and mechanical aids.

**Hooded flight** is flight in an aircraft for the purpose of training, during which the pilot is prevented outside visual reference of the ground by means of a hood or canopy.

The following requirements and safety precautions will be observed in connection with **instrument flights:**

1. Each airplane shall be equipped for instrument flying.
2. Each airplane shall be equipped with two-way radio and capable of communicating with ground stations along the route to be flown.
3. CAA flying regulations will govern on all instruments flights.

The following requirements and safety precautions will be observed in connection with **hooded flights:**

1. Requirements listed in **instrument flights** above.

2. On all hooded flights the pilot will be accompanied by a passenger who has, at all times, outside visual reference, who can readily communicate with the pilot, who is competent to act as instructor or to act as safety observer and warn the pilot of the proximity of aircraft or other danger, and who is capable of following on a map the course flown. The responsibilities of the above passenger may be assumed by the personnel of an accompanying airplane. In the latter case, both airplanes must be equipped with two-way radio to facilitate intercommunication and radio and visual contact with the hooded airplane will be maintained by the accompanying airplane at all times.

3. A rated airplane pilot will always accompany the pilot receiving hooded flying training, unless the pilot has been given the test described in PIF No. 1-9 and found qualified. **Student instrument team missions, where neither student has yet been rated as prescribed by PIF No. 1-9, will be permitted at AAF flying schools when necessary for the completion of the regular instrument flying training.**

4. Instructor pilots will be responsible for the observance of safety precautions during take-offs and landings, and when the pilot is enclosed by the hood. They will be responsible for compliance with military and CAA flying regulations.

5. The hood or enclosure for the pilot undergoing training will be of a type approved by the Commanding General of the Army Air Forces.

6. When service type airplanes are used, the hood will be installed in the cockpit normally occupied by the pilot. When training type airplanes are used, the hood installation may be in either cockpit.

7. In service type airplanes all take-offs and landings will be made by the individual occupying the pilot's cockpit. In training type airplanes, take-offs and landings will be as determined by the instructor pilot.

8. Landings by the pilot when hooded are prohibited except in connection with special instrument landing training at stations where ground equipment, an instrument-landing airplane, and qualified personnel for instruction are available. This paragraph will not be interpreted as prohibiting the practicing of approaches to the landing field while hooded.

9. Take-offs may be practiced by the pilot when hooded, provided he is accompanied by an instructor pilot who has satisfied himself regarding the pilot's competency.

10. Airplanes will not be flown with the hood closed at altitudes lower than 3,000 feet when the passenger accompanying the airplane as a safety observer is not a rated pilot or when the responsibility of the safety observer is assumed by personnel of an accompanying airplane.

### Hooded Flight Restrictions

The various phases of hood flying will be restricted to the following types of airplanes.

**Air work** may be performed in any type of airplane provided a safety observer is carried or an accompanying airplane with a safety observer is employed.

**Take-offs or landings** may be performed in:

1. Primary and basic trainers.
2. Basic combat types.
3. Corps and Division observation aircraft.
4. Bombardment and cargo airplanes with pilot and co-pilot in side-by-side seating arrangement.

### Red and Green Hoods

What the Students See

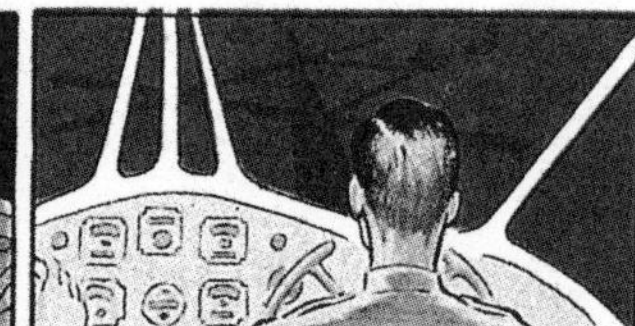

What the Instructor Sees

The recently developed red and green hood has one outstanding advantage for daylight instrument flying training. While it limits the student's view to the inside of the cockpit, it allows the instructor to sit next to him and yet to have an unimpeded view outside as well as inside the cockpit.

The equipment consists of transparent green sheeting that is attached to the cockpit windows, and red goggles that are worn by the pilot. **As the green sheeting and the red lenses in the goggles are complimentary, it is impossible to see through the combination.** Wearing the goggles, the student can clearly see everything within the cockpit but nothing beyond it. While the student is flying on instruments, the instructor, without goggles, can see through the green sheeting and protect the plane from outside hazards.

REFERENCE: AAF Regulations 60-7, dated December 9, 1942, and AAF Regulation 60-4, dated September 3, 1943.

Official military flying time and time spent in training devices is recorded on AAF Forms 1, 1G, 5 and 5S. Form 1G is used for Glider pilots.

For detailed information on Form 1 study AAF Reg. 15-1 carefully, and see the sample on PIF 1-13-1. Your Form 5 is accomplished from the information you enter on Form 1. Check it regularly, however, to be sure you are receiving proper credit.

**To Compute Time**

For the purpose of computing flying time, an aerial flight begins when the airplane starts to move forward on the takeoff run, and ends when enough momentum is lost so that the airplane either stops completely or power must be applied to continue the motion. **It does not include taxiing.**

As pilot, you enter on Form 1 or 1G the time in flight, the duty symbol, and the symbol indicating the type of flight for each person accompanying you on each flight. After landing, be sure you have made all necessary entries before you leave the airplane.

The following information is a guide to use of duty symbols. For situations not covered here refer to AAF Regulations 15-1 and 15-5. Read them in conjunction with each other.

### ★ FIRST PILOT TIME (P)

First Pilot Time is given all qualified pilots at the controls of the airplane and actually flying in the capacity of first pilot. In addition, it is given to any qualified pilot who is at one set of controls of a dual control airplane giving any type of dual instruction, such as transition training, instrument training, or check time, etc., to another qualified pilot. If your co-pilot, or other rated pilot, takes your place as first pilot, he will be given the credit for First Pilot Time for the time he is actually in full control of the ship. Only one pilot in an aircraft may claim First Pilot Time for the same time interval.

### ★ CO-PILOT TIME (CP)

Credit Co-pilot Time to all qualified pilots flying in the capacity of co-pilot at co-pilot controls, in airplanes equipped with dual controls, side-by-side arrangement only.

### ★ COMMAND PILOT TIME (C)

Use this symbol for any person holding an effective Command pilot rating if he is directing the flight operations of one or more aircraft but not flying as pilot or co-pilot. This time may not be credited in single-seater aircraft, nor for more than one individual in the same aircraft for the same time interval.

★ **ACTING COMMAND PILOT TIME (CA)**

Credit Acting Command Pilot Time to any qualified pilot not holding Command pilot rating but performing the duties of Command pilot.

★ **QUALIFIED PILOT DUAL TIME (QD)**

Use this symbol for any qualified pilot who is receiving any type of dual instruction such as transition training, instrument training, or check time. Qualified Dual Time may be logged only in aircraft having dual controls, either side-by-side or tandem; such time may be logged only for time actually spent at the controls.

★ **FLYING INSTRUCTOR TIME (S)**

Use this symbol for all qualified pilots for the time spent instructing non-rated student pilots. Any Instructor Time (S) appearing on the Form 1 will be credited on the Form 5 both as Instructor Time (S) and First Pilot Time (P).

★ **STUDENT FIRST PILOT TIME (SP)**

Use this symbol if you are a student pilot engaging in a solo flight. You may be credited with solo time, even if the instructor is present but not participating in any way except as passenger.

★ **STUDENT DUAL TIME (SD)**

This time will be recorded for all student pilots flying with and receiving dual instruction from an instructor in the same aircraft.

★ **STUDENT CO-PILOT TIME (SC)**

Student Co-pilot Time will be recorded for student pilots flying in the capacity of co-pilot, under the same conditions required for Qualified Co-pilot Time.

★ **NON-PILOTS AND NON-PILOT STUDENTS**

Time symbols used for non-pilots and non-pilot students will be found on the inside front cover of all Forms 1 and 1G.

★ **DUTY CHANGE ALOFT**

Whenever an occupant of an airplane changes his duty position in flight he will enter the time spent in his first position under that symbol, then enter the new duty symbol in the next column on the Form 1.

## ★ SYMBOL FOR TYPE OF FLIGHT

The following symbols indicate the type of flight performed. They will be entered on the Form 1 along with the number of hours and minutes flown under each condition. The symbol N covers night flying; I is instrument flight; NI indicates instrument flight at night; the symbol AI is used for flight in actual instrument conditions, and the symbol NAI will be used for flight in actual instrument conditions at night. Contact flight requires no symbol. Each time you change the symbol for type of flight enter it for each occupant except passengers.

Actual instrument time will be logged for flight conducted "in the soup," that is, when the attitude of the plane cannot be controlled by visual reference to the ground, clouds, or in any way except actual reference to your instruments. **Flight over a solid overcast, for example, would constitute instrument flight (I), but could not be logged as actual instrument time (AI). Actual instrument time cannot be logged unless the aircraft is being flown manually.**

Officers, warrant officers, flight officers, and enlisted men of the AAF and officers, warrant officers, and enlisted men of the AUS assigned to air observation of the Field Artillery, who have demonstrated the required qualifications may, under prescribed regulations, be rated by the Commanding General, AAF, or by such officer or officers he may designate, as:

1. Command pilot
2. Senior pilot
3. Pilot
4. Service pilot
5. Liaison pilot
6. Senior balloon pilot
7. Balloon pilot
8. Glider pilot
9. Senior aircraft observer
10. Aircraft observer
11. Technical observer

### Flying Officers

A flying officer is an officer who has received an aeronautical rating as a pilot of aircraft as an aircraft observer, or as any other member of a combat crew.

During the present war and for six months after its termination a flying officer as defined, includes Flight Surgeons, and commissioned officers or warrant officers while undergoing flying training.

### Command of Flying Units

Any flying officer on duty with the AAF who has an aeronautical rating as a pilot of service types aircraft is competent to command flying units.

Only an officer who has received an aeronautical rating as Command Pilot, Senior Pilot, Pilot, Senior Balloon Pilot, or Balloon Pilot, may command tactical units of the Army Air Forces where a comprehensive knowledge of flying on the part of the commander is necessary.

An officer who has received an aeronautical rating as Service Pilot may command air transport units or units equipped solely with liaison or glider aircraft.

An officer with an aeronautical rating as Liaison Pilot may command only organizations equipped solely with liaison aircraft.

An officer who has received an aeronautical rating as a Glider Pilot may command only organizations equipped solely with glider aircraft.

### Flight Officers

The Flight Officer Act enacted July 8, 1942, provides that an Aviation Cadet at the completion of training may, under regulations prescribed by the Secretary of War, if not selected for appointment as Second Lieutenant in the Army of the United States, be appointed a Flight Officer, with the pay and allowances of Warrant Officer, Junior Grade, on flying status. Flight Officers may, after three months, be appointed by selection to the grade of Second Lieutenant.

## REQUIREMENTS FOR RATINGS

### Command Pilot

Senior Pilots or Pilots may obtain a rating as Command Pilot, provided they have one of the following:

1. 15 years active military duty as a rated pilot with 2000 hours accredited flying time.
2. 10 years active military duty as a rated pilot with 3000 hours accredited flying time.
3. 20 years service as a rated Reserve or National Guard pilot with 2000 hours accredited flying time.
4. 15 years as a rated Reserve or National Guard pilot with 3000 hours accredited flying time.

### Senior Pilot

Pilots may obtain a rating as Senior Pilot provided they have 5 years service as a rated pilot with the air components of the military service and 1500 hours accredited flying time.

### Pilot

Anyone who is physically qualified may obtain a rating as Pilot, provided he has:

1. Successfully completed the necessary course of instruction at an AAF advanced pilot school, or
2. Held, on 2 July 1926, any rating as Pilot in the Army Air Corps, and in addition:
   a. Demonstrated by flight test, his ability to operate tactical or basic training type plane; and
   b. Has flown as pilot of aircraft of 400 or more horsepower, 25 hours alone, within the preceding 12 months; and
   c. Is recommended by a board of officers for such rating; or
3. Has passed his 18th birthday but not his 38th birthday; is not above the rank of Captain; holds the rating of Senior Service Pilot or Service Pilot, and in addition:
   a. Has flown as an aircraft pilot 500 hours or more, of which 400 hours were in aircraft of 400 or more horsepower; 300 hours alone, except when a copilot was required; 100 hours in planes of the AAF with engines of at least 400 horsepower, 50 hours within the preceding 12 months, and:
   b. Is recommended for such rating by a board of officers, as a result of examination of the applicant's flying records, showing he has the necessary qualifications and is ready for assignment to the combat duties appropriate for a pilot who has graduated from an AAF advanced flying school; and
   c. Demonstrates by a flight test conducted at an AAF station designated by the Commanding General, AAF, his ability to operate safely and efficiently, a tactical or advanced training type airplane in accordance with the procedure prescribed in AAF Reg. 50-7, Par. 7d.

4. A pilot physically qualified and recommended for such rating by a board of officers, who

a. Has graduated from a standard course of instruction for pilots in the armed forces of friendly foreign nations within the 12 months preceding the date of application; or

b. Has flown with pilots of the armed forces of friendly foreign nations 300 hours or more, of which: 200 hours were in airplanes with engines of 400 or more horsepower; 100 hours were alone, except when a copilot was required; and 50 hours were within the 12 months preceding date of application.

### Service Pilot

Anyone between the ages of 18 and 38 who is physically qualified, may obtain the rating of Service Pilot, provided he:

1. Has flown as an aircraft pilot 1000 hours or more, of which 200 hours were flown alone except when a copilot was required; 200 hours in airplanes of 200 or more horsepower or in PT Army type aircraft while a flight instructor at an AAF Contract Pilot school or in the CAA-WTS program, of which at least 100 hours were flown during the nine months preceding selection date and in addition:

a. Successfully completes a course of training conducted by the AAF Training Command for the training of Service Pilots;

b. Is recommended by a board of officers for such rating; or

2. Has flown as pilot of aircraft 1000 hours or more, of which 300 or more were flown alone except when a copilot was required; 300 hours in airplanes with engines of 400 or more horsepower, of which 50 hours or more were flown within the 12 months preceding the date of application; and

a. Demonstrates by a flight test at an AAF station designated by the Commanding General, AAF, his ability to operate safely and efficiently, multi-engine aircraft of tactical, cargo, or advanced training types in accordance with procedure prescribed in AAF Reg. 50-7, Par. 7d., and

b. Successfully completes the written professional examination prescribed in 50-7, Par. 7e.

c. Is recommended by a board of officers for such rating.

### Liaison Pilot

1. An enlisted man of the AUS assigned to the AAF, who is physically qualified for duty as Pilot, may obtain a rating as Liaison Pilot, provided he has successfully completed, since 1 July 1942, the prescribed course of instruction at an advanced AAF Liaison Pilot school, or an advanced AAF Contract Liaison Pilot school.

2. An officer, warrant officer, or enlisted man of the AUS assigned to air observation of the Field Artillery, who is physically qualified, may obtain a Liaison Pilot rating provided he has successfully completed the course prescribed at an AAF elementary Liaison Pilot school or at an AAF Contract school for elementary liaison training.

### Glider Pilot

Anyone physically qualified for duty as pilot may obtain a rating as Glider Pilot provided he:

1. Has successfully completed a prescribed advanced course of glider pilot training at an AAF special service school; or

2. Holds the rating of Command Pilot, Senior Pilot, Pilot, Senior Service Pilot, or Service Pilot, and in addition:

a. Has flown as pilot of tactical type gliders 3 hours or more, has made at least ten landings, and

b. Demonstrated by a flight test his ability to operate safely and efficiently at least a 15-place glider in various types of towed and free flights, and

c. Is recommended by a board of officers.

### Senior Aircraft Observer

Anyone holding the rating of Aircraft Observer and who has had at least 5 years service as a rated Aircraft Observer with the air components of the military services of the United States, and who has flown as a rated Aircraft Observer 500 hours or more, performing observation, reconnaissance, bombardment, or navigation missions, may be rated as Senior Aircraft Observer.

### Aircraft Observer

An individual who holds the rating of Command Pilot, Senior Pilot, Pilot, Senior Balloon Pilot, or Balloon Pilot may obtain the rating of:

1. Aircraft Observer, provided he has served as a regularly assigned member of a combat crew in observation or reconnaissance aviation units of the AAF or a Federally recognized National Guard squadron as a commissioned officer and a pilot for a period of not less than one year; and

a. Has flown at least 100 hours on such duty, at least 50 hours of which were as observer; and

b. Has qualified as expert aerial gunner or aerial sharpshooter; and

c. Is certified by his Commanding Officer as

competent to carry out the functions of an aircraft observer at the time of application.

A Command Pilot, Senior Pilot, Pilot, Senior Balloon Pilot, or Balloon Pilot may be rated Aircraft Observer (Navigator) or Aircraft Observer (Bombardier) if he has successfully completed a prescribed course in that category and is, in addition, qualified as expert aerial gunner or aerial sharpshooter and is certified by his Commanding Officer as qualified as an aircraft observer.

An individual who is physically qualified for duty as Aircraft Observer, and who, since 1 July 1940 has successfully completed the prescribed course in one of the following categories, may be rated as Aircraft Observer in that category:

Aircraft Observer (Navigator)
Aircraft Observer (Bombardier)
Aircraft Observer (Navigator-Bombardier)
Aircraft Observer (Radar-Night Fighter)
Aircraft Observer (Radar-RCM)
Aircraft Observer (Aerial Engineer)
Aircraft Observer (Flexible Gunner)

#### Technical Observer

Commissioned officers of the AUS who hold the rating of Command Pilot, Senior Pilot, Pilot, Senior Balloon Pilot, or Balloon Pilot, and whose principal flying duty should be the performance of technical duty incident to the operation of aircraft in flight, may be rated as Technical Observers.

### RERATING

Applications for rerating from Service Pilot and Senior Service Pilot to Pilot will be submitted to the aeronautical rating board of the station to which the individual is assigned. This board will not conduct a flight check, but will determine if the applicant possesses the other qualifications necessary for rerating.

The report will then be forwarded through command channels to the Headquarters of the air force or AAF command to which he is assigned.

That headquarters will act on the application; then if rerating is favored, forwards the application to the station designated by the Commanding General, AAF, to conduct flight checks for rerating; the applicant will be notified when he is to report for a flight check.

### APPLICATIONS

When an individual has determined that he is eligible for any aeronautical rating, he should forward his application through channels.

### FLIGHT STATUS

To receive flying pay, a qualified aircraft pilot who participates regularly and frequently in aerial flights must meet the following flight requirements:

1. During one calendar month perform 10 or more flights totaling at least 3 hours, or have a total of 4 hours in the air.

2. During 2 consecutive months, when the requirements above have not been met, perform 20 or more flights totaling at least 6 hours, or have a total of 8 hours in the air.

3. During 3 consecutive months, when the requirements above have not been met, perform 30 or more flights totaling at least 9 hours, or have a total of 12 hours in the air.

### SUSPENSION FROM FLYING DUTY

An officer unfit for flying for any reason other than an aviation accident, shall be suspended from flying by his Commanding Officer. **During such time he is not entitled to flying pay even though he has sufficient time to his credit to cover the period of suspension.** When the Commanding Officer believes that the suspended officer is again fit for flying, he shall revoke the suspension. Flying pay then accrues from and including the date of reinstatement, provided the flight requirements are met.

When an officer is suspended from flying because of sickness or injury incurred in line of duty, and the suspension is subsequently removed before the expiration of a 3-month period, the suspension does not affect flying pay provided the officer in such cases meets the flight requirements of the 3-month period.

An officer whose suspension from flying is the result of an aviation accident incurred in line of duty is excused from meeting flight requirements for a period not exceeding 3 months. If, at the time of the accident, he has flights and time enough to cover the month in which the accident occurred, he is paid for that month and the 3-month period will commence on the first of the following month.

REFERENCE: AR 95-60; AR 35-1480; AAF Reg. 50-7, and TM 14-250.

### Duty Required

The Flight Commander or senior member not incapacitated will, after an aircraft accident involving military personnel only:

1. Secure necessary medical attention or undertaker services.

2. Report to the Commanding Officer of the nearest Air Forces station: location of accident, nature and cause of accident, assistance required, extent of injury to personnel, action taken to care for injured, place at which injured can be reached, and any other pertinent information.

3. Secure all available evidence and testimony of witnesses, etc., which may serve to determine the cause of the accident.

4. See that the wreckage is not moved or tampered with in any way until an investigation has been made, except when necessary to assist persons injured or trapped in the wreckage, or in the interest of public safety.

5. Secure the necessary guard for the aircraft.

When an accident has occurred at or near an Army station, secure the guard from the C. O.

If the accident has occurred at a distance from any Army station, secure the guard from civilian sources at the local rate of pay for each caretaker. (The pay of civilian caretakers will be handled on AAF Form 15, and will be forwarded promptly to the home station of the airplane for payment.)

6. Take necessary action for the disposition of the airplane and the return of personnel to their stations.

7. In case of damage to private property the investigation will be made by the claims officer of the home station or the station nearest the scene of the accident. In reporting the accident ask that the claims officer be notified.

8. Upon return to home station, submit a full report to the Commanding Officer.

### *Public Relations Policy*

Make no statement to the press about any accident. The C.O. of the station or base nearest the scene of the accident will issue whatever information is to be released for publication. Don't talk to unauthorized persons about the accident nor the flight involved.

While in charge of wreckage not on Government property, do not interfere with representatives of the press who may take photographs or cover the news story, provided they do not disturb the wreckage. The use of material they gather or photos they take is covered by the censorship code and is not your responsibility.

REFERENCE: AAF Reg. 62-14. AAF Reg. 62-14A, AR 95-120, WD Circular 230.

# Forms

The pilot is responsible for the proper execution of a number of forms, and shares responsibility in others. In order that you may know how to fill them out properly and recognize what is of importance to you on forms that others fill out, these typical forms, properly executed, are presented.

You will enter each flight you make, with all necessary details, on a Form 1. A book of Form 1's belongs in every airplane.

DATE: 6-15-44 | STATION: PATTERSON FIELD, OHIO | GROUP NO. AND TYPE: 38th BOMB Gp. (H) | AIRCRAFT MODEL: B 17 G

CREW CHIEF OR AERIAL ENGINEER: T/Sgt Edward J. Marcus | SQUADRON NO. AND TYPE: 71st BOMB SQ. | AIRCRAFT SERIAL NO.: 43-7353

| PERS. CLASS (1) | – PRINT PLAINLY – NAME – RANK – ORGANIZATION (2) | USE AS DIRECTED LOCALLY (3) | ALWAYS ENTER DUTY SYMBOLS. WHEN APPLICABLE ENTER N - NIGHT OR I - INSTRUMENT. ENTER TIME FLOWN THEREUNDER: DUTY (4) | N OR I (4) | DUTY (5) | N OR I (5) | DUTY (6) | N OR I (6) | DUTY (7) | N OR I (7) | FLIGHT DATA: TERMINALS AND MISSION (8) | (9) |
|---|---|---|---|---|---|---|---|---|---|---|---|---|
| 00 | HENDERSON, DAVID H., 056321, CAPT., MITCHELL FIELD; 29th BOMB Gp. 14th BOMB SQ. | | P 1:05 | N | P 1:00 | NI | CP :55 | NAI | : | | FROM: MITCHELL FLD. L.I., N.Y. | 03:15 |
| 01 | GALLOWAY, CHAS. R., 0576425, CAPT., MITCHELL FLD; 29th BOMB Gp. 14th BOMB SQ | | CP 1:05 | N | CP 1:00 | NI | P :55 | NAI | : | | TO: PATTERSON FLD. OHIO | 06:15 |
| 38 | LEDBETTER, JOHN W., 18175688 T/Sgt MITCHELL FLD.; 29th BOMB Gp. 14th BOMB Sq. | | E 1:05 | N | E 1:00 | NI | E :55 | NAI | : | | MISSION: T; NO. OF LANDINGS: 1 | 3:00 |
| 00 | WHARTON, CHAS. T., 02412. Lt. Col. PATTERSON FLD. | | P 3:35 | | : | | : | | : | | FROM: PATTERSON FLD. OHIO | 08:00 |
| 01 | RUTHERFORD, DONALD L., 0354781, MAJ. PATTERSON FIELD | | CP 3:35 | | : | | : | | : | | TO: SALINA KANS. | 11:35 |
| 38 | HANKS, JOHN K., 14056490 T/Sgt. AMARILLO FLD., AAF. TEX.; AAF CFTC 21st SCH. SQ. | | E 3:35 | | : | | : | | : | | MISSION: A; NO. OF LANDINGS: 1 | 3:35 |
| | | | P 2:50 | | P 1:05 | I | : | | : | | FROM: SALINA KANS. | 13:00 |
| | | | CP 2:50 | | CP 1:05 | I | : | | : | | TO: PATTERSON FLD. OHIO | 16:55 |
| | | | E 2:50 | | E 1:05 | I | : | | : | | MISSION: A; NO. OF LANDINGS: 1 | 3:55 / 7:30 |
| 00 | NEWLON, FRANK C., 04698 CAPT. PATTERSON FLD., OHIO | | P 1:05 | N | : | | : | | : | | FROM: PATTERSON FLD. | 20.05 |
| 00 | ALDRIDGE, HOWARD R. 06064, Lt. RANDOLPH FLD. TEX.; C/S SING. ENG. Gp. | | CP 1:05 | N | : | | : | | : | | TO: LOCAL PATTERSON FLD. | 21:10 |
| 38 | GOODWIN, WILLIAM C., 37587991 T/Sgt. PATTERSON FLD., OHIO | | E 1:05 | N | : | | : | | : | | MISSION: A; NO. OF LANDINGS: 1 | 1:05 |
| 38 | FRANCISCO, WELLING P., 14056811, T/Sgt. PATTERSON FLD., OHIO | | R 1:05 | N | : | | : | | : | | FROM: | : |
| 18 | MITCHELL, DONALD M., 0065911, MAJ. AC | | X 1:05 | | : | | : | | : | | TO: | : |
| | | | : | | : | | : | | : | | MISSION: NO. OF LANDINGS: | : |

WAR DEPARTMENT A. A. F. FORM NO. 1 2-2-42 | FLIGHT REPORT - OPERATIONS | CHECKED: LEGIBLE AND CORRECT W.H.W OPER. CLERK | TRANSCRIBED: TOTAL FLIGHT TIME ENTERED ON FORM 1A F.S.C. CREW CHIEF | TOTAL FLIGHT TIME 11:35

The crew chief ordinarily fills out the top lines. Check the entries as soon as you enter the cockpit. Fill in data required on each member of the flight, and as much data on the flight as possible before takeoff.

**Each time you land, and before you leave the cockpit complete the Form 1.** For instance, if you are on a cross country flight, fill in the information on each leg of the trip. Don't wait until you reach your destination and then try to remember all the details of time credit and duty status. **Every time you come to a stop on the ground you have a date with your Form 1.**

Be sure that all enlisted men on flying status are properly credited with the right duty symbol, so they will receive credit for their time.

Use column 3 as directed locally. Your C.O. may direct you to write your organization there, instead of as shown.

For details on time credit and duty symbols see PIF 1-10.

WAR DEPARTMENT
A. A. F.
FORM NO. 1 A
2-2-42

## FLIGHT REPORT - ENGINEERING

**INSPECTION STATUS**

| | DATE OF OR HOURS DUE | INSPECTED TODAY BY | INSPECTED TODAY STATION |
|---|---|---|---|
| PREFLIGHT | 8-26-43 | K | F.A.D. |
| DAILY | 8-26-43 | M | F.A.D. |
| 25 HOURS | 148:00 | | |
| 50 HOURS | 150:00 | | |
| 100 HOURS | 200:00 | | |
| | | | |

**SERVICING AT STATION OF TAKE-OFF**
(CHECK IMMEDIATELY BEFORE TAKE-OFF)

| SERVICE | FUEL (GALLONS) SERVICED | FUEL IN TANKS | OIL (QUARTS) NO. 1 SERVICED | NO. 1 IN TANKS | NO. 2 SERVICED | NO. 2 IN TANKS | NO. 3 SERVICED | NO. 3 IN TANKS | NO. 4 SERVICED | NO. 4 IN TANKS | RADIATOR CHECKED |
|---|---|---|---|---|---|---|---|---|---|---|---|
| 1ST | 1150 | 2300 | 15 | 100 | 20 | 100 | 22 | 100 | 18 | 100 | |
| 2ND | | | | | | | | | | | |
| 3RD | | | | | | | | | | | |
| 4TH | | | | | | | | | | | |
| 5TH | | | | | | | | | | | |

**INSPECTION OF AUXILIARY EQUIPMENT**

| EQUIPMENT | SYMBOL | INSPECTED BY | STATION |
|---|---|---|---|
| BOMBARDMENT | — | | |
| GUNNERY | — | | |
| ~~CHEMICAL~~ | | | |
| COMMUNICATIONS | B | Sgt. Burns | F.A.D. |
| ~~PHOTOGRAPHIC~~ | | | |
| NAVIGATION | — | | |
| | | | |

STATUS TODAY: 1. / 2. / 3. / 4.

EXPLANATION: LEFT MAIN TANK FUEL GAUGE INACCURATE

**EXCEPTIONAL RELEASE**

WHEN THE "STATUS TODAY" IS INDICATED BY A RED SYMBOL, AND AN "EXCEPTIONAL RELEASE" HAS NOT BEEN GRANTED BY AN AUTHORIZED ENGINEERING OFFICER, THE PILOT OF THE AIRCRAFT WILL SIGN THIS RELEASE BEFORE FLIGHT.

RELEASED FOR FLIGHT: 1. John Doe 2. 3. 4.

REMARKS: PILOTS AND MECHANICS - SEE INSTRUCTIONS INSIDE FRONT COVER.

#1 Left Brake Weak - John Doe, 1st Lt. A.C.
#1 Serviced by Sgt. H. Ross - F.A.D. - 8/26/43

**AIRCRAFT AND ENGINE TIME RECORD**
(ENTER IN HOURS AND MINUTES)

| ENGINE | NO. 1 | NO. 2 | NO. 3 | NO. 4 |
|---|---|---|---|---|
| HOURS TO DATE | 143:55 | 143:55 | 143:55 | 143:55 |
| HOURS TODAY | 5:00 | 5:00 | 6:00 | 5:00 |
| TOTAL | 148:55 | 148:55 | 148:55 | 148:55 |
| OIL CHANGE DUE | H.T. | H.T. | H.T. | H.T |
| CUNO. CLEANING DUE | 150:00 | 150:00 | 150:00 | 150:00 |

| AIRCRAFT | |
|---|---|
| HOURS TO DATE | 143:55 |
| HOURS TODAY | 5:00 |
| TOTAL | 148:55 |

Form 1A will be executed by the crew chief, and should be in the airplane when you get in it. Check it carefully for the fuel and oil service, and watch for any **red diagonal.** This mark indicates that something is not functioning properly, but that the airplane is flyable. You must sign the release to show that you have noted the fault **before the flight.**

**If there is a red cross—do not fly the airplane.** After the flight, list anything needing attention.

It is recommended that symbols prescribed for use on Form 41B be used on Form 1A under heading "Inspection of Auxiliary Equipment." Crew chiefs and pilots should see that this is done.

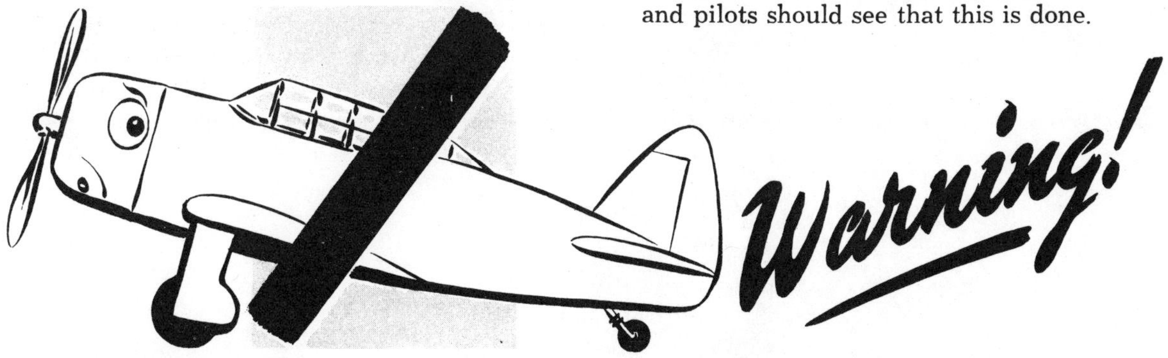

WAR DEPARTMENT
AAF FORM NO. 5
APPROVED DEC. 7, 1942

## INDIVIDUAL FLIGHT RECORD

(1) SERIAL NO. 0-101010 (2) NAME DOE (LAST) JOHN (FIRST) ALBERT (MIDDLE) (3) RANK Captain (4) AGE 1911
(5) PERS. CLASS 01 (6) BRANCH Air Corps (7) STATION Patterson Field, Ohio (ATTACHED FOR FLYING)
(8) ORGANIZATION ASSIGNED AF (AIR FORCE) ASC (COMMAND) (WING) 18 ADG (GROUP) Repair (SQUADRON) (DETACHMENT)
(9) ORGANIZATION ATTACHED (AIR FORCE) (COMMAND) (WING) (GROUP) (SQUADRON) (DETACHMENT)
(10) PRESENT RATING & DATE Pilot 3-14-41 (11) ORIGINAL RATING & DATE Pilot 3-14-41
(12) TRANSFERRED FROM (13) FLIGHT RESTRICTIONS
(15) TRANSFERRED TO (14) TRANSFER DATE
(17) MONTH December 19 43

(16) DO NOT WRITE IN THIS SPACE

| PERS CLASS | RANK | RTG. | A. F. | COMMAND | WING | GROUP NO. | GROUP TYPE | SQUADRON NO. | SQUADRON TYPE | STATION | MO. | YR. |
|---|---|---|---|---|---|---|---|---|---|---|---|---|
| : | | | : | : : | : : | : : | : : | : : : | : : | : : : | | |

| DAY 18 | AIRCRAFT TYPE, MODEL & SERIES 19 | NO. LANDINGS 20 | FLYING INST. (INCL. IN 1ST PIL. TIME) S 21 | COMMD. PILOT C CA 22 | CO-PILOT CP 23 | QUALIFIED PILOT DUAL QD 24 | FIRST PILOT DAY P 25 | FIRST PILOT NIGHT P N OR NI 26 | RATED PERS. NON-PILOT 27 | 28 | 29 | NON-RATED OTHER ARMS & SERVICES 30 | NON-RATED OTHER CREW & PASS'GR 31 | SPECIAL INFORMATION INSTRUMENT I 32 | NIGHT N 33 | INSTRUMENT TRAINER 34 | PILOT NON-MIL. AIRCRAFT OVER 400 H.P. 35 | UNDER 400 H.P. 36 |
|---|---|---|---|---|---|---|---|---|---|---|---|---|---|---|---|---|---|---|
| | | | | | | | | | | | | | | | | :45 | | |
| | | | | | | | | | | | | | | | | 1:00 | | |
| | | | | | | | | | | | | | | | | 1:00 | | |
| 3 | | | | | | | | | | | | | | | | 1:00 | | |
| 5 | | | | | | | | | | | | | | :55 | | | | |
| 12 | | | | | | | | | | | | | | | | | | |
| 15 | | | | | 2:15 | | 2:00 | | | | | | | | | | | |
| 16 | C-54 | 2 | | | | | 2:00 | | | | | | | | | | | |
| 18 | C-54 | 2 | | | | | 2:00 | | | | | | | | | | | |
| 19 | C-54 | 3 | | | | | 4:05 | | | | | | | :45 | | | | |
| 20 | C-54 | 3 | | | | | 5:35 | | | | | | | | | | | |
| 21 | C-54 | 5 | | | | | 2:50 | | | | | | | 1:05 | | | | |
| 23 | C-54 | 2 | | | | | 1:15 | | | | | | | | | | | |
| 31 | C-54 | 1 | | | | | 2:40 | | | | | | | | | | | |
| 31 | C-54 | 3 | | | | | 1:10 | | | | | | | | | | | |
| 31 | C-54 | 1 | | | | | | | | | | | | | | | | |
| COLUMN TOTALS | | | | | 2:15 | | 23:35 | | | | | | | 2:45 | | 3:45 | | |

Certified Correct
Frank S. Boone
2nd Lt., Air Corps

| | (42) TOTAL STUDENT PILOT TIME | (43) TOTAL FIRST PILOT TIME | (44) TOTAL PILOT TIME |
|---|---|---|---|
| (37) THIS MONTH | | 23:35 | 25:50 |
| (38) PREVIOUS MONTHS THIS F. Y. | | 130:00 | 130:00 |
| (39) THIS FISCAL YEAR | | 153:35 | 155:50 |
| (40) PREVIOUS FISCAL YEARS | 319:40 | 281:10 | 600:50 |
| (41) TO DATE | 319:40 | 434:45 | 756:40 |

DO NOT WRITE IN THIS SPACE

| CARD NO. 1 AIRCRAFT 19 | NL 20 | 21 | 22 | 23 | 24 | 25 | 26 | CARD NO. 2 27 | 28 | 29 | 30 | 31 | CARD NO. 3 32 | 33 | 34 | 35 | 36 |
|---|---|---|---|---|---|---|---|---|---|---|---|---|---|---|---|---|---|
| | | | | | | | | | | | | | | | | | |

756:40 HOURS

Form 5 will be executed for you. Your responsibility is merely to check that you have received proper credit for your flights, and for your time in the Link Trainer.

WAR DEPARTMENT
AAF FORM NO. 15
(REVISED 1 JAN. 44)

WAR DEPARTMENT
ARMY AIR FORCES

INVOICE

DATE 24 March 1944

THE UNITED STATES, DR.
TO Dallas Aeronautics Inc.

ADDRESS (*Write plainly name and address of firm or individual from whom supplies or services were obtained*)
Arlington Airport, Arlington, Texas

| Airplane Number | ARTICLES OR SERVICES | Quantity | Unit | Unit Price | Total |
|---|---|---|---|---|---|
| 44-2354 | (1) Aircraft gasoline, 100 Octane | 200 | Gal | 18 | 36 00 |
| | (2) Aircraft Engine oil, Grade 120 | 25 | qt. | 12 | 3 00 |
| | | | | TOTAL | |

**WHEN** it is necessary to procure emergency services or supplies you will execute an AAF Form 15, a supply of which will be found in an envelope (AAF Form 15A) in your airplane. Mailing envelopes are also included. You must prepare the form in quadruplicate. Use pencil (indelible if possible), ink, or a typewriter. All copies must be signed by both the dealer and yourself at the time of purchase.

Give the quadruplicate copy to the dealer. Immediately mail the original and remaining two copies to the contracting officer of the airplane's home station.

VENDOR'S CERTIFICATE

STATE TAX 0.05 LOCAL TAX NONE FEDERAL TAX 0.02 ~~is~~ (is not) included.
Strike out words not applicable. Always insert unit tax for fuel and oil, percentage or other basis, if any tax is included for other commodities or services.

I certify that the above bill is correct and just; that payment therefor has not been received; that all statutory requirements as to American production and labor standards, and all conditions of purchase applicable to the transactions have been complied with; and that state or local sales taxes are not included in the amounts billed.

| NAME OF FIRM | BY (*Print or Type*) | TITLE | SIGNATURE |
|---|---|---|---|
| DALLAS AERONAUTICS Inc. | M. L. SMITH | SALES MANAGER | M. L. Smith |

PURCHASER'S CERTIFICATE

Pursuant to authority vested in me, I certify that the supplies enumerated above or on attached list have been received in good condition and in quantities as stated; that the services enumerated have been satisfactorily performed. That the supplies or services were purchased in an emergency for the maintenance, operation, or protection of government equipment and were necessary for the public service. I further certify that the aircraft for which the supplies or services were purchased to maintain and operate, is engaged on an authorized cross country flight or is assigned at the point of purchase under proper authority.

| PRINTED NAME | ASN | GRADE | STATION | SIGNATURE |
|---|---|---|---|---|
| E.D. JONES | 0-484533 | CAPT. A.C. | BOLLING FLD, D.C. | E. J. Jones |

**CAUTION:** BE SURE TO INCLUDE BOTH THE SERIAL NUMBER OF THE AIRPLANE AND THE NAME OF ITS HOME STATION IN FILLING OUT THE FORM

AAF Form 23 is the clearance form. It is used for all flights except those listed in AAF Reg. 15-23, which may require different forms or be governed by other regulations. For more detailed information on clearances and the use of the Form 23 see PIF 2-2.

You are responsible for information entered in Sections "B" and "D" and the first line of Section "F." Other sections will be filled in by operations, weather office, and Clearing Authority.

The original copy (white) is retained by operations. Section "F" is detached and given to you. After you have filled out the first line give it to the crew chief just before departure. Retain the carbon copy (yellow).

Upon landing complete the "ARRIVAL REPORT" (Section "G") at bottom of the yellow copy and give the form to the crew chief of the alert crew which meets your airplane.

**Pilots must furnish information for sections marked by RED arrows**

**Others make entries in sections indicated by BLUE arrows**

**OPERATIONS**

**WEATHER OFFICE**

**CLEARING AUTHORITY**

**ARRIVAL REPORT**

**First Line Only**

WAR DEPARTMENT
AAF FORM NO. 23
(REVISED AUG. 15, 1943)

ARMY AIR FORCES
AIRCRAFT CLEARANCE

A OPERATIONS OFFICE PATTERSON FIELD
ADDRESS FAIRFIELD, OHIO
DATE 10 DEC. 1943

B PILOT'S NAME ANDERSON, M. L. RANK MAJ. HOME STATION PATTERSON ORGANIZATION 478 BASE HQ AIRCRAFT NUMBER 41-11804

NAME, INITIALS, RANK, HOME STATION OF OTHER OCCUPANTS
MC NAIR, M. E., MAJ.
CANNOY, M. L., CAPT. DPK
GREEN, M. E., 1ST LT. DPK
SHARPE, M. J., 2ND LT. DPK
WELTY, H. R., M/SGT DPK
JETT, F. L., T/SGT DPK
DLD

LIST ADDITIONAL PASSENGERS ON SEPARATE SHEET

C WEATHER DATA
EXISTING LOCAL C M8⊕3F-K-
EXISTING ROUTE CC M9⊕1S-R-K- LV E4⊕3R-F- SO +16⊕10⊕6R-F- NA Z5⊕3R-F-
JA E20⊕4K- UO E25⊕6K- ZH M30⊕ DL ⊕/25⊕6K-
DESTINATION (LATEST) E 25⊕/⊕5K- TIME 0730C
ALTERNATE (LATEST) E25⊕8 TIME 0730C
FORECASTS (ESTIMATED FLIGHT TIME PLUS 2 HOURS)
ROUTE OVC 8 HND TO NA IMPVG TO 25 HND SCTD AT JA TO CLEAR AT DBY LWRNG TO 15 HND OVC AT FV. VSBY 1 MI IN SNW RAIN MXD TO NA IMPVG TO 5 MIS TO JA IN LGT SMOKE IMPVG TO 10 MIS AT DBY DCRSG TO 2 MIS VRBL IN SNW. WARM FRNT VCNTY OF NA COLD FRNT 15 MIS WEST OF DL. ICG IN ALL CLDS.
DESTINATION 15 HND OVC VSBY 2 MIS IN LGT SNW. ICGIC.
ALTERNATE HI THIN BRKN CIG UNL. VSBY 9 MILES.
WINDS ALOFT GIVE ALT. DIR. VEL. AS PILOT REQUESTS WINDS 4 THSD 160 DGRS 20 MPH TO NA. WNDS 2 THSD 260 DGRS AT 15 MPH TO FV. TO DBY. WINDS AT 2 THSD 230 DGRS 15 MPH
AAF FORM 23A REQUIRED ☒ NOT REQUIRED ☐ FORECASTER D L KELLEY M/SGT TIME 0745C
ALTIMETER SETTINGS LOCAL 30.15 DESTINATION 30.08 ALTERNATE 30.11 RESET ALTIMETER BEFORE APPROACH

D FLIGHT PLAN (PILOT COMPLETES) RADIO CALLS 804 TYPE OF AIRCRAFT B24D PILOT (LAST NAME ONLY) ANDERSON POINT OF DEPARTURE DPK
1 ALT 6000 ROUTE AWYS IFR ☒ TO NA
2 ALT 6500 ROUTE DRT IFR ☒ TO JA
3 ALT CFR ROUTE AWYS CFR ☒ TO FV
AIRPORT OF FIRST INTENDED LANDING FV (MUNI) TRUE AIR SPEED 200 TRANSMITTING FREQUENCIES 4495 KC
PROPOSED TAKE OFF TIME 0800C EST. TIME ENROUTE 5:20 ALTERNATE AIRPORT BARKSDALE HOURS OF FUEL (CRUISING) 12 INSTRUMENT RATING WHITE FLIGHT PRIORITY D
REMARKS: SHOW FIXES WHICH WILL BE REPORTED WHILE ON INSTRUMENT FLIGHT. CC-LV-SO-NA-JA-UO-ZH-DL-FV
AM #6 NA-DRT JA-. RED #10 FV
TOWER FREQUENCIES DESTINATION 201 KC ALTERNATE 396 KC WEATHER CODE RECEIVED ☐ YES ☒ NO MILEAGE TO DESTINATION 1020 DEST. TO ALTERNATE 220
PILOT'S SIGNATURE M. Anderson, Maj. AC ☐ COMMAND PILOT ☐ SENIOR PILOT ☐ CONTRACT PILOT OF CARGO AIRCRAFT ☒ PILOT

E FLIGHT CLEARANCE AUTHORIZATION
SUBMITTED TO M.B. TIME 0750 BY J.S.
TIME APPROVAL RECEIVED 08:05 CONTROL INSTRUCTIONS RECEIVED (SEE ATTACHED ATC TRAFFIC SHEET)
INSTRUCTIONS AND APPROVAL TRANSMITTED TO PILOT OR TOWER BY: LK/BI ACTUAL TAKE-OFF TIME 08:10
CLEARING AUTHORITY J. A. WOODRUFF, COL., A. C. OPERATIONS IDENTIFICATION NO. 130
K.V. Laughton Col. A.C. CLEARANCE OFFICER

F PILOT COMPLETE FIRST LINE BELOW PRESENT TO LINE CREWMAN BEFORE TAKE-OFF
DEPARTURE RECORD
PILOT (LAST NAME ONLY) ANDERSON AIRCRAFT TYPE B24D AIRCRAFT NUMBER 41-11804 LINE CREWMAN WILL COMPLETE SECOND LINE AND DELIVER TO OPERATIONS OFFICE.
DATE OF DEPARTURE 10 DEC. TIME ACTUAL FUEL 2300 GAL GROSS WEIGHT 55070
LINE CREWMANS SIGNATURE Wm. Lamming

G THIS COPY TO BE GIVEN TO PILOT
ARRIVAL REPORT
Pilot will complete "Arrival Report" and present to line crewman meeting the arriving aircraft, for his information. Line crewman will then forward to the operations office. Home station will be notified in the event of an overnight stop.
DATE AND TIME OF ARRIVAL 12/10/43 1315 R. O. N. ☐ YES ☒ NO GAS 1000 OIL SERVICE REQUIRED DESIRED DATE AND TIME OF DEPARTURE 12/10/43 1500 DEPARTING FOR: KELLY FLD.
WHERE PILOT CAN BE REACHED AT THIS STATION Depot Supply REMARKS LINE CREWMANS SIGNATURE [illegible]

Consult AAF Regulation 15-23 for specific instructions on filling out Form 23. Certain flights for Air Transport Command, student training flights, and flights within local flying areas may require another form or a different procedure. It is your responsibility to know in what category your flight falls.

You are responsible for information entered in Sections "B" and "D" and the first line of Section "F." Other sections will be filled in by operations, weather office, and Clearing Authority.

The original copy (white) is retained by operations. Section "F" is detached and given to you. After you have filled out the first line give it to the crew chief just before departure. Retain the carbon copy (yellow).

Upon landing complete the "ARRIVAL REPORT" (Section "G") at bottom of the yellow copy and give the form to the crew chief of the alert crew which meets your airplane.

WAR DEPARTMENT
AAF FORM NO. 23
(REVISED AUG. 15, 1943)

ARMY AIR FORCES
AIRCRAFT CLEARANCE

A OPERATIONS OFFICE
ADDRESS PATTERSON FIELD
FAIRFIELD, OHIO
DATE 10 DEC. 1943

PILOT'S NAME
B ANDERSON, M. L. — RANK MAJ. — HOME STATION PATTERSON — ORGANIZATION 478 BASE HQ — AIRCRAFT NUMBER 41-11804
NAME, INITIALS, RANK, HOME STATION OF OTHER OCCUPANTS
MC NAIR, M. E., MAJ.
CANNOY, M. L., CAPT. DPK
GREEN, M. E., 1ST LT. DPK
SHARPE, M. J., 2ND LT. DPK
WELTY, H. R., M/SGT DPK
JETT, F. L., T/SGT DPK
DLD
LIST ADDITIONAL PASSENGERS ON SEPARATE SHEET

C WEATHER DATA
EXISTING LOCAL C M8⊕3F-K-
EXISTING ROUTE CC M9⊕1S-R-K- LV E4⊕3R-F- SO +16⊕10⊕6R-F- NA Z5⊕3R-F-
JA E20⊕4K- UO E25⊕6K- ZH M30⊕ DL ⊕/25⊕6K-
DESTINATION (LATEST) E 25⊕/⊕5K- TIME 0730C
ALTERNATE (LATEST) E25⊕8 TIME 0730C
ALTIMETER SETTINGS: LOCAL 30.15 — DESTINATION 30.08 — ALTERNATE 30.11 — RESET ALTIMETER BEFORE APPROACH
FORECASTS (ESTIMATED FLIGHT TIME PLUS 2 HOURS)
ROUTE OVC 8 HND TO NA IMPVG TO 25 HND SCTD AT JA TO CLEAR AT DBY LWRNG TO 15 HND OVC AT FV. VSBY 1 MI IN SNW RAIN MXD TO NA IMPVG TO 5 MIS TO JA IN LGT SMOKE IMPVG TO 10 MIS AT DBY DCRSG TO 2 MIS VRBL IN SNW. WARM FRNT VCNTY OF NA COLD FRNT 15 MIS WEST OF DL. ICG IN ALL CLDS.
DESTINATION 15 HND OVC VSBY 2 MIS IN LGT SNW. ICGIC.
ALTERNATE HI THIN BRKN CIG UNL. VSBY 9 MILES.
WINDS ALOFT (GIVE ALT, DIR, VEL AS PILOT REQUESTS) WINDS 4 THSD 160 DGRS 20 MPH TO NA. WNDS 2 THSD 230 DGRS 15 MPH TO DBY. WINDS AT 2 THSD 260 DGRS AT 15 MPH TO FV.
AAF FORM 23A REQUIRED ☒ NOT REQUIRED ☐ FORECASTER D L KELLEY M/SGT TIME 0745C

D FLIGHT PLAN (PILOT COMPLETES) RADIO CALLS 804 — TYPE OF AIRCRAFT B24D — PILOT (LAST NAME ONLY) ANDERSON — POINT OF DEPARTURE DPK
1 IFR ☒ ALT 6000 ROUTE AWYS TO NA
2 IFR ☒ ALT 6500 ROUTE DRT TO JA
3 CFR ☒ ALT CFR ROUTE AWYS TO FV
4
AIRPORT OF FIRST INTENDED LANDING FV (MUNI) — TRUE AIR SPEED 200
PROPOSED TAKE OFF TIME 0800C — EST. TIME ENROUTE 5:20 — ALTERNATE AIRPORT BARKSDALE — TRANSMITTING FREQUENCIES 4495 KC — HOURS OF FUEL (CRUISING) 12 — INSTRUMENT RATING WHITE — FLIGHT PRIORITY D
REMARKS: SHOW FIXES WHICH WILL BE REPORTED WHILE ON INSTRUMENT FLIGHT. CC-LV-SO-NA-JA-UO-ZH-DL-FV
AM #6 NA-DRT JA-. RED #10 FV
TOWER FREQUENCIES DESTINATION 201 KC ALTERNATE 396 KC — WEATHER CODE RECEIVED ☐ YES ☒ NO — MILEAGE TO DESTINATION 1020 DEST. TO ALTERNATE 220 — PILOT'S SIGNATURE M. Anderson, Maj. AC — ☒ PILOT

E FLIGHT CLEARANCE AUTHORIZATION
SUBMITTED TO M.B. TIME 0750 BY J.S.
TIME APPROVAL RECEIVED 08:05 — CONTROL INSTRUCTIONS RECEIVED (SEE ATTACHED ATC TRAFFIC SHEET) — OPERATIONS IDENTIFICATION NO. 130
CLEARING AUTHORITY J. A. WOODRUFF, COL, A. C.
INSTRUCTIONS AND APPROVAL TRANSMITTED TO PILOT OR TOWER BY: LK/BI — ACTUAL TAKE-OFF TIME 08:10 — CLEARANCE OFFICER K.V. Laughton, Col. A.C.

F PILOT COMPLETE FIRST LINE BELOW PRESENT TO LINE CREWMAN BEFORE TAKE-OFF
DEPARTURE RECORD
LINE CREWMAN WILL COMPLETE SECOND LINE AND DELIVER TO OPERATIONS OFFICE.
PILOT (LAST NAME ONLY) ANDERSON — DATE OF DEPARTURE 10 DEC. — AIRCRAFT TYPE B24D — AIRCRAFT NUMBER 41-11804 — ACTUAL FUEL 2300 GAL — GROSS WEIGHT 55070
LINE CREWMANS SIGNATURE Wm. Lamming

Pilots must furnish information for sections marked by arrows

First Line Only

ARRIVAL REPORT

G THIS COPY TO BE GIVEN TO PILOT — ARRIVAL REPORT — Pilot will complete "Arrival Report" and present to line crewman meeting the arriving aircraft, for his information. Line crewman will then forward to the operations office. Home station will be notified in the event of an overnight stop.
DATE AND TIME OF ARRIVAL 12/10/43 1315 — R. O. N. ☐ YES ☒ NO — GAS 1000 OIL — SERVICE REQUIRED — DESIRED DATE AND TIME OF DEPARTURE 12/10/43 1500 — DEPARTING FOR: KELLY FLD.
WHERE PILOT CAN BE REACHED AT THIS STATION Depot Supply — REMARKS — LINE CREWMANS SIGNATURE Ed Wamgbuchler

Standard Form No. 1051
Form approved by Comptroller General, U. S.
August 10, 1933

* SAMPLE *

No. ______

# FLIGHT CERTIFICATE AND SCHEDULE

Patterson Field, Fairfield, Ohio (Place), March 31 (Date), 1943

I HEREBY CERTIFY that during the period March 1, 1943, to March 31, 1943, I performed the flights listed on this schedule under orders involving flying issued by Chief of Air Corps, (Competent authority) P.O. #60 dated 3-14-41, effective 3-14-41, (Date of reporting and entering on duty) copy of which is filed herewith or with the accounts of F. S. McCauley, Colonel, F.D., disbursing officer, for the period ended October, 1942.

Eiler H. Smith (Signature of flyer)
EILER H. SMITH,
Captain, Air Corps, (Rank or rating)

| 1 DATE | 2 FLIGHT No. | 3 PERIOD IN THE AIR Hrs. | Min. |
|---|---|---|---|
| March 16 | 2 | 4 | 15 |
| Total | 2 | 4 | 15 |

I CERTIFY that Eiler H. Smith, Captain, Air Corps, (Name and rank or rating of flyer) during the period above mentioned fulfilled the flying requirements prescribed by Executive order in force during said period, under conditions specified therein and in the flying orders referred to, and that this certificate is made after checking the flight log book or record of said flyer with the aircraft log books or records of the aircraft in which he made the flights listed in the schedule, which is certified to be correct. I further certify that the above-named officer is a designated Pilot and has performed the duties as such for the period covered by this voucher. For the Commanding Officer:

Suspended from flying duty ______, 19__ A A Sheppard (Signature)

Suspension revoked on ______, 19__ A. A. SHEPPARD, Major, Air Corps, (Rank)
Post Operations Officer.

AAS/mmn

The flight certificate will be executed for you by the proper authority and turned over to you in duplicate each month. You will sign the original and attach both original and copy to your pay voucher.

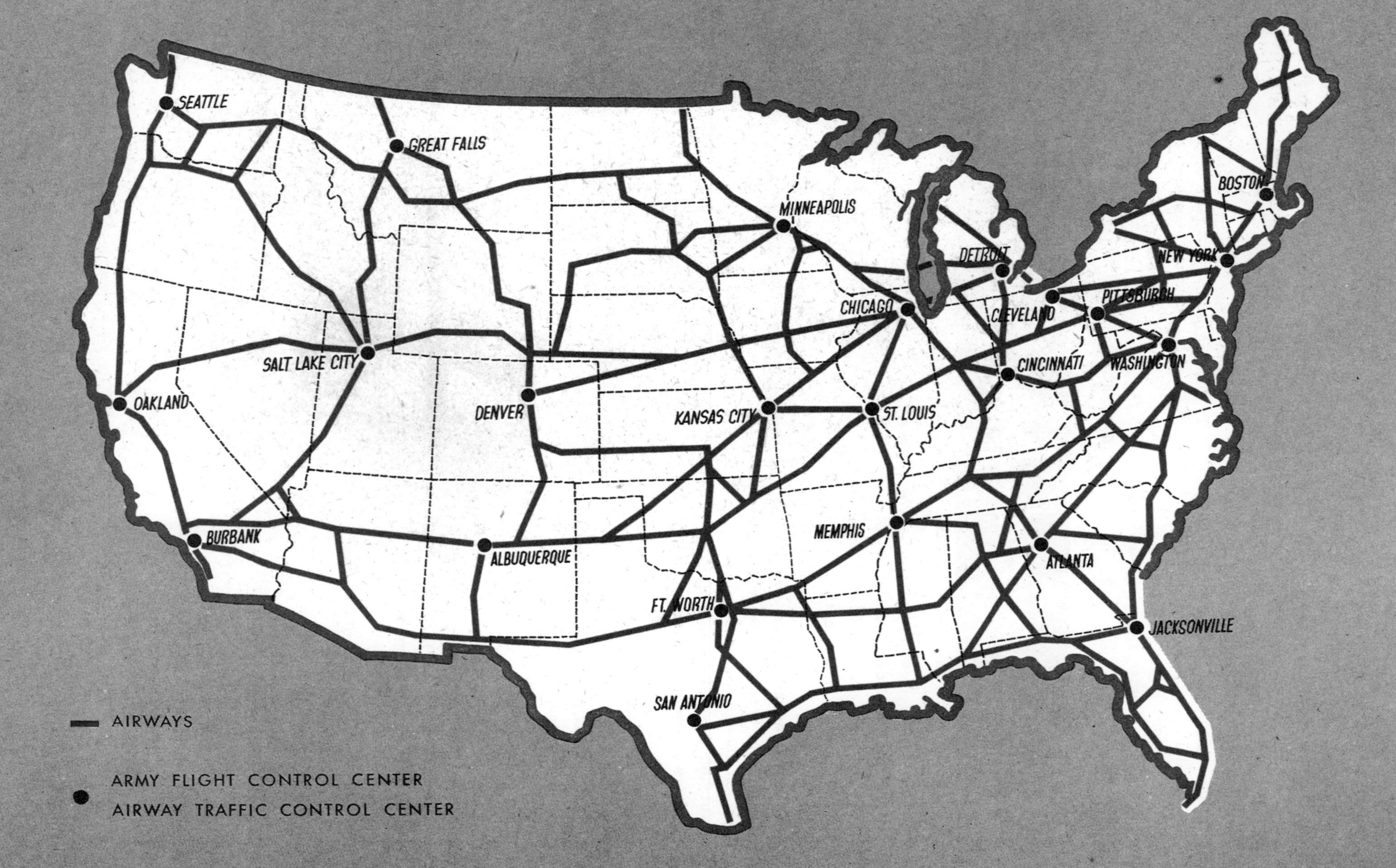


# SECTION 2 FLIGHT

# AAF FLIGHT CONTROL

Army Flight Control is an agency of Headquarters, Army Air Forces. It has one major responsibility within the continental limits of the United States: to facilitate all Army point-to-point flights, thus increasing the number of completed missions, with the greatest possible safety.

At present, Flight Control does this through:

1. Pilots' Advisory Service.
2. Clearance of Form 23 Flight Plans at fields where no AAF Operations exist. (PIF 2-2)
3. Approval of all changes in Form 23 Flight Plans made while in flight.

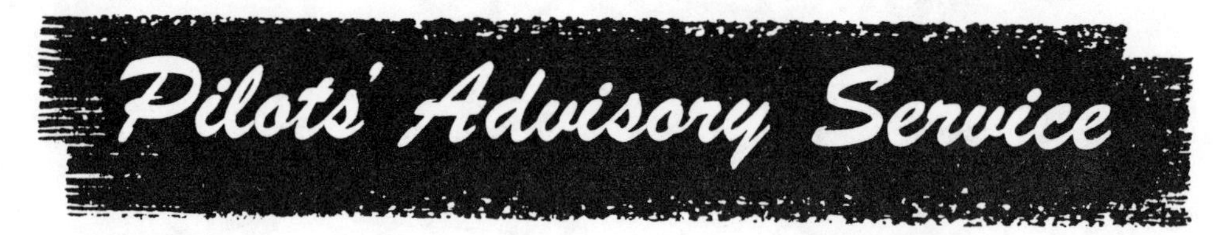


Pilot's Advisory Service is a nationwide service available to all Army and Navy pilots on point-to-point flights. Army Flight Control Centers operate alongside all CAA Airway Traffic Control Centers in the United States.

Your flight plan is transmitted from your point of departure to the Flight Control Center in your area. Trained officers, on duty 24 hours a day:

1. Trace your flight from takeoff to landing.
2. Analyze known weather and traffic conditions: anticipate weather hazards which might affect your flight as you have planned it.
3. Advise you of such hazards and conditions.
4. Answer your specific requests for information or suggestions concerning your flight.

If unforeseen weather, other flight hazards, or field conditions at your destination or alternate develop, Flight Control will advise you. If you're guarding the frequency of the nearest communications station you will hear a series of code dits. Then, "Army Flight Control advises" followed by the message.

1 NOTE ON FORM 23 THE STATIONS YOU WILL TUNE TO EN ROUTE.

2 FOLLOW YOUR FLIGHT PLAN.

3 MAKE POSITION REPORTS FREQUENTLY.

4 KEEP AN ALERT LISTENING WATCH.

5 REQUEST INFORMATION FROM ARMY FLIGHT CONTROL AT FIRST SIGN OF TROUBLE.

# Air Traffic Rules

1. **General:**

**a. Reckless Operation.** An AAF pilot will not operate aircraft in a reckless or careless manner, or so as to endanger friendly aircraft in the air, or friendly aircraft, persons, or property, on the ground.

**b. Proximity to Other Aircraft.** No aircraft will be flown closer than 500 feet to any other aircraft in flight, except when two or more aircraft are flown in duly authorized formation. On authorized formation flights, aircraft will not be flown closer to each other than the distance of one-half the wing-span of the largest aircraft concerned.

2. **Airspace Reservation.** An AAF pilot will not operate aircraft within any airspace reservation set aside by order of the President of the United States.

3. **Air Traffic Control Procedures.** AAF pilots will comply with air traffic control procedures established by competent authority and Standard Airport Traffic Control Procedures as approved by Army, Navy, and CAA, effective 1 January 1944.

4. **Over-the-top Flight.** No flight of aircraft, unless conducted under instrument flight rules, will be made over broken clouds or stretches of solid overcast, unless the attitude of the aircraft and its flight path can at all times be controlled by visual reference to the ground or water and unless ascent or descent can be made to or from the flight level without approaching closer than 2,000 feet horizontally to the cloud formation.

5. **Right-of-way.** The following rules will govern aircraft right-of-way:

**a. Order.** Aircraft in flight will have right-of-way in the following order:

**Balloons,** fixed or free (an airship not under control is classed as a free balloon).

**Gliders.**

**Airships.**

**Airplanes,** including rotorplanes.

**b. Crossing.** When two aircraft are on crossing courses at approximately the same altitude, the aircraft which has the other on its left will have right-of-way, and the other aircraft will give way.

**c. Approaching head-on.** When two aircraft are approaching head-on, or approximately so, and there is danger of collision, each will alter its course to the right so that they will pass each other at a distance of at least 500 feet.

**d. Overtaking.** An overtaken aircraft will have right-of-way and the overtaking aircraft will keep clear of the overtaken aircraft by altering its own course to the right.

**e. Landing.** Landing aircraft while maintaining a straight approach course for the last 1,000 feet before crossing the airport boundary will have right-of-way over all other aircraft, except aircraft landing in distress.

**f. Distress landing.** An aircraft in distress shall have right-of-way in attempting to land.

**g. Duty to give way.** When landing, or maneuvering in preparation to land, it will be the duty of the aircraft at the higher altitude to avoid the aircraft in the lower altitude.

6. **Right side traffic.** Aircraft or formations of aircraft operating along an airway will keep to the right of the radio range course projected along the airway, or if no radio range course is projected along the airway, will keep to the right of the center line of the airway, except:

**a.** When impracticable for reasons of safety;

**b.** When otherwise instructed or authorized by an airway traffic control center.

7. **Landing Area Rules:**

**a.** No aircraft will be taxied until the person in charge has ascertained that there will be no danger of collision with any objects in the immediate area. Aircraft being taxied within 100 feet of the parking area (except in established taxi lanes) or within the parking area will have a man to one wing tip to direct and assist in the taxiing operation. When this is impractical, due to the size or type of aircraft, a man will precede the aircraft at such distance that he is at all times within the view of the person taxiing the aircraft.

Station commanding officers are authorized to establish taxi lanes which will be clearly marked. Taxiing speed and procedures at any station will be as directed by the commanding officer or other competent authority and will be such as to insure reasonable and safe operation. Aircraft operation and movement on the ground will be governed by the provisions of AAF Reg. 62-10.

**b.** Landings and takeoffs will be made on the runway or landing strip most nearly aligned with the wind or when winds are light in the direction indicated by a controlled T or similar device, if available, unless local or operating conditions make landings and takeoffs in any direction inadvisable or unless exceptions are authorized by an air traffic control tower operator.

**c.** No takeoffs will be commenced until the pilot has ascertained that there is no danger of collision with any object in the takeoff path.

**d.** No turn will be made after takeoff until the airport boundary has been reached and the pilot has attained an altitude of at least 500 feet and has ascertained that there will be no danger in turning into the path of a following aircraft, unless exceptions are authorized by an air traffic control operator.

**e.** During and after a takeoff or when approaching, the pilot of an aircraft equipped with a functioning radio receiver will guard the air traffic control tower frequency until the air traffic control operator signs off.

8. **Flight in range approach channel, unless on an approved flight plan:**

**a.** Flight in a range approach channel will be below 1,500 feet except to enter the channel on the right side as determined by the direction of flight along such channel, and then continue along the channel in normal cruising flight.

**b.** Flight below 1,500 feet within the range approach channel will be confined to crossing the channel at an angle of not less than 45°, and to those maneuvers incidental to takeoffs and landings.

9. **Aircraft departing from or arriving at a controlled airport.**

Aircraft departing from or arriving at a controlled airport will take precedence over other air traffic within the control zone of such airport.

10. **Acrobatic flight.** No pilot will perform acrobatics:

**a.** At any height whatsoever over a congested area of any city, town, or settlement, or over any open-air assembly of persons.

**b.** Within the area included within the radius of five miles of the center of any airport except when such airport is within an established local flying area; or within the limits of a civil airway except within that portion lying adjacent to, but outside of, a range approach channel.

**c.** At any place other than unrestricted areas.

**d.** Unless the maneuvers can be completed and the aircraft under complete control at or above 1,500 feet (altitude above the surrounding terrain).

**e.** At any place, unless the visibility is at least 3 miles and the ceiling at least 3,000 feet, and unless the pilot has first ascertained that there is no danger of collision with other aircraft.

11. **Light and signal rules.** Between sunset and sunrise all aircraft in flight, or stationary, and likely to cause a traffic hazard, will show the regulation position lights.

12. **Fog signals.** Unless military necessity requires otherwise, an aircraft on the water in navigation lanes in fog, mist, or heavy weather will signal its presence by a sound device emitting a signal for about five seconds at one-minute intervals.

13. **Distress signals.** The following signals, separately or together, will, where practicable, be used in case of distress. **a.** The international signal, SOS by radio: In radio-telephony, the spoken expression MAYDAY (corresponding to the French pronunciation of the expression "m'aider"). When, owing to the rapidity of the maneuvers to be accomplished,

an aircraft is unable to transmit the intended message, the signal PAN not followed by a message retains such meaning. **b.** The international code flag signal of distress, NC. **c.** A square flag having a ball, or anything resembling a ball, above or below it.

14. **Aircraft on water.** Seaplanes on the water will navigate according to the laws and regulations of the United States governing the navigation and operation of watercraft, except as otherwise specifically authorized.

## FLIGHT RULES

15. **Aircraft approaching and circling:**

**a.** Aircraft approaching and circling for a landing or making contact flight below 1,500 feet above the ground or water within three (3) miles horizontally of the center of an airport or landing area will circle the airport or other landing area sufficiently to observe other traffic, unless the pilot receives other instructions from the air traffic control operator. All circles, whether approaching for a landing or after takeoff, will be made to the left, unless the pilot receives other instructions from the air traffic control operator, or unless in the interest of safety a different procedure has been prescribed for the particular airport or landing area.

**b.** Aircraft approaching for a landing will, unless impracticable, or unless exceptions are authorized by an air traffic control operator, maintain a straight approach course for the last 1,000 feet before crossing the airport boundary.

16. **Minimum altitudes of flight:**

**a.** Except during takeoff and landing, aircraft will not be operated:

(1) Below the following altitudes:

(a) 1,000 feet above any building, house, boat, vehicle, or other obstructions to flight.

(b) At an altitude above the congested sections of cities, towns, or settlements to permit an emergency landing outside of such sections in the event of complete power failure.

(c) 1,000 feet above any open-air assembly of persons.

(d) 500 feet above the ground elsewhere than as specified above.

(2) Within 500 feet of any obstruction to flight.

**b.** Any maneuver may be conducted at such altitude above the ground or water as is necessary for its proper execution in places other than specified above, when such maneuver is required to accomplish an ordered tactical flight, engineering, or training mission.

17. **Contact flight.** Pilots of military aircraft flying under contact conditions, except when engaged (1) in practicing takeoffs and landings under contact flight rules at an altitude of not more than 1,500 feet above the airport from which such practice is conducted, or (2) in flight in established local flying areas, will:

**a.** File a flight plan as defined in paragraph 41 with an airway traffic control center. In case of a formation flight, only one flight plan need be filed, which plan will indicate the number of aircraft making the flight.

**b.** Except for necessary ascent and descent, maintain flight altitudes on civil airways as nearly as existing conditions permit, as follows:

Eastbound flights on green and red civil airways and northbound flights on amber and blue civil airways will be conducted at the odd 1,000 foot levels (such as 5,000, 7,000, and 9,000 feet). Westbound flights on green and red civil airways and southbound flights on amber and blue civil airways shall be conducted at the even 1,000 foot levels (such as 4,000, 6,000, and 8,000 feet).

Do not cross a civil airway at an angle of less than 45° unless otherwise instructed.

18. **Proximity to cloud formation.** Aircraft will not be flown closer than 500 feet vertically to an overcast or cloud formation nor closer than 2,000 feet horizontally to a cloud formation.

19. **Weather minimums.** The following weather minimums will govern flights made in accordance

with contact flight rules, provided that, in the interest of safety, an airway traffic control center may restrict or suspend contact flight operation within the airway traffic control area of such center:

**a. Within control zones.** Aircraft will not be flown within a control zone (or within 3 miles horizontally of the center of a military airdrome) unless the ceiling is at least 1,000 feet and the visibility is at least 3 miles: **Provided,** that an airport traffic control-tower operator on duty in an airport control tower may authorize flight at altitudes of 1,000 feet or less above the ground or water when the visibility is less than 3 miles but not less than 1 mile.

**b. Outside control zones.** Aircraft will not be flown at or below 1,000 feet above the ground or water unless the visibility during the hours of daylight is at least 1 mile and during the hours of darkness is at least 2 miles and will not be flown above 1,000 feet above the ground or water, unless the visibility (forward) is at least 3 miles at flight altitude.

**c. Ferry Flights of Military Necessity.** Notwithstanding any of the foregoing provisions of this section, AAF aircraft making ferry flights, in case of military necessity, may, in compliance with all other contact flight rules:

(1) Take off from, or fly over airports where the local visibility conditions are below the minimum hereinbefore established for operations under such flight rules, provided the visibility elsewhere along the civil airways to be flown is equal to or above such minimum.

(2) Be cleared to fly above an overcast in the clear, provided: (a) The pilot holds a currently effective instrument rating. (b) The aircraft is equipped with a radio receiver capable of tuning through the 200 to 400 kilocycle range. (c) Flight on top of the overcast does not exceed one hour. (d) Visibility at flight altitude is at least two miles. (e) The ceilings at the points of departure and destination are unlimited, and the weather forecasts indicate that the ceiling at the point of destination will remain unlimited until the arrival of the flight thereat.

**d.** Requests for authority to deviate from minimums specified in **a** and **b** above when deemed necessary because of peculiar local conditions, will be submitted to the Commanding General, AAF, Attention: Chief of Flying Safety.

20. **Weather changes.** If weather conditions below the minimum prescribed in paragraph 19 are encountered en route, a landing will be made at the nearest airport at which weather conditions are equal to or better than those prescribed in paragraph 19, or the flight may be altered so that it may be made in weather conditions as good as, or better than, such minimums, unless such flight can and does proceed in accordance with the instrument flight rules prescribed in "INSTRUMENT FLIGHT RULES."

## Instrument FLIGHT RULES

21. **Pilot.** A pilot may, and will be encouraged to, engage in instrument flight provided he holds currently effective Instrument Pilot Certificate AAF Form 8 (white) or AAF Form 8A (green). Currently effective Certificates of Competency, signifying qualification to engage in flight under instrument conditions, issued by the Royal Air Force Transport Command, or the Royal Canadian Air Force, and held by pilots of such air forces, will be considered as equivalent to Instrument Pilot Certificate AAF Form 8 (white). (As to pilots of the Royal Canadian Air Force, see paragraph 8, AAF Regulation 61-2B.)

22. **Weather Minimums.** The following weather minimums will govern:

**a.** For holders of Instrument Pilot Certificate, AAF Form 8A (green):

No takeoff will be made when the ceiling is less than 200 feet or the visibility is less than one-half mile, except that Commanding Officers of OTU, RTU, Tactical and Training Units, may authorize takeoffs at minimums less than 200 feet and one-half mile visibility, provided such takeoffs are under the direct supervision of competent and duly authorized instructors of the training organization. (Training under these conditions is considered necessary because tactical pilots will be required to take off under actual weather conditions of less than 200 feet and one-half mile visibility during combat missions.)

No landing weather minimums are applicable to pilots who hold a currently effective Instrument Pilot Certificate, AAF Form 8A (green), and clearance will depend upon the judgment and be the responsibility of such pilot.

**b.** For holders of Instrument Pilot Certificate, AAF Form 8 (white):

(1) No takeoff will be made when ceiling is less than 500 feet or visibility is less than one mile.

(2) No clearance will be issued for a flight to a destination at which ceilings or visibilities are reported to be less than those established by standard Instrument Approach Procedures, TO 08-15-3, for the particular station cleared for and forecast to remain steady or improving until the expected arrival thereat, and in no circumstances will any clearance be authorized for flight to a destination where visibility is reported or forecast as being less than one mile. In case no weather minimums for the particular station to be cleared for are published in TO 08-15-3, the following minimums will apply:

| DAY | | NIGHT | |
|---|---|---|---|
| **Ceiling** | **Visibility** | **Ceiling** | **Visibility** |
| 800 | 1 | 1000 | 2 |
| 700 | 1½ | 800 | 3 |
| 600 | 2 | | |
| 500 | 3 | | |

**c.** In the case of Army contract air transportation service, weather minimums will be those set forth in the air carrier's operations manual and approved by the Air Transport Command. At stations where weather minimums have not been previously submitted by the air carrier and approved by the Air Transport Command, the minimums will be those specified in subparagraph **b,** above.

23. **Flight rules.** In addition to general flight rules or any special air traffic rules which apply, instrument flight rules will govern the following:

**a.** Flight in closer proximity to cloud formation than 500 feet vertically to an overcast or cloud formation, or closer than 2,000 feet horizontally to a cloud formation.

**b.** Flight in weather conditions worse than those described in paragraph 22.

24. **Aircraft Clearances and Flight Plan.** Aircraft clearances and flight plans will be governed by AAF Reg. 15-23.

25. **Alternate airport.** No takeoff of aircraft will be made unless:

**a.** The alternate airport named in the flight plan has a landing area suitable for the aircraft to be used, and

**b.** Weather reports and forecasts indicate that the weather conditions at the alternate airport will remain at or above the minimum specified in **c** or **d** of this paragraph, until the arrival of the aircraft thereat.

**c.** The alternate airport is equipped with a radio directional aid to air navigation in operation and there is at such alternate airport a ceiling of at least 2,000 feet and a visibility of at least three miles if an overcast exists; or a ceiling of at least 1,500 feet and a visibility of at least three miles if broken clouds exist; or

**d.** If the alternate airport is not equipped with a radio directional aid to air navigation, there is at such alternate airport an unlimited ceiling and a visibility of at least three miles.

**e.** In the case of Army contract air transportation service, weather minimums for alternate airports will be those set forth in the air carrier's operations manual and approved by the Air Transport Command. At stations where weather minimums have not been previously submitted by the air carrier and approved by the Air Transport Command, the minimums will be those specified in subparagraphs **c** and **d,** above.

**f.** The aircraft is provided with fuel and oil sufficient, considering the wind and other weather conditions forecast for the flight:

(1) To complete such flight to the point of the first intended landing, and thereafter.

(2) To fly to and land at the alternate airport designated in the approved flight plan, and thereafter.

(3) To fly, at normal cruising consumption, for a period of 45 minutes.

26. **Communication contacts.** The pilot will maintain a continuous listening watch on the appropriate radio frequency and will contact and report as soon as possible to the appropriate communications station the time and altitude of passing each radio fix or other check point established within the civil airways or specified in the flight plan.

27. **Communications failure.** In the event of failure of aircraft two-way communication equipment or in the event that the pilot does not receive radio signals sufficient to permit him to maintain instrument navigation, one of the following procedures

will be observed:

a. Proceed in accordance with contact rules.

b. Effect a landing at the nearest suitable airport at which favorable weather conditions exist.

c. When sufficient radio signals are received, proceed according to flight plan, including any amending instructions issued and acknowledged en route.

28. **Crossing an airway.** Unless otherwise instructed, a civil airway will not be crossed at an angle of less than 45° to such airway.

29. **Flight altitude.** Exclusive of taking off from or landing upon an airport or other landing area, aircraft making an instrument flight will not be flown below 1,000 feet above the ground or water nor within 1,000 feet of a mountain, hill, or other obstruction.

30. **Let-down procedures.** Descent through an overcast to contact flight condition at other than a range station for which a let-down procedure has been established, will not be attempted unless the pilot can specifically identify his position by means of navigational aids and has definite knowledge of the terrain.

31. **Airport control tower.** An airport control tower is an establishment equipped to allow the operator thereof to control air traffic in the immediate vicinity of an airport.

32. **Airway traffic control area.** An airway traffic control area is an area within the limits of an airway designated by the Administrator of Civil Aeronautics, or by competent military authority, over which an airway traffic control center exercises control.

33. **Airway traffic control center.** An airway traffic control center is a station operated by the Administrator of Civil Aeronautics, or by the military forces, for the purpose of controlling air traffic on established airways.

34. **Airway communications station.** An airway communications station is an airway radio, teletype, or other communications station operated by the Administrator of Civil Aeronautics, or by the military forces.

35. **Alternate airport.** An alternate airport is an airport, other than the point of first intended landing, specified in the flight plan, and to which the flight may proceed in case of emergency.

36. **Ceiling unlimited.** A ceiling is considered unlimited when clouds cover less than one-half of the sky or when the base of the clouds is more than 9,750 feet above the point of observation.

37. **Control airport.** A control airport is an airport where, in the interest of safety, air traffic is controlled from the ground by means of radio communication or visual signals.

38. **Contact flight.** Contact flight is flight conducted in such a manner that ground and water within gliding distance of the aircraft can at all times be used for visual reference.

39. **Control zone.** A control zone is the airspace above that area on the surface of the earth within 3 miles horizontally of the center of an airport designated by the Administrator of Civil Aeronautics as a control airport, and within one-half mile of a line extending from the center of such airport to the radio range station established for the purpose of directing air traffic to such airport.

40. **Cruising altitude.** Cruising altitude is a flight altitude, measured in feet above sea level, proposed for that part of a flight from point to point during which a constant altitude will be maintained.

41. **Flight plan.** A plan of flight filed with an Army Airway Control Center, or the Administrator of Civil Aeronautics, in such manner and containing such information as may be required by Section D of AAF Form 23.

42. **Instrument flight.** That portion of any flight which cannot be made by visual reference to ground or water within gliding distance of the aircraft, or which approaches closer than 500 feet vertically to the base of an overcast, or 2,000 feet horizontally to a cloud formation.

43. **Radio fix.** A radio fix is a geographical location on an airway above which the position of an aircraft in flight can be accurately determined by

means of radio only, such as: a cone of silence marker, Z type marker, fan type marker, or intersection of radio range "on course" signals.

44. **Radio range station.** A radio range station is a radio station from which radio signals are emitted for the purpose of assisting an aircraft to maintain a course.

45. **Range approach channel.** A range approach channel is the airspace above the ground or water below 17,000 feet above sea level located within 2 miles of either side of the center of the "on course" signal of any designated leg or legs of a radio range station serving a control airport, and extending along such leg or legs from such radio range station for a distance of 15 miles.

46. **Visibility.** Visibility is the greatest mean horizontal distance toward at least 50 per cent of the horizon at which conspicuous objects can be readily identified by the naked eye.

47. **Zone of intersection.** A zone of intersection is that part of a civil airway which overlaps and lies within any part of any other civil airway.

48. **Parking area.** Those portions of airport aprons, ramps, lines, or other spaces, so designated by the commanding officer and reserved for the open-air parking of friendly aircraft.

## NONAPPLICATION OF AIR TRAFFIC RULES

49. These air traffic rules will not apply in the following cases:

**a.** When special circumstances render nonobservance necessary to avoid immediate danger, or when such nonobservance is required because of stress of weather conditions which could not reasonably have been foreseen, or for other unavoidable causes. Such nonobservance, including the emergency making the nonobservance necessary, the results accomplished by such nonobservance, and when regular observance was resumed after the emergency had passed, will be reported in writing in full detail within 24 hours to the operations officer at the point of first landing after such nonobservance, which report will be transmitted by the operations officer without delay to the commanding officer of the pilot's home station. The commanding officer's remarks will be placed upon such reports, and the report forwarded to the Commanding General, AAF, Attention: Chief of Flying Safety.

**b. When military necessity requires.** If the military necessity is such as to endanger other aircraft operating on airways, the appropriate airway traffic control center will be notified.

**c. When in the interests of safety.** C.O.'s are authorized to prescribe more stringent regulations for any airport under their control.

**d. When Necessary to Carry out Training Directives.** Aerial reviews and parade flight formations, unless in the form of a tactical combat maneuver or demonstration, will not be deemed as necessary to carry out training directives, and will not be authorized. The provisions of this sub-paragraph will not be invoked to authorize deviations from the provisions of paragraphs 1 and 2 above. When deviation from the provisions of these regulations is deemed necessary for flying training activities, commanding officers will be responsible for: (1) the coordination of such training with all other activities concerned in the local area involved, (2) the notification of the appropriate airway traffic control center and (3) an entry, "Authorized Deviations" being made in the Remarks column of Section D, "Flight Plan", on the AAF Form 23 for the flight.

REFERENCE: AAF Regs. 60-16, 6 March 1944; 60-16A, 15 April 1944; 60-16B, 1 May 1944; 60-16C, 16 May 1944.

# CLEARANCE FOR AIRCRAFT

**Procedures governing the clearance of aircraft are contained in AAF Reg. 15-23, along with directions for filling out your individual clearance, the AAF Form 23. A sample of this form is shown in PIF 1-13-6.**

Contact Flight Rules (CFR) and Instrument Flight Rules (IFR) published in AAF Reg. 60-16 (See PIF 2-1) and rules published in AAF Reg. 15-23 will govern all aircraft clearances.

All pilots will use Form 23 for clearance on all flights except those specifically exempted by AAF Reg. 15-23.

### Weather Data and Flight Plan

Step by step procedure in filling out Form 23 is shown in the accompanying illustrations.

If there is no qualified weather personnel on duty, check the available weather information, insert a statement in Section C that you have done so, and initial same.

If you are a pilot with clearing authority, you may either present your Form 23 to the weather office in the usual manner, or simply initial Section C, signifying that you have current weather information.

After completion of Section C complete Section D by indicating whether your flight will be CFR or IFR.

If your flight is to be on instruments you must pick

## THE STORY OF THE FORM 23

1. OPERATIONS MAKES FIRST ENTRIES IN SECTION A OF YOUR FORM 23

2. YOU MAKE ENTRIES IN SECTION B AND FILL IN "AIRPORT OF FIRST INTENDED LANDING" IN SECTION D

3. WEATHER OFFICE MAKES ENTRIES IN SECTION C, "WEATHER DATA"

4. YOU THEN COMPLETE SECTION D, "FLIGHT PLAN"

5. CLEARING AUTHORITY SIGNS FORM

when regular observance was resumed after the emergency had passed, shall be reported in writing in full detail within 24 hours to the operations officer at the point of first landing after such nonobservance, which report shall be transmitted by the operations officer without delay to the commanding officer of the pilot's home station. The commanding officer's remarks will be placed upon such reports, and the report forwarded to Hq. AAF, Chief, Office of Flying Safety, Nissen Building, Winston-Salem, N. C.

**b. When military necessity requires.** If the military necessity is such as to endanger other aircraft operating on airways, the appropriate airway traffic control center will be notified.

**c. When in the interests of safety.** C.O.'s are authorized to prescribe more stringent regulations for any airport under their control.

**d. When necessary to carry out training directives.** When deviation from the provisions of these regulations is deemed necessary for flying training activities, commanding officers will be responsible for (1) the coordination of such training with all other activities concerned in the local area involved; and (2) for the notification of the appropriate airway traffic control center; and (3) the report of such deviation and the necessity therefore, in writing to the Commanding General, Army Air Forces.

# COMPARISON: *Contact and Instrument Rules*

| ITEM | CONTACT | INSTRUMENT |
|---|---|---|
| FLIGHT PLAN | Form 23, no alternate airport, radio, nor altitude necessary. | Form 23, including alternate airport radio equipment, and valid instrument card. (Green or White) |
| CHANGES IN FLIGHT PLAN | If two-way radio installed, advise nearest radio station. If no radio, land at nearest airport. | Contact nearest radio station to obtain ATC permission. In emergency, alter plan and inform as soon as possible. |
| FLYING ON AIRWAYS | To the right and away from "On Course" signal. | To the right, and close to "On Course" signal. If coming in to land, fly on "On Course." |
| FUEL REQUIREMENTS | Sufficient for proposed flight under any condition of wind, weather, and load. | Sufficient to make the alternate airport, with enough extra to sustain flight for 45 minutes. |
| LISTENING WATCH | Not necessary, but in general practice on airways and their approaches if radio installed. | At all times. |
| DISTANCE FROM BASE OF CLOUDS | 500 ft. below ceiling. | No rule. |
| ALTITUDE, MAXIMUM | None unless specified on Flight Plan. | As specified on Flight Plan. |
| ALTITUDE, MINIMUM | 500 ft. (1,000 ft. over congested areas, danger areas, & penal institutions.) | 1,000 ft. above terrain. 1,000 ft. distant from any mountain or obstacle. |
| VISIBILITY 1. In a Control Zone | 3 miles, (day or night). | No rule. |
| 2. On airway; not in a Control Zone | 1 mile (day), 2 miles (night). | No rule. |
| 3. On airway, not in a Control Zone but over 1,000 ft. alt. | 3 miles at flight altitude, (day or night). | No rule. |

### Aircraft Clearance

The pilot as well as the Operations Officer is concerned with AAF Form No. 23 (Aircraft Clearance). This form is made up of three distinct parts, each going to different persons and having a distinct function or use in the clearing and landing of an airplane. These parts are the Aircraft Clearance, Departure Record, and Departure and Arrival Report. The Departure Record is a stub at the bottom of the original Aircraft Clearance. The Departure and Arrival Report is a duplicate of the Aircraft Clearance, except that the lower part of the form contains the Arrival Record.

### AAF Regulations

These regulations direct this form be used by all activities of the Army Air Forces in clearing pilots for other than local flights, and for local flights when weather conditions require an instrument clearance. This requirement may be augmented by any additional requirements made by the Operations Officer (acting for the C.O.) of a specified field. Some Operations Officers require that AAF Form 23 be used to clear all flights, regardless of type.

### Weather Entry

The weather entry is completed by the weather forecaster in space provided on Form 23. It should contain information concerning existing classifications of the en route weather and a forecast of conditions to be expected for the duration of the flight plus two hours. Included in this forecast should be a forecast for the alternate route.

The Weather Officer is not authorized to tell the pilot he cannot be cleared. Clearing authority rests solely with the Operations Officer (acting for the Station C.O.) and the Weather Officer acts only as an adviser.

### Flight Rules and Weather Minimums

The Flight Rules and Weather Minimums as determined by the Civil Aeronautics Board and by Army Air Forces Stations will apply in the preparation of the Form 23.

### Authority for Flight Clearance

Command pilots and senior pilots are authorized to approve clearance for individual flights of units which they accompany and which are directly under their flight controls. Pilots of lower rating will be cleared only by the Operations Offcer, or his authorized representative, at the Army Air Forces station concerned.

### Flight Clearance by Radio (Sometimes Known as "Air Clearance")

Command pilots and senior pilots for individual flights, and for flights of units which they accompany and which are directly under their flight control, are authorized, while in flight, to approve clearance by radio.

an alternate airport at which weather conditions meet alternate airport weather requirements. You must also have enough fuel to (1) fly to airport of first intended landing, then (2) proceed to alternate airport if first airport is closed, and (3) remain aloft over alternate for at least 45 minutes.

### Clearance

Clearances will normally be valid for one hour from proposed takeoff time unless cancelled by competent authority.

**Command Pilots,** who for purposes of clearance will be considered to possess a green Instrument Pilot Certificate (AAF Form 8A) are authorized to:

1. Act as their own clearance authorities for IFR or CFR flights.
2. Clear individuals for CFR flights which they accompany or which are under their control.

**Senior Pilots, Senior Service Pilots, and Civilian Contract Pilots of Air Transport Command** may act as their own clearance authorities, or may clear individuals on flights they accompany, or which are under their direct control, for CFR flights. These three classes of pilots may, if they hold valid white Instrument Pilot Certificate (AAF Form 8), clear only themselves for IFR flights.

**Pilots holding valid green cards** may clear themselves on CFR or IFR flights.

**In all other cases pilots can be cleared only by the authorized clearing authority.**

You cannot be cleared for instrument flight if you don't hold a valid white certificate.

Even if you hold a white certificate, but it shows total pilot time less than 500 hours, you cannot be cleared for instrument flight if ceiling or visibility is at or near instrument flight minimums.

**No matter what your rating nor the extent of your clearing authority, don't taxi nor take off until authorized by the tower.**

### No AAF Operations

Before departing from a location where there is no AAF operations office you must file a flight plan with an AAF Flight Control Center, using any Army approved communications facility. Army Flight Control Centers are located at all CAA Airway Traffic Control Centers. Unless you are your own clearance authority, you must wait for Army Flight Control's approval before taking off. **All pilots are urged, however, to wait for any further information from Army Flight Control.**

If approved communications facilities are not available, you may take off under CFR flight conditions only, and fly to the nearest practicable point where they are available. You may then request a change of flight plan in flight, or land and file a flight plan in the usual manner.

### Formation Flights

If you are the Flight Commander of a formation flight, make out a Form 23 for your own airplane. Put the words "Flight Commander" after your signature, and on the form or an attached list enter: model and serial number of each plane in the formation; name, rank, serial number of each crew member in the flight; name, rank of other occupants. Then brief pilots and see that they have proper maps and instrument ratings. Certify to this action under "Remarks."

FILL IN SECTION F, "DEPARTURE RECORD," HAND IT TO LINE CREWMAN. HOLD DUPLICATE YELLOW FORM, "ARRIVAL REPORT."

DON'T TAKE OFF UNTIL TOWER GIVES YOU FINAL GO-AHEAD

AFTER ARRIVAL FILL IN G (YELLOW FORM), "ARRIVAL REPORT," HAND TO CREWMAN

# RADIO CHANGE IN FLIGHT PLAN

Any deviation from the original flight plan you have filed on your Form 23 requires a new clearance by radio. Note: Changing your altitude or route under CFR conditions is not considered a deviation.

*Unless you are a pilot with clearing authority, here is what you must do:*

Request approval for your intended change **before you make it,** using the nearest available radio facility (consult radio facility charts).

Report (1) your position, (2) details of the change you wish to make, (3) remaining hours and minutes of fuel aboard, and (4) estimated time en route from current position to destination.

The reply you get, after your request has been processed and approved by the proper agencies, will begin with either the words "Army Flight Control approves - -" or "Airway Traffic Control approves - -". This is all you need to remember, as the approval you receive is final unless specifically stated otherwise.

**If you are a pilot with clearing authority** and flying under CFR conditions, you can make the change without approval. However, you must report your change and give same information as outlined above. **All pilots, regardless of their rating, will make no change in flight plan when flying under IFR conditions without first obtaining approval.**

In case of an emergency, you will use your own good judgment and take whatever action is necessary. You are required by regulation, however, to report such action as soon as possible through the nearest available radio communications facility.

## How Approval Is Granted

The following information, although you need not remember it in flight nor let it confuse you in any way, is given so that you can better understand how radio requests for changes in flight plan are handled.

A pilot in flight is subject to two types of clearances by radio. One is concerned with **traffic separation** on the airways and is always subject to the approval of CAA Airway Traffic Control Centers. The other is in the nature of a **dispatch clearance** and is granted (or refused) by Army Flight Control.

These two agencies are always located in the same office and coordinate closely. Your request through your nearest available radio facility is relayed by interphone to an Airway Traffic Control Center where the two agencies process it. Airway Traffic Control is concerned only with change of flight plan on the airways under IFR conditions. However, Army Flight Control is concerned with all changes of flight plan, on or off the airways, under CFR or IFR conditions. The approval involved is something that will be determined for you automatically on the ground, depending on the nature of your intended change. Thus when you hear either "Army Flight Control approves - -" or "Airway Traffic Control approves - -" you need not worry whether or not the proper agency has received your request.

## Number of Passengers

The number of persons carried in Army aircraft will not exceed that for which the aircraft is designed, except when necessary in cases of emergency:

1. Involving great physical hardship or possible loss of life.
2. In the performance of urgent military duties.

In the absence of specific instructions from higher authority, the character of the emergency will be determined by the commanding officer of the Army Air Forces station where the flight originates, except when the emergency arises elsewhere than at such a station, in which case its character will be determined by the senior Army Air Forces officer on duty with the flight.

## Travel by Army Aircraft

Passenger space on routine interstation flights of Army aircraft may be utilized for authorized travel of members of the Army of the United States on official duty for the purpose of reducing the expense to the Government of the travel involved or to save time.

Under no consideration will this type of travel impose an added burden in the way of additional flights or the alteration of previously planned flying schedules.

Any exception to these regulations will be made only after a presentation of the case and action thereon by the Commanding General, Army Air Forces.

Commanding officers of AAF stations are authorized to grant exceptions in case of civilians who are to be flight-tested for a military aeronautical pilot rating under the provisions of AAF Reg. 50-7. No release is required.

## On Leave or Furlough

Military, Naval, Marine Corps, and Coast Guard personnel, while on leave of absence, furlough, or on detached service, may be permitted by commanding officers of AAF stations or higher authority to ride as passengers on flights in Army aircraft when such flights are incident to a regularly scheduled mission, and **provided such transportation does not involve additional expense to the Government.**

## Passengers on ATC Aircraft

AAF Reg. 75-3 establishes special regulations for the transportation of passengers on aircraft operated by the Air Transport Command. All aircraft of the Air Transport Command engaged in air transport operations, whether flown by civilian crews under contract or by military crews, are military aircraft and as such are authorized to transport only those passengers who are in possession of military orders expressly authorizing or directing travel by military aircraft or those whose travel will facilitate the prosecution of war activities.

Passengers will be carried on such aircraft only when priority for transportation has been established by the Commanding General, ATC, Washington, D. C., or his authorized representative.

Releases will be required from such passengers as would be required to sign releases in case of travel in other military aircraft.

### No Release Required

Commanding officers of AAF stations or higher authority in the chain of command are authorized to permit the persons listed below to ride as passengers in Army aircraft under their control under proper orders and clearance **without signing a release.**

1. Military personnel of the Army, Navy, Marine Corps, Coast Guard, Women's Army Corps, and commissioned officers of the U. S. Coast and Geodetic Survey.

2. Civilian employees of the War Department, of other Government agencies, of Government contractors, and technical advisors to military authorities engaged in activities for the Army which require such flight.

3. Students, foreign or otherwise, undergoing instruction with or under supervision of the AAF, when such flights are a part of the authorized course of instruction.

4. Red Cross personnel when serving with the armed forces of the United States in the field either at home or abroad.

5. Subject to the approval of the Commanding General, AAF, members of the Cabinet and certain other officials of the Government, members of Congress, and governmental employees.

### Release Required

Commanding officers of AAF stations or higher authority in the chain of command are authorized to permit the persons listed below to ride in Army aircraft under their control, **only, in the case of civilians, upon signing the injury or death release.**

1. Military and civilian personnel of foreign nations in cases involving war activities, including projects carried out under the Lend-Lease Act.

2. On flights which will not extend beyond the local flying area: wives, mothers, and children of military personnel of the U. S. who hold aeronautical ratings and who are required in the performance of their duties to participate in regular and frequent flights. Such children must be ten years of age or over and have the permission of a parent. These flights must not exceed two in number for any person during any calendar year.

3. Subject to the approval of the Commanding General, AAF, representatives of the pictorial and news disseminating agencies on flights which will benefit the War Department. Such flights may be conducted within the continental limits of the U. S. and its territorial possessions.

4. The commanding general of any theater, or any department, base command, defense command, or task force, **outside the continental limits of the United States** may authorize any person to ride as passenger in Army aircraft under his control when this action is necessary or desirable in the Government interest.

5. AAF members of the U. S. Military missions in Latin America are authorized to carry as passengers, on local flights, within the jurisdiction of the mission, distinguished nationals of the country in which they are stationed, on the approval of the U. S. ambassador or minister. At the discretion of the chief of the mission, members of the armed forces of the country in which they are stationed and students who are under their instruction may be carried as passengers only after signing a release.

### Release Required, When Practicable

Commanding Officers of AAF stations or higher authority in the chain of command may permit any person to ride as passenger in Army aircraft, in case of emergency involving catastrophe or possible loss of life, when other means of transportation is not available. **Civilians will be required to execute releases, when practicable.**

REFERENCE: Army Reg. 95-90, AAF Reg. 50-7, AAF Reg. 60-20, AAF Reg. 75-3.

***IMPORTANT***

**Whenever you clear from a station or airport where there are no AAF facilities, you are required to leave with the local operations officer, airport manager, or other suitable person a list of all passengers, showing name and home station of each.**

# AIR SPACE RESERVATIONS AND FLIGHT HAZARDS

Before making a flight, study the map in the Operations Office and ascertain whether your projected flight plan will bring you close to any areas shaded or marked as shown below. If it does, consult the **Weekly Notice To Airmen** and current **Notams** for specific details and familiarize yourself with the location and limits of the area as well as the conditions that apply within it.

These are the types of shadings and warnings you will find:

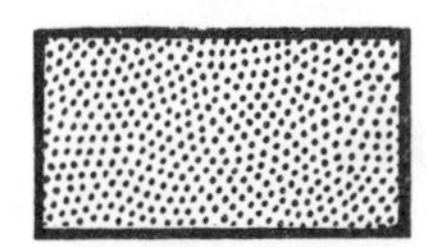

Never Fly Over Prohibited Areas

**1. Prohibited Areas (Air Space Reservations) are areas from which all aircraft, except those engaged in actual defense missions, are prohibited.** At present there are two such areas established in the United States by executive order: **Washington, D. C. and Hyde Park, New York.** The former may be avoided by flying close on course when near the Capitol.

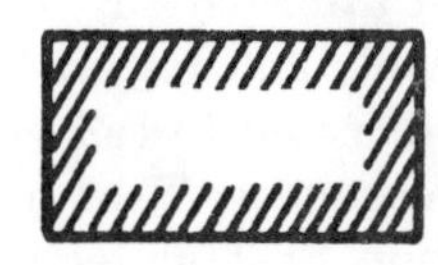

Keep Sharp Lookout in Caution Areas

**2. Caution Areas** are areas in which under CFR there are **visible** hazards. These areas should be avoided if practicable.

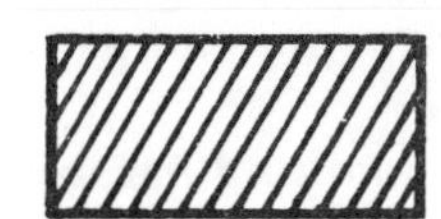

Get Authority to Cross Danger Areas

**3. Danger Areas** are reservations in which there is an **invisible** hazard such as might exist over an artillery or bombing range. No flight of aircraft shall be conducted in this area without specific authority issued by the agency having jurisdiction over the danger area.

Tunneled Airway

**4.** Where a **Tunneled Airway** is indicated on charts, through traffic must utilize airspace between 1500 and 3500 feet above ground.

Red Flag

**5. Red Flagged Items, or Warnings.** This symbol indicates an airport flooded or otherwise closed; range frequency changed; range not operating or course swinging badly; field lights out of commission; special flight restrictions; any decommissioning of a navigational aid.

Orange Flag

**6. Orange Flagged Item, or Caution.** Indicates airport under partial construction; field soft; control inoperative; etc.

NOTE: When entering Vital Defense Areas on contact flight, it is desirable that you contact the nearest radio station and report your position and receive pertinent instructions.

# HAZARDS TO AIR NAVIGATION

Commanding Officers of AAF stations at which there is flying activity, are charged with reporting local conditions affecting flying safety to the proper authorities, so that the information will reach pilots as quickly and easily as possible.

Some of the hazards which you may encounter include: unsafe conditions of landing fields or runways due to construction, repairs, snow, floods, etc.; installation or changes of lighting systems; failure in servicing facilities; congested training areas; aerial, bombing, and gunnery ranges; barrage balloon, searchlight, and certain construction areas; changes in operation of control towers and Army operated radio range or radio beam facilities.

Chief source of information on the existence of hazards or their removal through corrective action is "Weekly Notice to Airmen." Further information can be found in "AAF Radio Facility Charts," "AAF Radio Data and Flight Information," CAA "Danger Areas in Air Navigation", and regional maps. Local operations offices are directed to call your attention to them in issuing clearances. Teletype and Army Airways Communication System facilities are also used to bring them to your attention, and by using Pilots' Advisory Service you can get current news of hazards in the area.

The system used in marking hazards at air fields is discussed in full in "MARKING OBSTRUCTIONS," PIF 2-11-1.

REFERENCE: AAF Regulation 63-1, dated January 1, 1943.

**Caution**

Don't ignore the precautions taken to safeguard you in flight. Learn everything you can, in advance, about what you will be flying over or the field you will be landing on.

# LOW FLYING

### Over Cities

In flying over cities and towns, observe all safety precautions possible to eliminate danger to persons and property on the ground.

Except in taking off from fields located in or near populous areas and in climbing immediately thereafter, always maintain sufficient altitude to glide to a landing field in the event of engine failure.

When a mission prescribed by proper authority requires low-altitude flying, the above instructions may be disregarded; but every precaution will be taken to insure perfect functioning of engines, instruments, etc., and the amount of low flying will be strictly limited to that which is necessary to accomplish the mission.

### Over Crowds

No flying by any type of aircraft will be done at any altitude over or in the immediate vicinity of stadiums, ball parks, fair grounds, or other localities where crowds are gathered, except those around the borders of flying fields witnessing flying demonstrations. When flying from a field where there is a large crowd of people, take precautions to maintain such position and altitude as will insure the safety of persons on the ground. **Never dive lower than 1,000 feet over or in the direction of crowds of people.**

### Over Water

Buzzing water is a particularly dangerous violation of the regulation against unauthorized low flying. It is extremely difficult to judge exactly the distance you are flying above water, particularly when the surface of the water is calm.

An elementary fact about vision is that your judgment of distance is determined largely by the relative size of objects. You make the judgment subconsciously and it is based upon experience accumulated during a lifetime.

A smooth body of water, however, contains no patterns or objects which you can use as points of reference. It is, therefore, easy to misjudge your clearance over water by as much as fifteen or twenty feet—enough to let a propeller strike the water.

The tip of a propeller moving at high speed does not knife through the water, it hits violently, and causes an immediate loss of power; the plane is thrown out of control, and a major catastrophe is almost inevitable.

---

**Remember**

**Unauthorized low flying over cities and towns, diving over crowds, buzzing of any kind, and flying through or over Prohibited Areas are subject to disciplinary action.**

---

PIF 2-7-1
May 1, 1943

# CREW REQUIREMENTS FOR MULTI-ENGINED AIRPLANES

Multi-engined aircraft, the cockpits of which are arranged for side-by-side seating of pilot and co-pilot, ordinarily will be operated with a co-pilot. However, commanding officers may authorize the operation of such aircraft with a minimum pilot crew of one pilot, provided the pilot is accompanied by a crew chief or aerial engineer who is thoroughly familiar with the mechanical operation of the airplane.

Two-engined aircraft with a single pilot cockpit and two-engined advanced training type aircraft may be operated by the pilot only, when in the opinion of the commander concerned other crew members are not required to perform the mission.

Commanders may authorize changes in these requirements in cases of military necessity.

REFERENCE: AAF Regulation 55-5.

# Hand Signals for Taxiing

**THESE SIGNALS, PRESCRIBED BY AAF REGULATION 62-10, DATED 2 AUGUST 1943, REPLACE ALL FORMER TAXI SIGNALS. THEY WILL BE USED BY CREWS OF AAF, USN, USMC, RAF, RCAF, AND RN.**

**A FLAGMAN** with checkered flag will meet aircraft on any landing space where the nature of traffic demands it. He will direct the pilot toward the taxi signalman who will stand with both arms extended full length above his head.

COME AHEAD

To signal turns, signalman will beckon "Come Ahead" with hand on the same side as the wing to be brought around, and point with other hand at wheel to be braked.

**THE SIGNALMAN** will direct taxiing from a position forward of the left wing tip of the airplane, where the pilot can see him easily all the time.

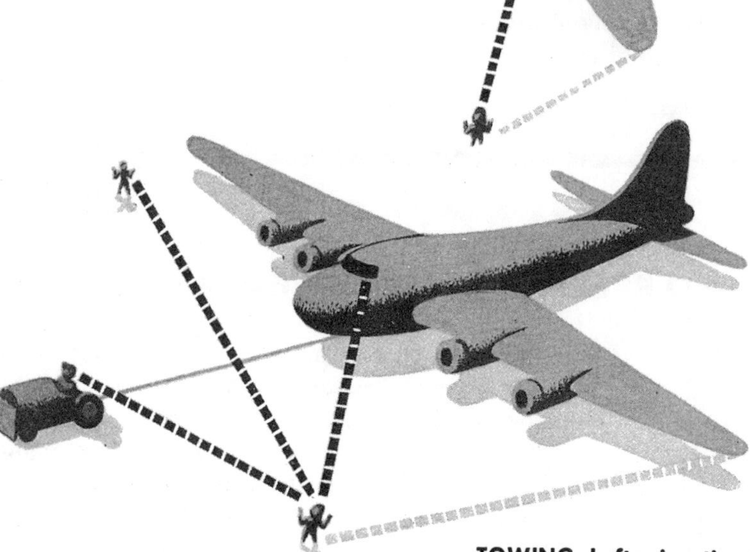

**TOWING.** Left wing tip signalman gives all signals to tractor driver.

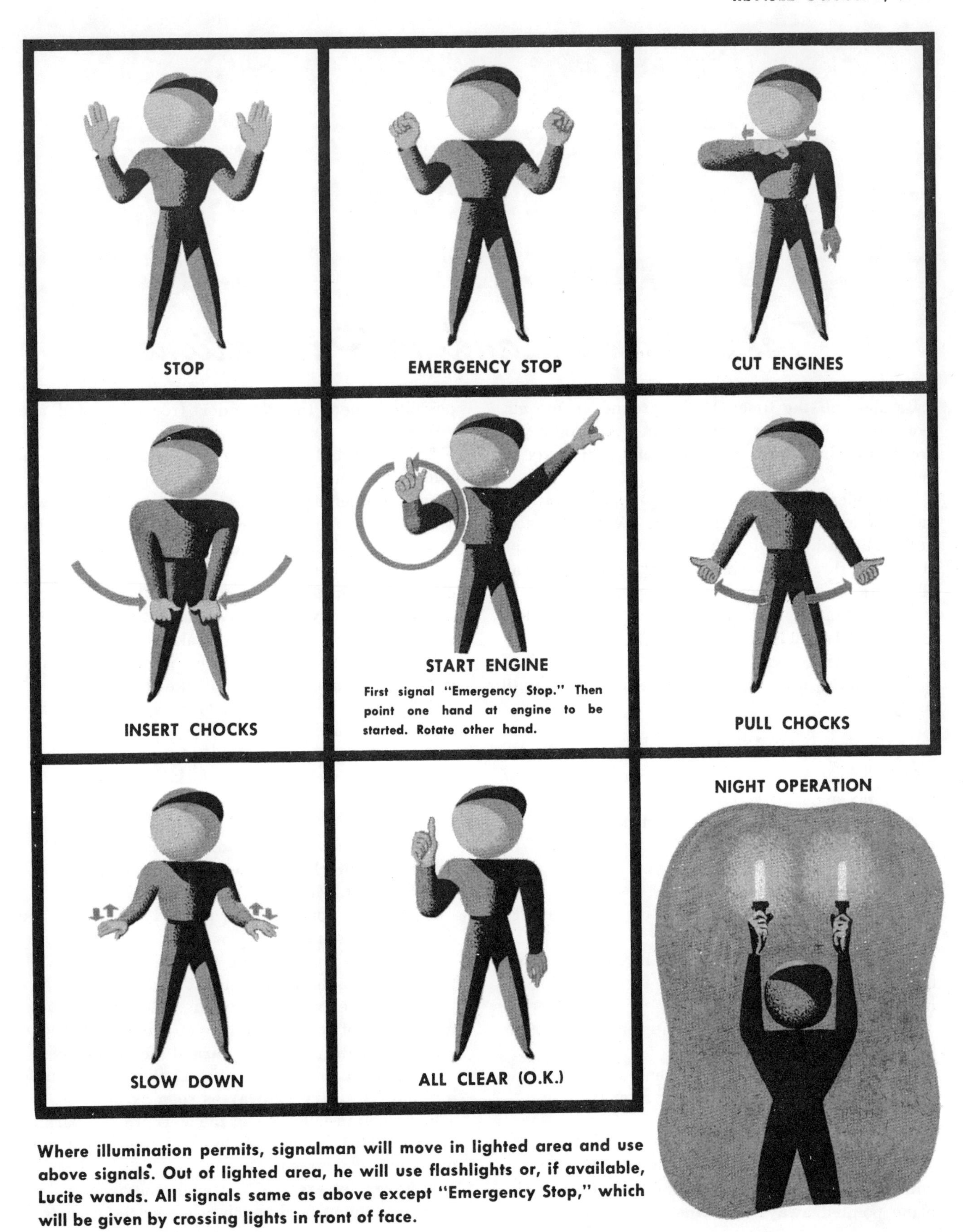


Where illumination permits, signalman will move in lighted area and use above signals. Out of lighted area, he will use flashlights or, if available, Lucite wands. All signals same as above except "Emergency Stop," which will be given by crossing lights in front of face.

# Takeoff and Landing Precautions

**Be alert all the time:** On the ground, in the air, every minute you are in an airplane. As a pilot you are responsible for **your airplane and everybody on board.**

**To be alert all the time** develop a simple routine habit of looking **automatically in the right places** under all conditions. If you maneuver your airplane around its three axes while looking around you can eliminate all blind spots.

Any maneuver that may result in excessive loss of altitude **demands extra alertness.**

**If tower orders you to clear the runway, taxi off unless checks are complete. Never take off until ready.**

### Taxiing

1. Before leaving ramp contact control tower for proper taxiing instructions.
2. **Taxi slowly** and on hard surface areas unless otherwise advised.
3. Don't drag your brakes. If you overheat them they may lock in the air.
4. Use your upwind engine if taxiing multi-engine airplanes crosswind.
5. **Take nothing for granted.** If you can't see, fishtail right and left far enough to **see everything in front of you.**
6. If your radio fails, watch the control tower for instructions by biscuit gun.
7. Remember, over fifty percent of all airplane accidents occur **before takeoff or after landing.**

### Takeoff

1. **Use your checklist prior to takeoff.** Run up your engine or engines with the tail of your airplane pointed **away from other equipment,** and, if possible, where the ground is free from sand and gravel.
2. Don't stand with your engine or engines running for a long period of time. If you must wait for takeoff any length of time, **turn off your engines.**
3. Stay clear of the prevailing runway until the tower says "clear for takeoff."
4. Observe airfield conditions and all boundary obstructions in the proposed takeoff lane.
5. **Look backward, above, and forward** before pulling onto the runway for takeoff.
6. **Use all the runway.** There is never too much runway if an engine fails on takeoff.
7. Be sure your crew is properly stationed and ready for takeoff.
8. Adjust your seat to the proper position. **Check and double check your seat lock.**

### Entering Traffic and Landing

1. Contact tower for landing instructions.
2. **Use your checklist before entering traffic.**
3. **Be alert.** Maintain traffic altitude.
4. Be sure your crew is properly stationed and ready for landing.
5. **Land on the first one third of the runway.** If this is impossible, **go around.** It's no disgrace.
6. Don't relax after landing. Remember, your mission is not complete until you have **parked your airplane on the ramp and turned off the engines.**
7. **Be alert.** Stay within the bounds of safe taxiing, but, **clear the runway as soon as possible.**
8. Follow instructions of the alert crew in parking your airplane. Advise the control tower that your mission is completed.
9. **Except in cases of extreme emergencies** don't land on a flying field which is under the process of

construction, until that field has been declared officially open by the proper authorities. Do not construe this order as preventing the landings, when necessary, of aircraft carrying inspection personnel on official business.

### Precautions on Muddy or Snow-Covered Fields

When airplanes are operated from muddy or snow-covered fields, the landing gear and tail wheel sometimes clog with mud or wet snow which freezes at the lower temperatures encountered after takeoff. When the gear is retracted, there is danger of its freezing in the retracted position. To avoid this retract and lower landing gear several times after takeoff. If it seems likely that the gear will freeze in the retracted position even after the above precautions have been taken, leave the gear lowered.

### Landing with Gear Retracted

Remember that landing wheels-up always results in more or less damage to the airplane, so try everything else first. Then, if the gear won't function properly, and you have fuel enough, try to make the wheels-up landing at a depot or station where there are facilities for extensive repairs.

If you see that you must land with the gear up, and have time, take the following precautions:

1. If carrying bombs, release them in "SAFE" over uninhabited area, at an altitude above 500 feet.
2. If carrying bomb bay tanks, drop them over a suitable area. Leave the bomb bay doors open long enough to get rid of gas fumes.
3. **Close bomb bay doors before landing.**
4. Though flares are not likely to ignite when landing wheels-up, you may drop them from 2000 feet, or at any altitude over suitable terrain.

In emergencies, when you have to land on questionable terrain, or on a field too short for a landing run, extend flaps and land wheels-up.

Exception: In making a forced landing with a B24 the recommended procedure is to **land with gear extended on any type of terrain.**

In case of engine failure with single-engine aircraft on takeoff, if there is not runway enough left, retract wheels and land straight ahead.

### Wheels-up Landing on Paved Runways

Experience has shown that **it is better to make a wheels-up landing on a paved runway** (where available) than on unpaved terrain. A belly landing on soft ground may result in nosing-up, or digging in and stopping abruptly, which may endanger personnel and cause severe damage to the airplane.

### Runway Shoulders

**Stay on the runways.** Shoulders along runways and taxi-strips have been treated to withstand occasional emergency use; but they are not constructed for hard usage, such as parking, taxiing, or landing.

The lights along the outer edge of the runway pavement indicate the full area to be used for all operations. On fields where no lights have been installed stripes have been painted six inches wide and parallel to the outer edge of the pavement. **The areas outside these stripes are for emergency use only.**

REFERENCES: AAF Regulation 60-6; Technical Order 00-25-6; and Technical Order 01-1-10.

# Standard MEMORY CHECKS

USE THE PILOT'S CHECK LIST IN THE COCKPIT. IN ADDITION, IT IS EXCELLENT SAFETY INSURANCE FOR EVERY PILOT TO MEMORIZE THE STANDARD CHECKS GIVEN BELOW.

## BEFORE TAKEOFF

**C** Controls: free and easy.

**I** Instruments and switches: check from left to right.

**G** Gas: proper tank "ON," fuel pressure, mixture control.

**F** Flaps: proper takeoff setting.

**T** Trim tabs: set for takeoff.

**P** Propeller at "INCREASE RPM."

**R** Run-up.

**THROTTLE BRAKE** Firm enough to prevent slipping.

**TAILWHEEL LOCK** Locked when plane is lined up for takeoff.

## BEFORE LANDING

**G** Gas: to fullest tank.

**U** Under-carriage: down, locked, and checked.

**M** Mixture: full rich.

**P** Propeller: "INCREASE RPM" to correct landing position.

**FLAPS** Proper setting for landing.

# MARKING OBSTRUCTIONS

Obstructions in the vicinity of landing fields are marked so that they will be easy for pilots to see as they approach or take off. A standard system of marking is now used by the Army, Navy, and CAA for all uncamouflaged airports in the United States.

Permanent structures are painted orange and white for daytime recognition, marked by red lights so that they are easily spotted at night. Temporary hazards on landing fields are indicated by yellow markers by day and also lighted at night.

Watch for obstructions and hazards when you approach a field, particularly a field new to you. Recognize the markings and appreciate obstruction hazards.

## Standard Markings

Radio towers, masts, flag poles, smoke stacks, water standpipes and other similar structures are painted in alternate bands of orange and white. Mounted water tanks are painted in a checkerboard design of the same colors. The tops of tanks and standpipes are divided into eight segments, which are painted alternate orange and white.

Where hangars or other buildings are situated within a normal landing area or landing approach, so that they constitute a hazard, the roofs are painted with an orange and white checkerboard design so that they stand out clearly.

## Marking by Lights

From sunset to sunrise, and at other times when visibility is poor, obstructions are marked with bright red lights so that they are easy to spot. The lights outline the obstructions from every angle of approach.

Guy cables for structures in the immediate vicinity of landing fields will be marked with a danger cone at the midpoint of each cable.

## Hazardous Areas on Airfields

Holes, soft spots, construction, or other conditions which make any area of a landing field unsafe for use, are marked with yellow flags or pyramids during the day and with red lights at night. The unsafe portions are outlined; and if the area is large, it is also marked with a cross of flags or lights placed in the center of the area.

A large cross of flags, colored material or red lights at each end of a runway, marks it as unsafe for use at the time.

## Along Airways

Any structure which projects above the surrounding terrain so that it can be considered a serious hazard, is marked according to the same system of orange and white markings and red lights, whenever it is possible to do so. However, often it is impractical to mark some structures, so allow altitude sufficient to clear all obstructions and variations in the terrain while in flight.

## Exceptions

Although every attempt is being made to standardize marking and lighting, local conditions and camouflage or blackout requirements make deviations necessary. Until you are familiar with local marking procedure, use extra caution while in flight.

REFERENCE: Army Regulation 95-35, AAF Regulation 62-5, and Technical Order 19-1-20

# DOMESTIC RUNWAY MARKINGS

**Runway and landing strips on uncamouflaged airfields in the zone of the interior are marked to aid in the control of traffic and to help the pilot. Runways are marked in black or white, the color having the greatest contrast to the runway surface. Taxiways are marked in yellow.**

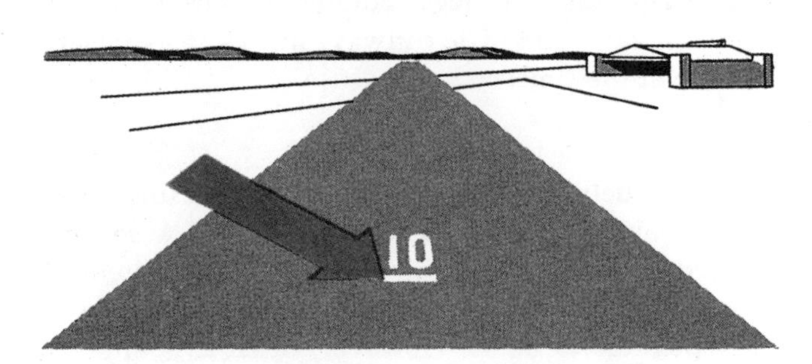


**EACH RUNWAY IS NUMBERED**

The numbers indicate, in units of ten degrees, the approximate magnetic bearing of the runway. A large "R" for Right and "L" for Left, designates parallel runways.

**RUNWAY LENGTH SHOWN BY VERTICAL BARS**

Up to 4500 feet, each vertical bar represents a thousand feet, a half bar, 500 feet. Each five thousand feet is indicated by four vertical bars with a cross bar intersecting the verticals. Length over 5000 feet is indicated to the nearest thousand by additional vertical bars.

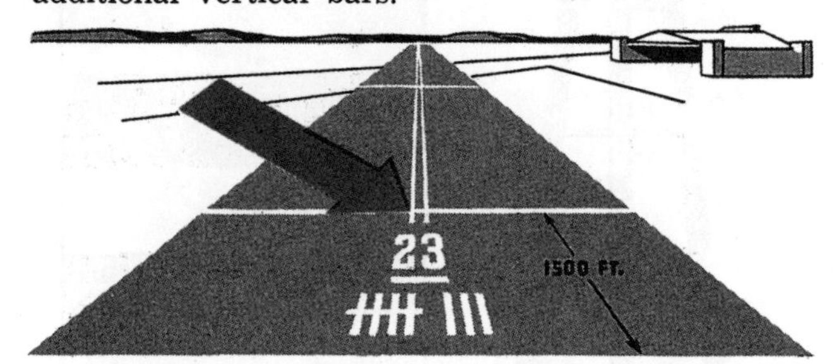


**BANDS RUN LENGTH OF EACH RUNWAY**

They extend down the center to within 25 feet of the numerals on each end. A band is painted across the runway 1500 feet from each end.

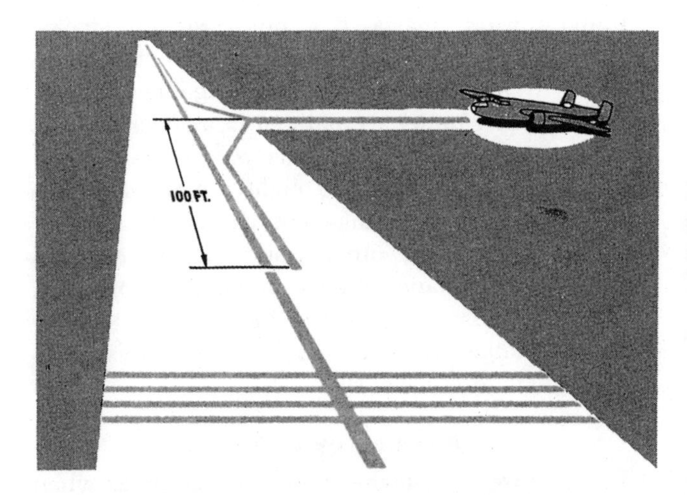


**YELLOW STRIPES FOR TAXIWAYS**

Taxiways are marked by a single yellow center stripe, with guide stripes for turns into hardstands, aprons, etc. A yellow cross band warns of approach to within 100 feet of a runway. Where taxiways intersect runways, the yellow taxiway stripe extends to the adjacent runway stripe.

**Camouflaged airfields in the zone of the interior** will be marked as authorized by the Commanding General, Army Air Forces.

**Airfields in the theater of operations** will be marked as authorized by the Theater Commander.

REFERENCE: Technical Order 00-25-7.

# CAMOUFLAGE



Lack of strict camouflage discipline at an airdrome can ruin the whole camouflage scheme. The simplest human activities leave marks on the earth which are visible from the air. Men around an airdrome leave distinctive marks, easily identified by aerial observers as being the "spoor" of an army flying field. To keep this "spoor" at a minimum everyone must obey strictly the rules laid down by camouflage authorities on the airdrome.

Camouflage is designed to reduce visibility and conceal identity. It is accomplished by disruption, disguise, and concealment.

Officers charged with the layout and execution of camouflage will do everything possible to protect you and your plane; but as a pilot you have a special responsibility. You are in command of your plane on the ground as well as in the air.

## Protect Your Aircraft

Just because your plane is normally taken care of by your crew chief after you have landed, don't forget that the safety of your plane is still your responsibility.

The necessity of thinking about your plane while it is parked on a continental airfield may seem unimportant to you, but when you set down at an airfield in a Theater of Operations, it will become vital. Then you will realize the importance of having trained yourself to think about your plane all around the clock.

**The enemy likes nothing better than to find a group of planes standing in a line, wing tip to wing tip—an ideal target!** If you have unthinkingly contributed to creating such a target you may suddenly find yourself without a plane. This means that upon landing in any theater of operations you must automatically take additional precautions in parking your plane—just in case the enemy starts bombing or machine gun strafing.

Here is a check list to remember:

1. See that your plane is parked a reasonable distance from runways; they invariably are primary objectives for bombing attacks.

2. See that it is parked at least 400 feet from any other plane; all other things being equal.

3. Park your plane near to trees, buildings or any other objects which make it less conspicuous and break up the distinguishing shadow cast by any airplane.

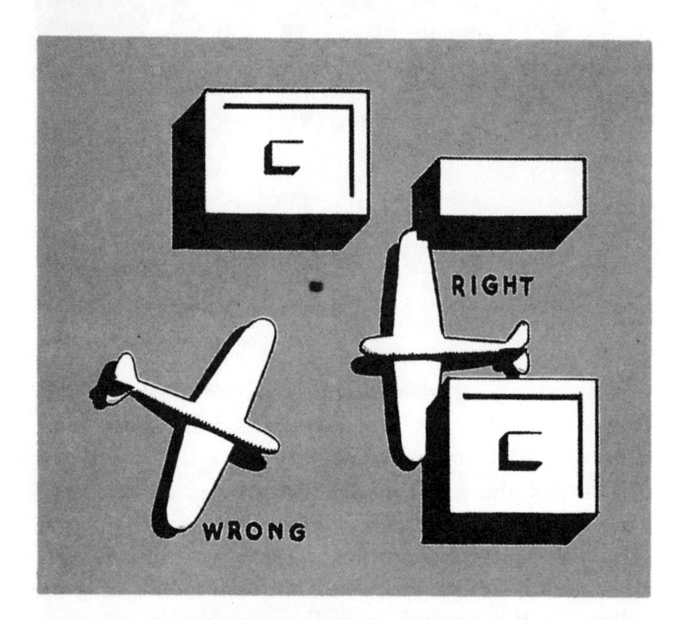


4. If it is in an open field, a plane parked along a hedgerow, at the junction between two or more fields, or near low scrubby bushes is more difficult to see from the air.

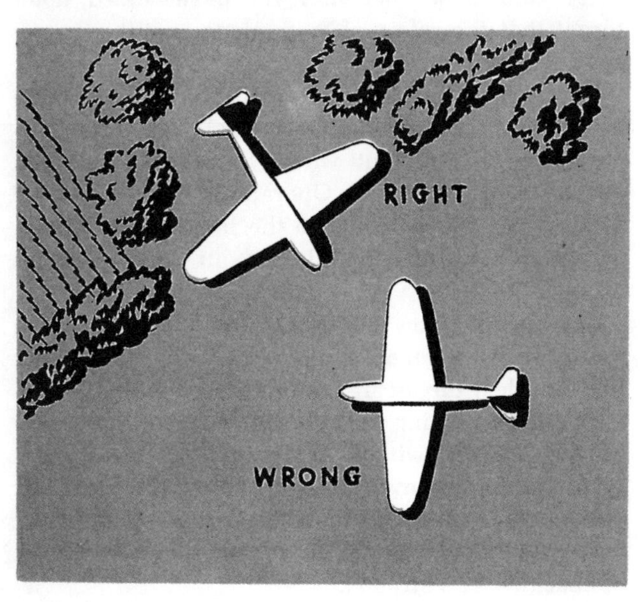

5. Anything thrown over the cockpit greenhouse, windows or Plexiglas nose will kill the reflections which might be a dead give-away.

6. Park it so you can taxi out quickly in an emergency.

**Don't park your plane close to and in line with other planes.**

When you are operating from a dispersed position on an airfield in a theater, always think about these precautionary measures:

1. Always taxi to your designated dispersal point immediately upon landing.

2. Don't leave that position until you know you can take off soon after reaching the runway.

3. Don't jam up behind other planes on a taxiway leading to the take-off position. The unexpected might happen, resulting in a string of dead ducks.

That is just the sort of target the enemy likes—where he can eliminate planes and combat crews with a minimum of effort.

Make it tough for the enemy if he does sneak in! Just remember these simple facts:

1. If your plane is in a dispersed position he will have more trouble finding it.

2. Even if he does bomb or strafe, the chances are good that you will still have a plane when the attack blows over.

3. Follow the instructions of your camouflage officer in all matters of dispersion and discipline.

4. Make your crew members conscious of this problem also.

5. If necessary ask your camouflage officer to request suitable nets from the Theater Service Center for further protections of your aircraft.

Continually train yourself so that it becomes second nature to **think about the safety of your plane on the ground as well as in the air.**

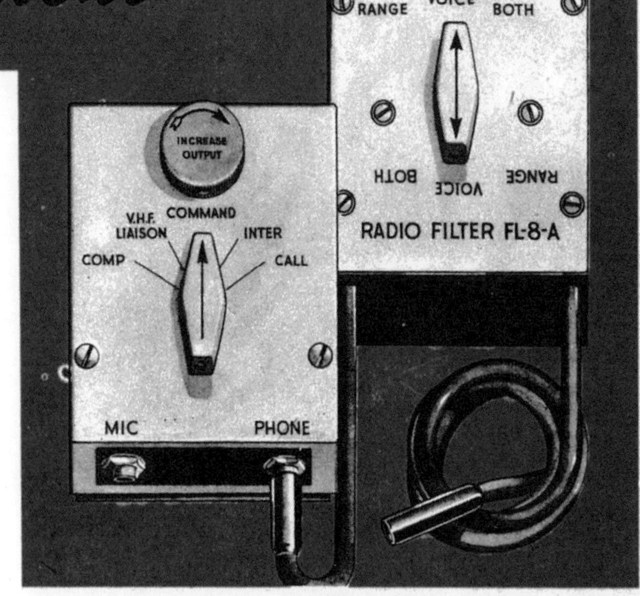


Jackbox and Range Filter

**Turn on radio equipment by switching on:**

1. **The airplane battery switches.**
2. **The master switch.**

## THE JACKBOX

The pilot and crew members use the jackbox to connect microphones and headsets to various pieces of radio equipment in the airplane. Following is a list of jackbox positions and their uses:

### Compass

The "COMP" position is used to listen to the radio compass receiver. You cannot transmit with jackbox in this position. The compass does not include transmission facilities.

### Liaison-VHF

Use "LIAISON-VHF" to transmit and receive with VHF equipment. To transmit, press microphone button-switch. To receive, release microphone switch.

If your jackbox does not have a "VHF" position, the liaison equipment will be connected to that position. Use the same procedure for operating both.

### Command

Operation in the "COMMAND" position is the same as in the "LIAISON-VHF" position, except that the medium frequency command equipment is controlled. If the jackbox has no "VHF" position, a separate "MED FREQ-VHF" or "274-522" switch is provided for switching the microphone circuit to either of the command sets.

### Inter

Use "INTER" when you want to communicate with any other crew member, whose jackbox is also on this position. Press microphone button-switch to talk. Release it to listen.

**CAUTION: Interphone operation is seriously impaired if more than one microphone switch is depressed at the same instant. Don't attempt to speak while someone else is using the system, unless your message is urgent.**

### Call

When you want to call another crew member to the "INTER" position, use "CALL" To operate you must hold the jackbox switch in the "CALL" position, and at the same time press the microphone switch. After calling, immediately return the jackbox switch to "INTER" while awaiting an answer, since "CALL" position blocks all other reception and gives precedence to the person calling. To answer, the person called switches to "INTER" and proceeds with normal interphone communication.

## RADIO RANGE FILTER

The radio range filter system is used by the pilot or copilot to select either voice or range signals, when both signals are being transmitted simultaneously by the radio range station. When the selector is on "RANGE" position you can listen only to the 1020 cycle A and N signals. When the switch is on "VOICE" the A and N signals are filtered out to facilitate reception of speech. When switch is on "BOTH" the filter is out of the circuit, and you can hear both speech and A and N signals.

**Remember, when the filter system is in "RANGE" position interphone operation is blocked. Never allow both pilot's and copilot's filters to be in this position at the same time.**

1. The microphone should fit firmly about the throat, with the buttons spaced equally on each side of your Adam's Apple and slightly above it.

2. Place the strap slightly higher than the microphone level to maintain this position.

3. Don't allow clothing to get between the buttons and your throat.

4. You will get better reception by placing the positioning clip between the buttons.

5. Be sure the proper sides of the buttons are against your throat.

6. Speak as loudly as you can, but don't shout.

### T-17 Hand Microphone

If your airplane is equipped with the T-17 hand-held microphone, observe the following steps for proper operation:

1. Hold the microphone squarely in front of your mouth, with your lips slightly touching the mouthpiece when speaking.

2. Speak as loudly as you can, but don't shout.

When flying at high altitudes always use the oxygen mask microphone. Always make sure the connecting plugs fit firmly.

## LIAISON EQUIPMENT

The liaison equipment includes a high-powered long range transmitter and receiver. It is operated primarily by the radio operator. In some airplanes the pilot can operate the liaison equipment by switching the interphone jackbox to "VHF-LIAISON" or "LIAISON" position. If it is necessary for you to use the liaison equipment, ask the radio operator to tune it to the proper frequency. Don't use the liaison transmitter to contact stations within the range of your command equipment, unless it is absolutely necessary.

## IFF RADAR EQUIPMENT

Operation of IFF Radar equipment is automatic, once it is turned on. "ON-OFF" switches are provided for the pilot and also the radio operator in airplanes which carry one. The "EMERGENCY" switch, under a springed guard, operates an emergency channel. This switch is usually safety-wired to the "OFF" position. Details of its operation are given at briefing.

Detonator Switches

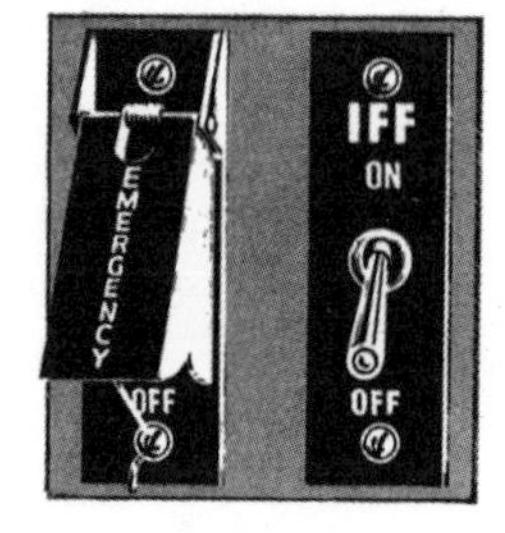


IFF Control Switches

**Warning**

Your IFF equipment contains two detonator push-buttons.

**If you must abandon your airplane in enemy territory, push both buttons simultaneously. They will set off a detonator which destroys the equipment internally, and without danger to you.**

There is also an intertia switch that sets off the detonator automatically, if you have to make a crash landing.

SCR- 274N Receiver Control Box

# COMMAND EQUIPMENT

Command equipment is of two classes; medium frequency and very high frequency. It is used chiefly by the pilot for voice communication from plane to plane, or from plane to ground when requesting operational instructions. Command equipment is used also to receive radio navigational aids.

### SCR-274N Medium Frequency Set

The SCR-274N command set is used for medium frequency command communication. Most airplanes are equipped with two transmitters and three receivers. They are remotely controlled. The two transmitters are usually tuned to the most frequently used channels. With three receivers available, the pilot can use one for receiving navigational signals, and the other two for monitoring stations along the planned route.

### Receivers

All three receivers are controlled from the same remote control unit. The operating frequency ranges of the receivers are found on the calibrated dials of the control unit. These dials are calibrated either in kilocycles or megacycles. A megacycle is 1000 kilocycles. The receivers are capable of receiving continuous wave or modulated signals. Their outputs are fed into an "A TEL" or "B TEL" line. The "A" line feeds into the "COMMAND" position of the interphone jackbox. The copilot may plug his headset into the "B TEL" jack on the control unit if it is ever necessary to receive two stations simultaneously. If the "B TEL" line is used, the "A"-"B" selector switch on the control of the receiver being used must be in "B" position.

**To operate receiver:**

1. Turn "CW-OFF-MCW" switch to the type of reception desired. For voice reception turn to "MCW" position.
2. Place "A"-"B" switch to desired channel. Remember the "A" channel feeds the "COMMAND" position of the interphone jackbox.
3. Turn the dial crank to the desired frequency.
4. Adjust the volume to a level comfortable in your headset.

### Transmitters

The remote control unit which operates the transmitters has a "TRANSMITTER SELECTION" switch to enable the use of four transmitters. Since there are, however, usually only two transmitters installed in an airplane you will use only number "1" and "2" positions. By turning to position "4" you can use the sidetone of the modulator unit for an **emergency interphone.**

SCR-274N Transmitter Control Box

**To operate transmitters:**

1. If the VHF command set is not connected to the "VHF-LIAISON" position of the interphone jackbox be sure the "MED FREQ-VHF" or "274-522" switch is in "MED FREQ" or "274" position.
2. Select the proper transmitter with the "TRANSMITTER SELECTION" switch.
3. Select the type of emission desired; "TONE-CW-VOICE".
4. Turn "TRANS POWER" switch "ON".
5. Allow the tubes about 10 seconds to warm up before pressing microphone push-button.

A key is mounted on the top of the transmitter control unit so that you can transmit either tone or CW. The normal transmitting range of the SCR-274 is approximately 25 miles. However, in the absence of atmospheric and local disturbances, you can obtain ranges up to 100 miles.

### SCR-522 Very High Frequency Set

The SCR-522 set is used for VHF (line of sight) command communication. The transmitters and receivers operate on four pre-tuned crystal controlled channels. You can select any one of these four receiver or transmitter channels from the pilot's compartment. Consult your communications officer for the functions of the four channels.

The "A-B-C-D" push-buttons turn the equipment on, and select the respective channels. The green lights alongside the buttons indicate the channel selected. The lever tab directly above these lights controls a dimmer mask. The "OFF" button turns the equipment off.

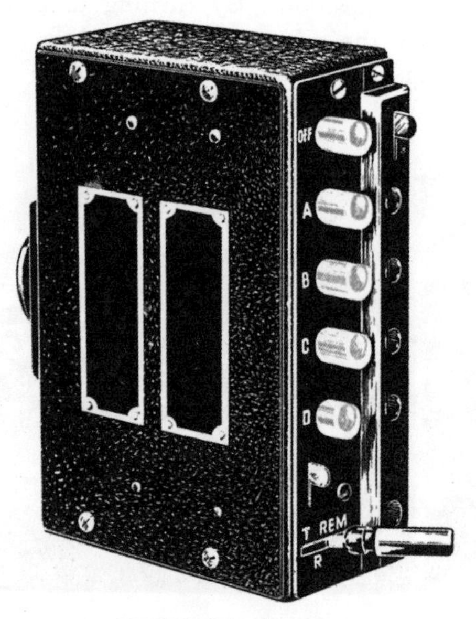


SCR-522 Control Box

**To operate the SCR-522:**

1. Press the desired channel push-button.
2. Place "T-R-REM" switch in "REM" position. The lever tab directly above the switch must be up to allow it to be placed in this position. The switch is usually safety-wired or permanently fastened to "REM". With the switch on "REM" the receiver will be in operation at all times, unless the microphone button is depressed. While the button is held down the set will be in "TRANSMIT" position. If there is no "VHF" position on your interphone jackbox be sure the "MED FREQ-VHF" or "274-522" switch is in "VHF" or "522" position. This connects the microphone to the input circuit of the transmitter. The distance range obtainable with the SCR-522 is limited to line of sight transmission (i.e. at an altitude of 5000 feet the line of sight distance is approximately 90 miles, at 10,000 feet 120 miles).

### Operating Tips, Command Sets

**274N Set:** Transmission on frequencies at which this set operates vary from month to month and from day to day, because of changes in atmospheric conditions. Signals are stronger in the winter than in the summer, and may vary from hour to hour on a summer day.

At certain times of the day signal fading is so rapid it produces severe distortion of voice signals. Signal fading increases with an increase in distance of transmission.

**522 Set:** Occasionally you will observe dead spots when the metal structure of another airplane gets between the antenna radiation pattern of the communicating airplanes. You can correct this by maneuvering the airplane so that the disturbing structure is removed from between the airplanes.

## RADIO COMPASS RECEIVER

You may use the radio compass for radio reception, for homing, or for taking bearings on a radio transmitting station. Two antennas are used with the receiver, one of them directional, the other non-directional. Complete controls for operating the radio compass are located in the pilot's compartment, and also in the navigator's or radio operator's compartments. As a pilot you are primarily concerned with homing, and using the radio compass receiver to receive navigational signals, when static or other interference prohibits the use of command receivers.

### Operation for Homing

1. Place the interphone jackbox switch in "COMP" position.

2. Place "OFF-COMP-ANT-LOOP" switch in "ANT" position.

3. Push "CONTROL" switch, hold in for a few seconds until green light comes on.

4. Select the proper frequency range either "200-400", "410-850", or "850-1750" kilocycles.

5. Turn the tuning crank to the frequency desired, rock it back and forth over this setting until you obtain a maximum reading on the "TUNING METER".

6. Place the "OFF-COMP-LOOP-ANT" switch on "COMP"

7. Maneuver the airplane until the radio compass indicator reads zero. The airplane is now headed for the transmitting station. Remember, when homing on a station you are heading for the transmitting towers, which are usually located outside of towns.

If the indicator points to the right of zero, the station is to the right. If it points to the left the station is to the left. The homing operation of the instrument is such that the airplane will arrive ultimately over the radio station, regardless of drift. The flight path, however, will be a curved line and coordination with ground fixes or landing fields will be difficult. By trial and error you can fly the airplane on a relatively straight line course by offsetting the heading to compensate for wind. A decreasing magnetic bearing indicates a wind from the left, an increasing bearing a wind from the right.

Radio Compass Control Box

Radio Compass Indicator

### Operation for Range Reception

1. Place "OFF-COMP-ANT-LOOP" switch to "ANT".

2. Tune in radio range station. Keep the interphone jackbox volume control turned fully clockwise, and the volume of the radio compass as low as possible.

If reception on "ANT" is poor because of precipitation static, switch the function switch to "LOOP". Rotate the loop antenna by means of the "LOOP L-R" switch for maximum volume in the headsets. Keep compass volume control as low as possible.

If the loop is in the null position when flying on a radio range course, the signal may fade in and out and be mistaken for a cone of silence.

Cone of silence indications are not reliable on loop type range stations when the receiver is in "LOOP" position. Directly over the station the signal may increase in volume to a strong surge instead of a silent zone.

**Never use "COMP" position for flying the radio range. The course might appear much broader than it actually is.**

**For details of radio compass operation see Technical Order 30-100B-1.**

## MARKER BEACON RECEIVER

The marker beacon receiver indicates visually to the pilot that the airplane is passing through the radiation field of a marker beacon transmitter. These transmitters are used to mark locations of radio range stations, to indicate range course intersections, boundaries, and position when making an instrument approach landing. The length of visual indication depends upon the type of marker station and the altitude of the airplane. A station marker indication lasts approximately one minute at 10,000 feet, if the airplane is flying at 150 mph.

Marker beacon transmitter locations and keying data are found in your Radio Facility Charts.

**To operate:** Turn on radio compass, as it furnishes the power for the marker beacon receiver.

## POLARIZATION ERRORS

The radio compass was designed to serve primarily as a navigational aid in flying. So long as it is used in this capacity and its limitations recognized, it is a useful and valuable device. Unfortunately, in actual flight, there are certain periods when the instrument's indications are not correct. This is caused by radio wave polarization errors. Failure to recognize these errors can throw you far off course, and make you mistrust the compass.

The principal polarization error is **night effect.** Other causes of faulty bearings are **mountain effect, shore-line effect, and magnetic disturbances,** such as those found in auroral zones of polar regions.

Ordinarily, for homing, the radio compass receiver depends on reception of vertically polarized radio waves. However, when these waves are reflected from the sky they may change polarity and become horizontally polarized. These horizontally polarized waves conflict with the vertically polarized waves and cause fluctuations in the reading of the radio compass indicator.

### Night Effect

Since radio waves are reflected in greater strength by night than by day, the opposition of horizontally opposed waves is stronger at night. Errors caused by this phenomenon are called night effect.

Polarization errors may flare up for a few seconds at intervals through the day, and cause the needle to hunt more than normally about the bearing. Real night effect causes hunting of more than 30 seconds' duration. Variations in the intensity of this hunting may be conveniently classified into two types.

In less severe form, the indicator hunts over a total angle of 15° or less around the true bearing; and often a bearing with an accuracy of 5° can be taken by computing a mean reading. In more severe cases the indicator moves constantly, usually through a wide angle and not around the proper bearing. Therefore taking an average of the fluctuations is not possible. You sometimes encounter short periods of nearly normal operation during these periods of extreme instability.

The times at which night effect begins vary considerably, even on the same station. Usually, the first and last disturbances appear during the periods just before sunset and just after sunrise. The errors increase with an increase in frequency, or in the distance of the airplane from the station.

### Remedies

Night effect recurs frequently. However, there are definite steps with which to combat it. First, recognize it by remembering that a period of fluctuation in the bearing indications lasting more than 30 seconds is a sure sign. Then try the following:

1. Use other methods of navigation to check the bearings.
2. Increase altitude.
3. Average the fluctuations if possible.
4. Select a station of lower frequency.
5. Remember that comparatively large errors are tolerable for purposes of homing, since accuracy increases as the distance diminishes.

### Other Effects

You may notice fluctuations when flying across coast lines when the radio waves cross the coast at acute angles. Errors may occur, also, when you are flying over certain mountainous regions, and, to a limited extent, through cold fronts.

## RADIO ALTIMETERS

There are several types of radio altimeters. Consult the operating instructions for the particular equipment installed in your airplane.

Instrument Landing Indicator

The SCS-51 Instrument Landing System provides the pilot with a straight line glide path and a localizer or on-course guidance. Some installations include marker stations to provide a further check on location of the airplane in relation to the airport. The airborne equipment consists of a localizer receiver, a glide path receiver, a visual indicator, and a control box.

### Visual Indicator

The visual indicator shows you the position of your airplane with respect to the localizer and glide paths. The vertical needle registers the blue and yellow localizer course. The horizontal needle registers the glide path course.

### Localizer Indications

The vertical needle of the indicator registers the area of the localizer transmitter signal in which the airplane is flying. If the airplane is in the blue-shaded area when making an approach, the needle swings to the left, indicating you are to the right of the runway. The yellow area indicates you are to the left of the runway. These areas do not necessarily indicate the proper landing direction, since the course is transmitted down the approach of front beam and the downwind or back beam. If the airplane is approaching from the back beam the indications are reversed. When they are reversed the needle swings further away from the center as you fly toward the direction indicated.

Contact the control tower for instructions if in doubt. By keeping the needle within a one-quarter scale deflection you are assured of landing within a safe portion of the runway.

### Glide Path Indications

The horizontal needle of the indicator indicates the position of the airplane in relation to the glide path signal.

If the airplane is above the desired glide path the needle points downward, indicating the direction in which you must fly the airplane to approach the desired glide path. If the airplane is below the glide path the needle points upwards.

The glide path needle is sensitive. It indicates a full scale deflection when the airplane is 0.3° above the glide path or 0.5° below it. You must align the airplane accurately on the glide path before nearing the field. Only minor corrections are allowable near the field.

The glide path (horizontal) needle points straight up when no signal is being received.

### Control Unit

The same unit controls both the localizer and glide path receivers. **To operate:**

1. Turn "ON-OFF" switch "ON".
2. Select the proper channel for the field you are approaching: "U", "V", "W", "X", "Y", or "Z"

SCS-51 Control Box

**For complete instructions on the SCS-51 system see Technical Order 30-100F-1.**

**The use of standard radio telephone procedures is the responsibility of all pilots.**

### Operating Tips

1. Transmit in a concise and business-like manner. Make only necessary official transmissions.
2. Speak slowly. Pronounce each word and number distinctly.
3. Know your message. Group your words so the idea is clear.
4. Don't hesitate. Don't say "uh" and "er".
5. Speak loudly enough to be heard above the surrounding noises.

### PHONETIC ALPHABET

When necessary to identify any letter of the alphabet or to spell a word, use the standard phonetic alphabet.

| Letter | Spoken as | Letter | Spoken as |
|---|---|---|---|
| A | Able | N | Nan |
| B | Baker | O | Oboe |
| C | Charlie | P | Peter |
| D | Dog | Q | Queen |
| E | Easy | R | Roger |
| F | Fox | S | Sugar |
| G | George | T | Tare |
| H | How | U | Uncle |
| I | Item | V | Victor |
| J | Jig | W | William |
| K | King | X | X-ray |
| L | Love | Y | Yoke |
| M | Mike | Z | Zebra |

Difficult words will be both spoken and spelled, for example: "Solved—I spell—Sugar Oboe Love Victor Easy Dog—Solved."

### Radio Telephone Terms

**"Roger"** means "Received your message."

**"Wilco"** means "Received your message and (where applicable) will comply." Use "Wilco" to acknowledge you will carry out landing, takeoff, or other instructions.

**"Say again"** means "Repeat."

**"I say again"** means "I will repeat."

**"Wrong"** means "That is incorrect. The correct version is --"

**"Correction"** means "An error has been made. The correct version is --"

**"That is correct"** is self-explanatory.

**"Wait"**, if used by itself, means "I must pause for a few seconds." If the pause is to be longer than a few seconds use "Wait. Out." If "Wait" is to be used to prevent another station from transmitting, follow it with "Out."

### Pronunciation of Numbers

When you transmit numbers by radio telephone use the following standard pronounciation:

| NUMERAL | SPOKEN AS | NUMERAL | SPOKEN AS |
|---|---|---|---|
| 0 | Ze-ro | 5 | Fi-yiv |
| 1 | Wun | 6 | Six |
| 2 | Too | 7 | Seven |
| 3 | Thuh-ree | 8 | Ate |
| 4 | Fo-wer | 9 | Niner |

### Statement of Numbers

Speak all numbers in serial form, as "wun ze-ro" for 10; "ate niner too" for 892.

An even hundred or thousand is spoken as "Hund-red" or "Thow-sand". An exception is made for ceiling and flight levels of ten, eleven and twelve thousand feet. For example:

| FLIGHT LEVEL | STATEMENT |
|---|---|
| 1,200 | Wun Thow-sand Too Hund-red |
| 12,000 | Twelve Thow-sand |
| 13,000 | Wun Thuh-ree Thow-sand |
| 15,500 | Wun Fi-yiv Thow-sand Fi-yiv Hund-red |

### Radio Call Signs

Call signs for airplanes are composed of the last three numbers of the airplane serial number. If further identification is necessary, the ground station will request four or more numbers.

### Establishing Communication

To establish (or reopen) communication make the initial call-up once, as illustrated in the following example. If you get no reply within 30 seconds make a second call-up, this time making it twice. Repeat this double call-up at 1 minute intervals until communication is established.

| Item | Example |
|---|---|
| Station called | "Scott Army Airways |
| Introduction | "this is |
| Station calling | "Army ate wun fo-wer. |
| Invitation to reply | "Over." |

After communication is established you can continue without further call-up, but each message should begin with the airplane's identification and end with the proper termination.

### Termination of Message

All messages will end in "Over" or "Out", depending on which is appropriate.

**"Over"** means "My transmission is ended. I expect a response."

**"Out"** means "This conversation is ended and no response is expected."

*For all frequencies see your*
**RADIO FACILITY CHARTS**

### Contacting Ground Stations

Various ground radio communications facilities are at the disposal of the pilot. They include Army Airways Communications System stations, Army or CAA radio range stations, and Army or CAA control towers (See Radio Facility Charts for listings). Examples of call-ups are:

| | |
|---|---|
| AACS stations | "Scott Army Airways" |
| Radio ranges | "Scott Radio" |
| Control towers | "Scott Tower" |

### AACS

Army Airways Communications System operates control towers, Army airways and radio navigational aids. AACS stations will relay traffic to any point, including CAA or Navy, via telephone, telegraph, teletype, or interphone.

### To Prearrange a Channel

Before departing from a base where an AACS station is located a pilot can file a radio message to notify airways stations along the route and at destination to stand by on any air-ground channel.

## RADIO DIALOGUE

### Takeoff Instructions

Contact the tower for taxiing and takeoff instructions before moving from the line or parking area. Towers should give: (1) Wind direction and velocity and direction of takeoff traffic. (2) Runway clearance. (3) Special instructions for local conditions. (4) Taxi clearance. (5) Takeoff clearance. (6) Altimeter setting. (7) Time. For example:

**Plane:** "Washington tower, this is Army one five seven, over."

**Washington:** "Army one five seven, this is Washington tower, over."

**Plane:** "Washington tower, this is Army one five seven, taxi clearance, over."

**Washington:** "Army one five seven, traffic northwest, cleared to runway three three. Taxi to front of terminal building, turn right and follow taxi strip. Altimeter three zero zero four, over."

**Plane:** "Army one five seven, wilco, out."

**Plane** (in takeoff position): "Washington tower this is Army one five seven, over."

**Washington:** "Army one five seven, this is Washington tower, over."

**Plane:** "Washington tower, this is Army one five seven, takeoff clearance, over."

**Washington:** "Army one five seven, cleared for takeoff, over."

**Plane:** "Army one five seven, wilco, out."

**Washington:** "Army one five seven, off at zero five, over."

**Plane:** "Army one five seven, roger, out."

**Remain tuned to the tower frequency for at least 5 minutes after departure, unless cleared to another frequency by the control tower.**

### Landing Instructions

When approaching a field, contact tower when approximately 10 minutes out, or in time to receive an answer before entering the control zone. Give your position, and stand by for landing instructions. Call tower again about 1 minute from the field and give position.

Clearance to enter traffic pattern is issued to the pilot when it is desired that you approach in accordance with current traffic patterns, and when traffic conditions are such that a clearance authorizing actual landing cannot be given. Information is given concerning landing direction and runway, so you may plan your entry into the traffic pattern. Don't confuse clearance to enter pattern with clearance to land. Landing instructions should include wind direction and velocity, altimeter setting, traffic information, landing sequence, observation of landing gear, and field conditions. For example:

**Plane:** "Waco tower, this is Army five nine two, over."

**Waco:** "Army five nine two, this is the Waco tower, over."

**Plane:** "Waco tower, this is Army five nine two, position one five miles northwest, time zero eight one five at four thousand. Request landing instructions, over."

**Waco:** "Army five nine two, roger. Cleared to enter right traffic pattern. Traffic south, runway one seven R. Make an overhead approach at one thousand five hundred indicated. Call tower on base leg, over."

**Plane:** "Army five nine two, wilco, out."

**Plane** (on base leg): "Waco tower, this is Army five nine two, on base leg, over."

**Waco:** "Army five nine two, wheels down and locked, cleared to land, over."

**Plane:** "Army five nine two, wilco, out."

**Plane** (now on ground): "Waco tower, this is Army five nine two, on ground, taxi instructions, over."

**Waco:** "Army five nine two, make a right hand turn at first taxi strip. Taxi north on ramp to center control tower, alert crew will park you, over."

**Plane:** "Army five nine two, wilco, out."

**Remain tuned to the tower frequency until you have taxied airplane to parking position and shut off the engines.**

### CAA Range Stations

Contact CAA radio range stations to report your position, request weather and traffic information, file flight plans, or make changes in flight plans. Report position to range stations in this sequence: (1) Plane's call. (2) Position. (3) Time. (4) Altitude. (5) Flight conditions, such as contact or instrument flight, on top, between layers (include visibility), icing, or turbulence. (6) Estimated time over next fix. For example:

**Plane:** "Denver radio, this is Army five zero eight, over."

**Denver:** "Army five zero eight, this is Denver radio, over."

**Plane:** "Denver radio, this is Army five zero eight. Crossing northeast leg Denver range one five miles northeast, time one six four zero at nine thousand on instruments. Estimate Cheyenne time one seven one five, over."

**Denver:** "Army five zero eight. Crossing northeast leg Denver range one five miles northeast, time one six four zero at nine thousand on instruments. Estimate Cheyenne time one seven one five, over."

**Plane:** "Army five zero eight, that is correct, out."

### Change of Flight Plan

Give: (1) Plane's call. (2) Position. (3) Time over position. (4) Altitude. (5) Flight conditions. (6) Details of change in flight plan. (7) Remaining hours and minutes of fuel aboard. (8) Estimated time en route present position to destination. (9) Alternate airport. (See PIF 2-3-1.)

**Plane:** "Washington radio, this is Army three zero eight, over."

**Washington:** "Army three zero eight, this is Washington radio, over."

**Plane:** "Washington radio, this is Army three zero eight. Over Washington, time one six one five at six thousand on instruments. Request change in flight plan, six thousand red airway two zero to Pittsburgh. Fuel supply three hours and three zero minutes. Estimate Pittsburgh one hour and one five minutes. Alternate Columbus, over."

**Washington:** "Army three zero eight, roger, wait, out."

**Washington:** "Army three zero eight, this is Washington radio, over."

**Plane:** "Washington radio, this is Army three zero eight, over."

**Washington:** "Army three zero eight, Washington Airway Traffic Control clears you to two five miles northwest Martinsburg, to cruise at six thousand. Contact Martinsburg radio for Pittsburgh clearance, over."

**Plane:** "Army three zero eight, roger, out."

Wartime use of airplanes often will not permit the use of radio to control the formations in which they are flying. The majority of formation training flights also require partial or total radio silence. Consequently, a system of signals is required in order that a commander may exercise proper control of his formation during flight.

Because each type of flying employs formations peculiar to itself and because variety in formations also is caused by non-uniformity of airplane types, it is not considered expedient to prescribe all signals to be used in formation flying.

To insure uniformity throughout the Army Air Forces in certain items of control common to all formations of airplanes, the following visual signals are prescribed:

| SIGNAL | SIGNIFICANCE |
| --- | --- |
| **FLUTTER AILERONS:** Repeated and comparatively rapid movement of ailerons.  |  **ATTENTION:** Used on the ground or in the air to attract attention of all pilots in the formation. Stand by for radio message or further signal. When on ground and in proper position to take off, this signal will normally mean "Ready to take off." |

**FISHTAIL OR YAW:** By rudder control during flight, move the tail of the airplane alternately and repeatedly right and left.

**OPEN-UP FORMATION:** Where applicable, this may be used to order a search formation.

**SERIES OF SMALL DIVES AND/OR ZOOMS:**

**PREPARE TO LAND:** An order to each pilot in the formation to prepare to land. In the absence of further signals the landing will be made in the normal landing formation of the unit, which should be predetermined. Any change in formation for landing will be ordered by supplemental signal by radio.

**DIP RIGHT (LEFT) WING:**

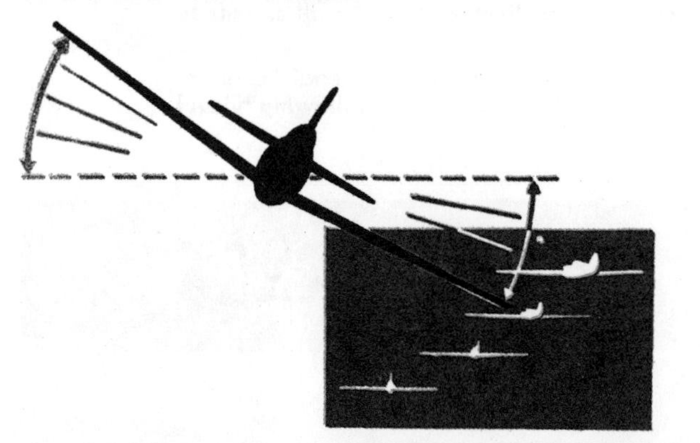

From any formation other than echelon go into echelon of flights to the right (left).

Being in an echelon of flights to the right (left), go into echelon of individual airplanes to the same side. Being in an echelon of individual airplanes, if wing is dipped on the side to which airplanes are echeloned, form echelon of flights to the same side. Being in an echelon of flights or individual airplanes, if wing is dipped on the side away from the echelonment, form same echelon to the opposite side.

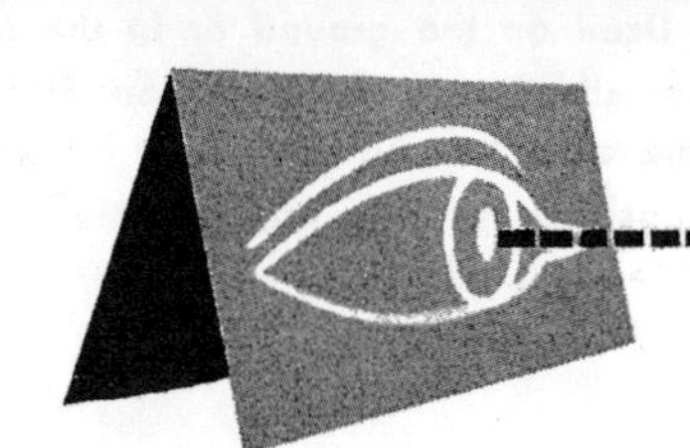

**KEEP YOUR EYE ON YOUR COMMANDER**

## ROCK WINGS:

Slow, repeated, rocking motion of airplane about longitudinal axis, by gradual use of ailerons.
Wing movement to be slower and of greater amplitude than in "Flutter of ailerons."

## ASSUME NORMAL FORMATION:

From any other formation, go into the normal closed-up formation for the unit concerned. This formation is to be prescribed in each group and/or squadron.

ROCK WINGS *Slowly*

Hold hand to headset and then swing hand up and down in front of face several times, as in "cease firing."

## RADIO OUT OF COMMISSION:

When necessary or desirable this signal may be given by an occupant of the airplane other than the pilot.

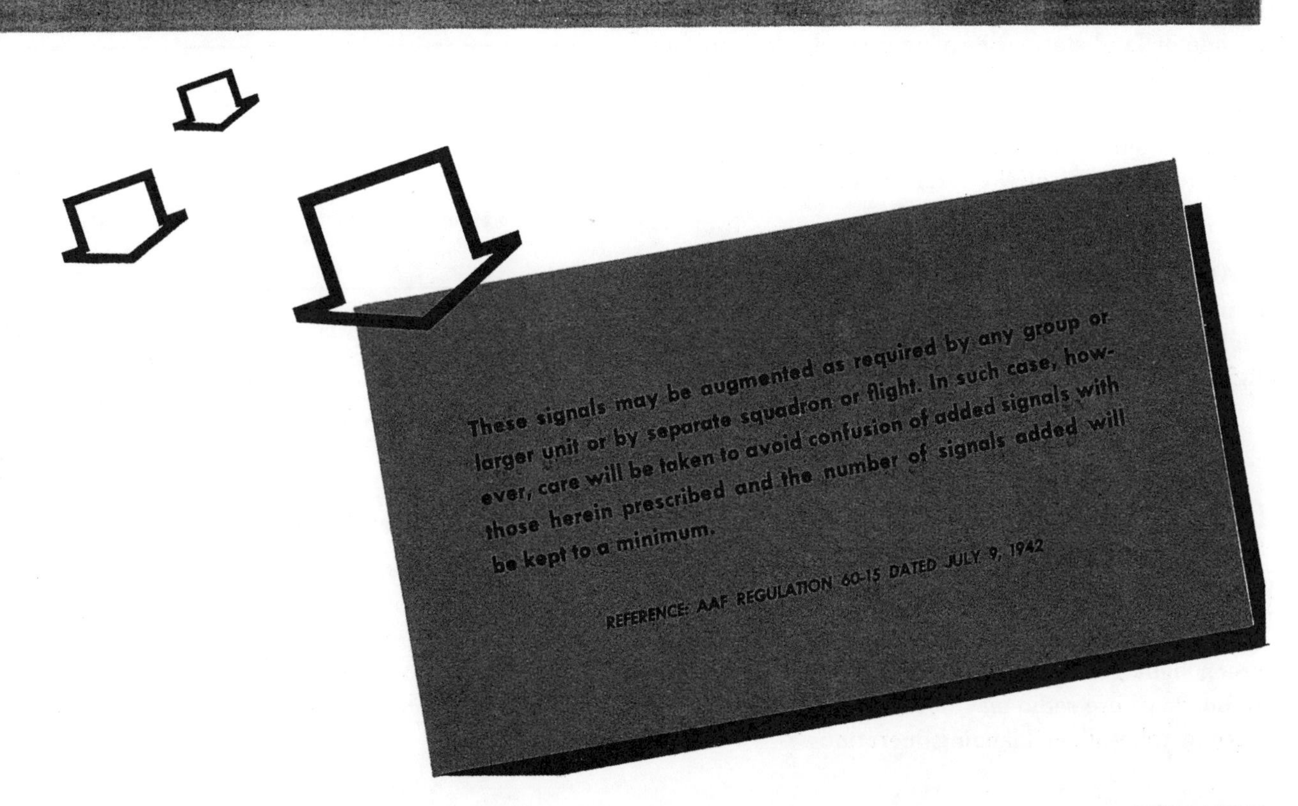

# Miscellaneous Signals

## Distress

The following signals, separately or together, will be used where practicable in case of distress:

The international signal, **SOS,** by radio.

Use **"Mayday"** if transmitting by voice.

The international code flag signal of distress NC.

A succession of white Very pistol lights fired at short intervals.

## Urgent

In radiotelephony the urgent signal consists of three transmissions of the expression **"PAN"** transmitted before the call.

The urgent signal indicates that the calling airplane has a very urgent message to transmit concerning its own safety or concerning the safety of another airplane or ship.

The urgent signal **"PAN"** also indicates that the airplane transmitting it is in trouble and is forced to land, but that it is not in need of immediate help. Follow this signal, so far as possible, by a message giving additional information.

The urgent signal has priority over all other communications, except distress communications, and all mobile or land stations hearing it must take care not to interfere with the transmission of the message which follows the urgent signal.

The urgent signal may be transmitted only with the authorization of the aircraft commander.

## Weather

At civil airfields, approach of unfavorable weather is indicated by day by a rotating tower beacon; by night, by flashing wind key lights or flashing tower light.

## Forced Landing

When forced to land at night at a lighted airport, signal by firing a red Very light or making a series of short flashes with navigation lights.

## Tower Light Gun

The control tower operator uses a directional red-green light gun to signal aircraft not equipped with radio, or whose radio may not be functioning, during taxiing, take-off, and landing operations.

## Light Gun Signals

| SIGNAL FROM | MEANING | | |
|---|---|---|---|
| Tower | Airplane in flight | Airplane taxiing | Airplane in take-off position |
| Green light | Clear to land | Continue taxiing | Clear for take-off |
| Flashing red light | | | Return to line when on ground |
| Red light | Do not land. Stay clear of field and continue circling | Stop immediately | Do not take off; wait |

The flag "Negat" (blue and white checkered) indicates control tower is inoperative.

## Signals From Aircraft to Tower

| SIGNAL | MEANING |
|---|---|
| Landing light on | Desire to land. (This signal should be acknowledged by control tower) |
| One flash of landing light | Acknowledge visual signal from ground |
| Series of flashes of landing light | If flood lights are off, turn on flood lights; if flood lights are on, turn off flood lights |

## Signal Lights

Signal lights are provided on some airplanes for showing that bomb bay doors are open, bombs are being dropped, and identification of friendly aircraft.

Handbooks of Flight Operating Instructions for specific airplanes contain instructions for use of signal lights.

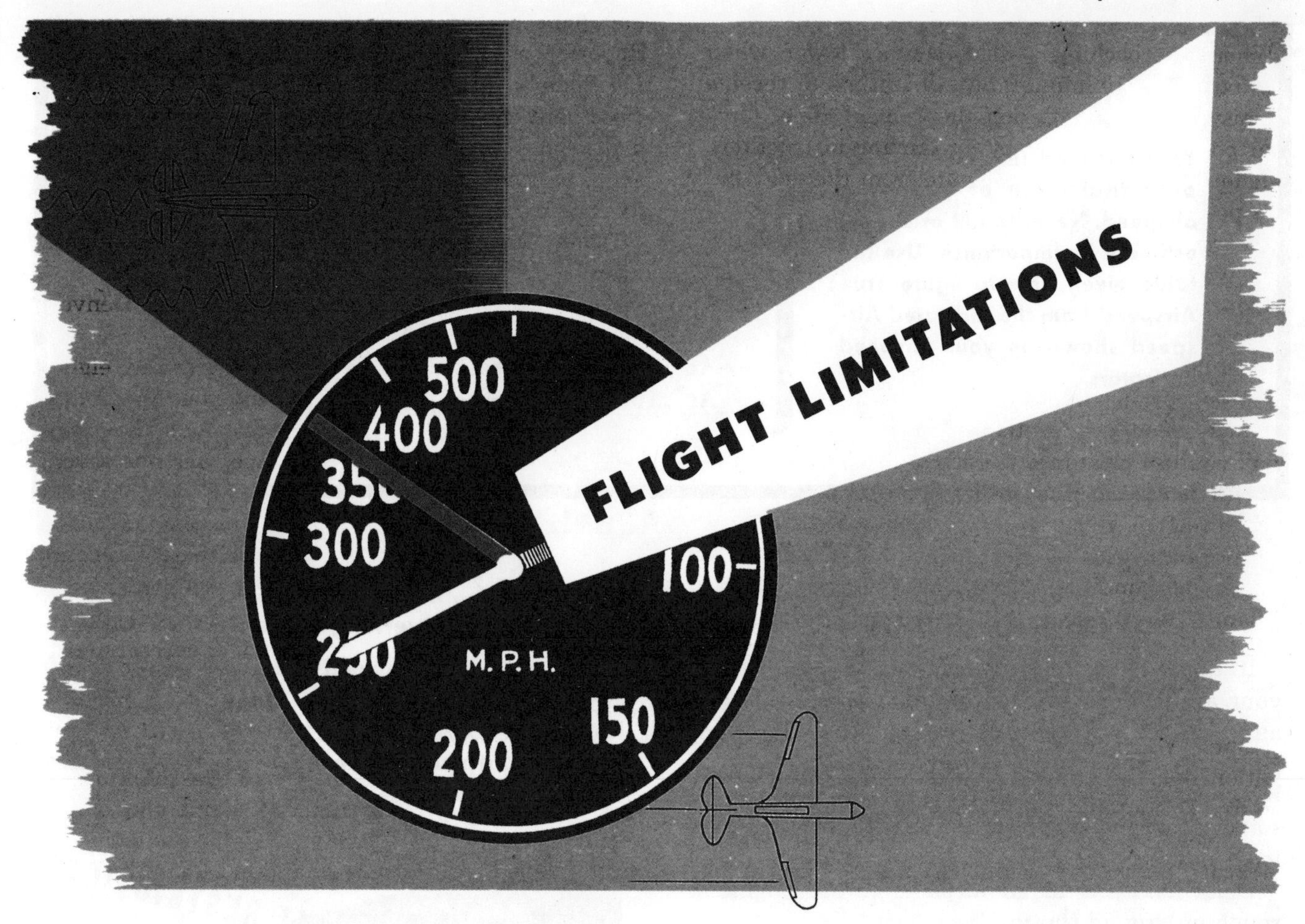


Don't exceed the red line speed on your airspeed indicator. If you do, structural or control failure may result.

The structure and flight characteristics of any type of aircraft limit its performance. In actual flight tests these limits are carefully determined, then each type is redlined at a point within its safety limits.

The red line has a definite purpose. It represents what the airplane was built to do. When you exceed the red line, you are demanding more performance of the airplane than it was designed to accomplish.

You will find that the red line is much nearer to cruising speed on heavy bomber and cargo aircraft than on lighter fighter and attack aircraft. A heavy bomber was designed to carry heavy loads for great distances, to fly at high altitudes, and to defend itself against fighters and flack. It can withstand safely approximately 5 Gs when empty, but not more than 2 or 3 Gs when fully loaded.

A fighter normally is built to withstand up to 12 Gs before permanent deformation or structural failure results. Its primary functions entail high speed and great maneuverability.

Factors influencing flight limitations are:

1. Type of work to be done. (Bomber—Fighter).
2. Design of the airplane. G forces it can withstand without permanent deformation or structural failure.
3. Size and load-carrying capacity of the plane.
4. Point of compressibility. (PIF 2-21-1.)
5. Special equipment (external fuel tanks, bomb racks, rocket tubes, etc.)

### Know Your Airplane

1. **Read all TOs and Pilot Operating Instructions that apply to your airplane.**
2. **Don't experiment with prohibited speeds or maneuvers** unless authorized to do so by competent authority for a specific purpose.
3. **Take it easy** in recovery from any unusual position or speed. Don't jerk or manhandle your airplane. No plane is built that can't be torn up by rough handling of controls, especially at high speeds or in unusual maneuvers.
4. **Use pressure, not movement** of the controls for smooth, uniform recovery.

**Your airspeed indicator is your only indication of your true airspeed. Never forget or underestimate its importance. Use the table given here to figure True Airspeed from the Indicated Airspeed shown on your airspeed indicator:**

**Figuring True Airspeed**

| | | | | | | | | |
|---|---|---|---|---|---|---|---|---|
| At | 5,000 | feet | increase | IAS | by | 7½% | to get | TAS |
| At | 7,500 | feet | increase | IAS | by | 10% | to get | TAS |
| At | 10,000 | feet | increase | IAS | by | 15% | to get | TAS |
| At | 15,000 | feet | increase | IAS | by | 25% | to get | TAS |
| At | 20,000 | feet | increase | IAS | by | 35% | to get | TAS |
| At | 25,000 | feet | increase | IAS | by | 50% | to get | TAS |
| At | 30,000 | feet | increase | IAS | by | 65% | to get | TAS |
| At | 35,000 | feet | increase | IAS | by | 80% | to get | TAS |

### Vibration and Flutter

If there is any flutter or unusual vibration in your airplane, reduce airspeed and engine speed to minimize it until you can land. In particular, any persistent oscillation in which relative motion can be observed between the wings and ailerons, or stabilizer and elevator, or fin and rudder, is likely to be dangerous.

Don't take off in an airplane if the engine is rough on ground run-up.

Small, high frequency vibrations existing over long periods of time may cause fatigue failures.

Factors likely to set up excessive vibrations in an airplane in use are:

1. Propeller—Unbalance, incorrect setting, improper mounting, etc.
2. Power Plant—Missing of one or more cylinders, improper adjustment, unsatisfactory mounting, etc.
3. Control Surfaces—Reduction in rigidity, loose hinge fittings, bearings, or balance weights, etc.
4. Control Systems—Excessive play in tab-control systems, insufficient tension in control cables, loose bearings, and attachments, etc.
5. Wings and Fuselage—Reduction in bending or torsional rigidity through damage, slack, or defects in the covering or truss systems.

### Loading Airplanes

**Use extreme care in loading airplanes.** Additional loads, other than those normally carried in the plane, alter the balance and stability. In some airplanes so burdened, the response to the controls may be sluggish and recovery from spins difficult or impossible.

### Conditions Governing Maneuvers Permitted

The permissible maneuvers and load factors for any airplane do not apply when the airplane is loaded in excess of its designed gross weight. **Never load your airplane in excess of its permissible gross-weight condition.**

## STALLS

Airplanes with high wingloadings require considerable altitude to recover from a stall.

Using aileron to lift the dropping wing increases the stall and a spin may develop unless the airplane is inherently stable against spinning.

To correct stall, **reduce the angle of attack and get flying speed by diving.** The most dangerous feature is that so much altitude must be lost to regain control.

Most maneuvers, including banks, increase the effective gross weight of an airplane. **Be sure you have enough flying speed to provide for the increased stalling speed created by banks or other maneuvers which impose greater than normal loads on your plane.**

Don't make sloppy turns; skids at low speeds will stall your plane.

Don't make side-slip landings, particularly with flaps down. Flap shadow, or wake, is likely to blanket vertical surfaces at low speeds, and reduce rudder control.

Watch for warnings of an approaching stall.

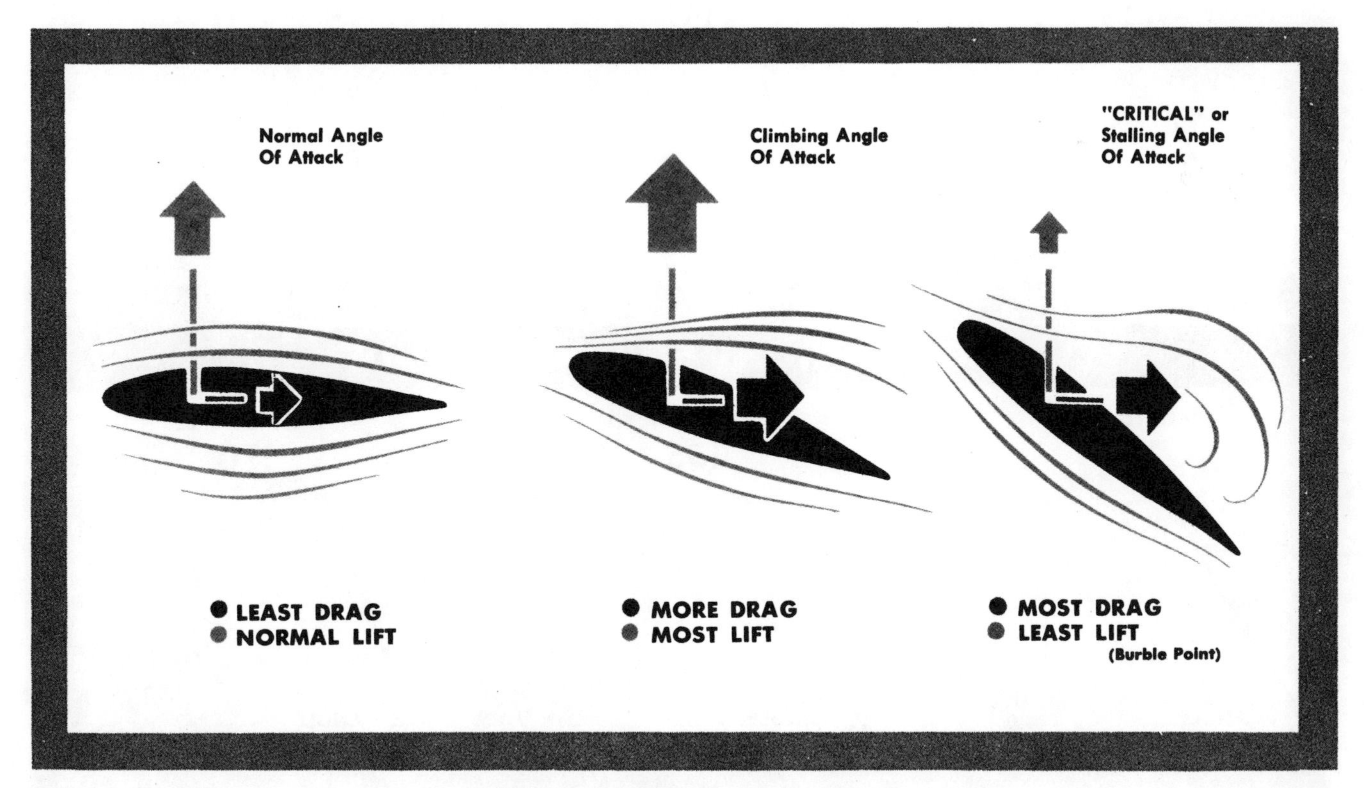


AERODYNAMICS OF A STALL

# HIGH-SPEED STALLS

You know that as the gross weight of an airplane goes up the stalling speed increases. When an airplane is turning in a banked attitude, centrifugal force has the same effect as increasing the gross weight, and the stalling speed of the airplane is thereby raised.

1. For example, if a B-17G airplane weighing 60,000 pounds is turned in a 60° bank, the pilot is in effect flying an airplane weighing 120,000 pounds and the stalling speed has been raised to 155 IAS. A 60,000 pound B-17G can be stalled at normal cruising speed merely by entering a steep bank, but the stall thus encountered is not a normal stall. The speed and high induced gross weight may cause a violent stall which endangers the airplane's structure. Stalling the airplane in a banked attitude may throw it into an inverted attitude from which several thousand feet will be required for recovery.

2. To show the degree of bank at which a B-17G stalls at 60,000 and 50,000 pounds for 135 IAS, 155 IAS, and 170 IAS, see the charts below:

3. Note that for a particular IAS the induced gross weight at the stall point is the same for both gross weights. As the stalling IAS is raised the induced gross weight at the stall point increases rapidly. The maximum induced gross weight which can be sustained without permanent deformation to the wing structure is 148,000 pounds. This corresponds roughly to a stalled bank at an IAS of 172 mph for any gross weight.

**INCREASED STALLING SPEED AND EFFECTIVE GROSS WEIGHT OF B17 AT VARIOUS ANGLES OF BANK**

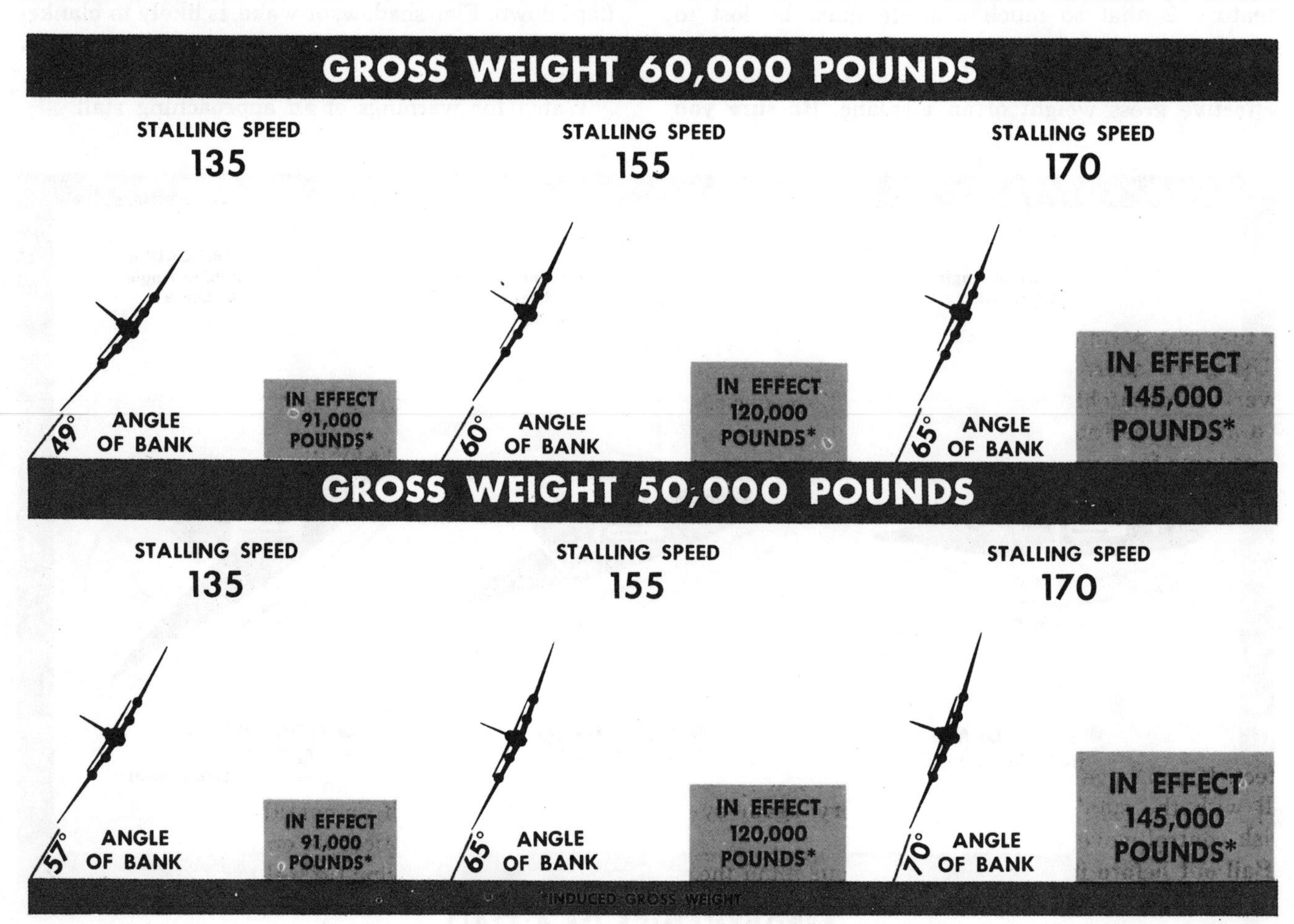

| Degree of Bank | Percent Increase in Normal Stalling Speed | Approximate Stalling Speed Based on 100 mph Normal Stalling Speed |
|---|---|---|
| 0° | 0 | 100 |
| 10° | .5 | 101 |
| 20° | 3.0 | 103 |
| 30° | 7.0 | 107 |
| 40° | 14.4 | 115 |
| 50° | 25.0 | 125 |
| 60° | 41.4 | 142 |
| 70° | 71.0 | 171 |
| 80° | 140 | 240 |
| 90° | Infinity | Infinity |

**The figures in the table show how banking an airplane with a stalling speed of 100 mph increases the stalling speed. The table may be used for any airplane by adding the percentage of increase in the second column to the known stalling speed of the airplane. Note particularly the sharp increase in steep banks.**

# SPINS

There is no infallible method for recovery from spins. In general, apply full rudder against direction of rotation. Follow by full forward stick until rotation stops.

**Move controls briskly to a position full against the spin.** Don't move controls slowly or cautiously, for that may permit the spin to continue indefinitely.

Using the ailerons as an additional means of recovery is debatable. In general, hold the ailerons in a neutral position unless otherwise instructed in procedures for specific airplanes.

### Precautions

**Don't spin an airplane that is restricted against spinning.** There are good reasons for the restriction.

**Don't get excited.** Mental confusion sometimes causes pilots to move controls opposite to intended procedure.

**Don't get impatient.** Don't overestimate the time element. Give the controls at least two turns to take effect. If you have sufficient altitude place controls full with the spin and attempt recovery again by brisk control movement.

**Bail out before it's too late.** Fix in your mind the altitude at which you're going to bail out if you get into a spin. Never attempt to stay with any tactical airplane in a spin closer than 5000 feet to the ground.

**Get plenty of altitude for recovery before starting an intentional spin in an airplane with doubtful spinning characteristics.** Your plane will lose a great deal of altitude during recovery.

**Caution**

**Complete any intentional spin in a tactical airplane at an altitude of at least 10,000 feet above the ground.**

### Factors Influencing Spin Recovery

**Position of wheels.** Generally, both spin and recovery are better with wheels up.

**Ammunition loads in wings.** Spin recovery usually is more sluggish with ammunition loads in wings. The effect is more noticeable on airplanes that normally spin with one wing tip well down; such a load may endanger recovery.

**Using the throttle** in an attempt to recover from a bad spin is effective at times. Use throttle for spin recovery only if normal recovery procedure is ineffective. Either short bursts or steady applied power may be necessary.

### Technique

The following techniques apply to most airplanes. **However, you must learn the specific procedure that applies to your airplane.**

Usually, if you hold the ailerons neutral and the elevator and rudder controls all the way with the spin, the spin will be more steady. Also, moving the controls from one extreme to the other (against the spin) gives a snappier and more forceful reaction for recovery.

Generally, use rudder first, then elevators because:

1. Reversed rudder checks the rate of rotation and causes the nose to go down.

2. Blanketing of the rudder by the elevator is usually less, and rudder is more effective when the elevator is up.

3. Hold the stick back while you hold the rudder opposite to the spin. As you slow down the rate of the spin, your elevators become progressively more effective.

4. The comparative effectiveness of rudder and elevator during spin recovery varies greatly. The elevator may be a more positive control than the rudder. Even then, don't change the sequence of control movement unless specifically instructed.

The first few turns of a spin are a transition period from straight and level flight to the final steady spin. During this time recovery often grows progressively more difficult. When spinning a strange airplane, use this period for testing the ease of recovery at various stages of the spin. In the first trial, for example, attempt recovery after 1/4 turn, and in succeeding trials, recover after 1/2, 3/4, 1, 1 1/2, 2, 3, etc., turns. Thus, you can learn the spinning characteristics more gradually.

Uncontrollable spins are not necessarily flat spins. In a flat spin the controls are probably less effective than in a steep (or normal) spin.

In general, an airplane spins differently in a right than in a left spin, because of rigging to overcome propeller torque and slipstream effects. An airplane may have satisfactory characteristics for a spin in one direction and unsatisfactory spin characteristics in the opposite direction.

Changes in weight and center-of-gravity position are likely to effect spinning characteristics. It is difficult to predict the effect of a definite center-of-gravity position on these characteristics.

**Don't spin an airplane intentionally when the CG position is rear of normal.** It is much more difficult, and sometimes impossible, to recover under this condition.

**Caution**

**Check your position in the cockpit before takeoff. Be sure that you get full travel and can apply full force all the way on all controls.**

## Remember these Tips

1. Be sure you can get full control movement and can apply force on the controls.
2. Don't get excited.
3. Don't be impatient. Leave the controls in recovery position long enough.
4. Fix in your mind beforehand the altitude at which you must bail out.
5. Get plenty of altitude before starting an intentional spin in tactical airplanes.
6. Don't spin with a rear of normal CG.
7. Don't spin an airplane restricted against spins.

REFERENCE: AAF Memorandum 62-4.

## COMPRESSIBILITY

When a fighter airplane approaches the speed of sound, it reaches a point where it loses its efficiency. Some or all of the following things happen:

1. Lift characteristics will be reduced or entirely destroyed.
2. Intense drag develops.
3. Stability and control characteristics change.
4. The tail buffets; the airplane may develop uncontrollable pitching and porpoising; it may develop uncontrollable rolling and yawing; or it may have a combination of these effects.
5. Severe vibration of the ailerons or of the entire plane may develop.
6. Trim changes in almost every case.

These are known as compressibility effects. The point at which an airplane enters compressibility varies with different models of aircraft; it varies on the airspeed indicator with changes of altitude, just as the speed of sound varies at different altitudes.

Compressibility is caused by a change in the physical behavior of air as it passes over any curved surface which is travelling at high speeds through the air mass. The continuity of the airflow breaks down; compression or shock waves develop on the wings or other surfaces and ahead of the airfoils; the air seems to split apart, shooting off at a tangent on both the upper and lower surfaces.

### Mach Number

The Mach (pronounced mock) number is the speed of an airplane divided by the speed of sound. The critical Mach number is the point at which a given airplane goes into compressibility. Because of design limitations, a plane actually goes into compressibility **before it reaches the speed of sound.** With some airplanes this point is reached at a speed 65% that of sound, others as high as 80%, depending upon the design of the model. Know the limiting speeds or critical Mach number of your plane before you attempt high speed maneuvers.

### Airspeed and Speed of Sound

True airspeed and indicated airspeed are identical at sea level. TAS is greater than that shown on the airspeed indicator as you reach higher altitudes. The higher you climb, the greater the discrepancy, until at 35,000 feet when your airspeed indicator registers 300 mph you're actually travelling more than 500 mph true airspeed.

The speed of sound is 760 mph at sea level and gradually drops to 660 mph at 35,000 feet. Your plane goes into compressibility when the percentage of TAS to speed of sound at any given altitude exceeds the critical Mach number.

### Avoid Speeds Within Compressibility Range

1. Know the critical Mach number or limiting airspeeds of your plane. Know TAS and IAS ratios at approximate altitudes where you intend to carry out maneuvers. Know the speed of sound at those altitudes.
2. Never attempt high-speed dives with external tanks until you are familiar with their effect.
3. Avoid violent use of ailerons in high-speed maneuvers and dives.
4. Keep rudder and elevator trim in relative position for level flight. Use trim tabs only if necessary. Make minute adjustments in trim only with great caution and, above all, observe the specific operating procedures for your airplane.
5. Never roll into a vertical dive above 30,000 feet, when it's possible to avoid it.
6. Use controls gently at high speeds and high altitudes. Do not attempt to recover from dives too quickly. Rough handling of the controls may cause entry into the compressibility range much more easily above 20,000 feet than below it.
7. Do not have a routine procedure for all fighter types as a group. Each type has its own peculiar high-speed dive and compressibility effects.

**Observe the Placard of Limiting Airspeeds posted in the cockpit of your airplane.**

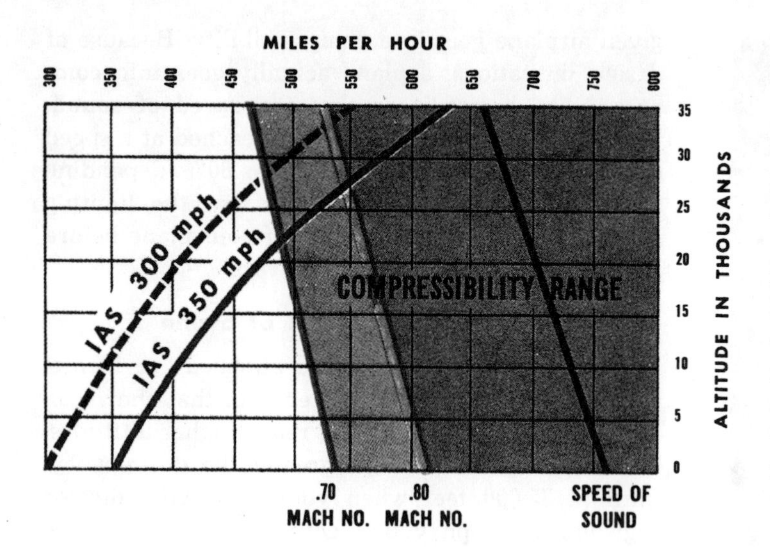


### Compressibility

The chart shows how easy it is to get into compressibility at high altitudes, with your airspeed indicator still well under the red line.

For example, if your airplane has a Mach number of .70 it will enter compressibility at 22,000 feet if the IAS is 350 mph (intersection of blue and red lines). With an IAS of 300 mph it would enter compressibility at 28,000 feet (dotted line and red line). Likewise shown are the points where it would enter compressibility if it had a Mach number of .80.

### Vertical Dive Recovery

**Don't dive too close to the ground.** Modern fighter airplanes require plenty of altitude for safe recovery from high-speed dives. Not only do you lose altitude rapidly; you must have a lot of air under you if you black out by exceeding the acceleration you can withstand in recovery.

The average pilot cannot withstand more than 4 G acceleration for periods (10 to 20 seconds) required for pullouts from high-speed vertical dives without blacking out. Accelerations higher than 4 G for extended periods will result in blacking out and loss of consciousness. **Don't get into a diving position which will require more than 4 G acceleration for safe recovery.** (See PIF 4-3-1.)

Unless you have an instrument showing acceleration, the best way to judge it is by the pressure on the seat of your pants. If you have ever blacked out, recall that pressure.

### Minimum Safe Altitudes

The chart on the opposite page shows the minimum safe altitude for recovery from vertical dives at various speeds, altitudes, and accelerations. These curves are based upon certain idealized conditions which make them applicable to all airplanes. However, they are not strictly accurate in all cases. For this reason, a safety factor of 25% is included.

**The altitude required for recovery from a vertical dive at constant acceleration is a function of true speed and not indicated speed. Different losses in altitude occur at different altitudes for the same indicated airspeed. This accounts for the difference in scales for the different altitudes.**

### Using the Chart on Next Page

The curves are plotted in terms of **indicated airspeeds** at several altitudes to facilitate their use in conjunction with the airspeed indicator.

**Remember altimeter readings usually lag several hundred feet behind the actual altitude in dives.**

The curves for accelerations higher than 4 G are given only to show the effect of variations in accelerations.

A chart, showing altitude required for recovery with an acceleration of 4 G, will usually be included in the operating instructions of all airplanes for which vertical dives are not prohibited.

### Caution

When diving at high speeds from high altitudes, avoid abrupt pullouts. Make special efforts to avoid them by proper handling of your trim tabs, control stick, and rudder. Such pullouts place dangerous strains on your airplane and your own body.

In high-speed dives bring your trim tabs back to approximately the neutral position when you reach an altitude of about 20,000 feet. Excessive use of trim tabs in these dives causes overload on the plane structure in excess of its designed load limits when you reach the lower altitudes where they become effective.

# MINIMUM SAFE ALTITUDE REQUIRED FOR PULL-OUT FROM A VERTICAL DIVE

## ALTITUDE AT START OF PULL-OUT

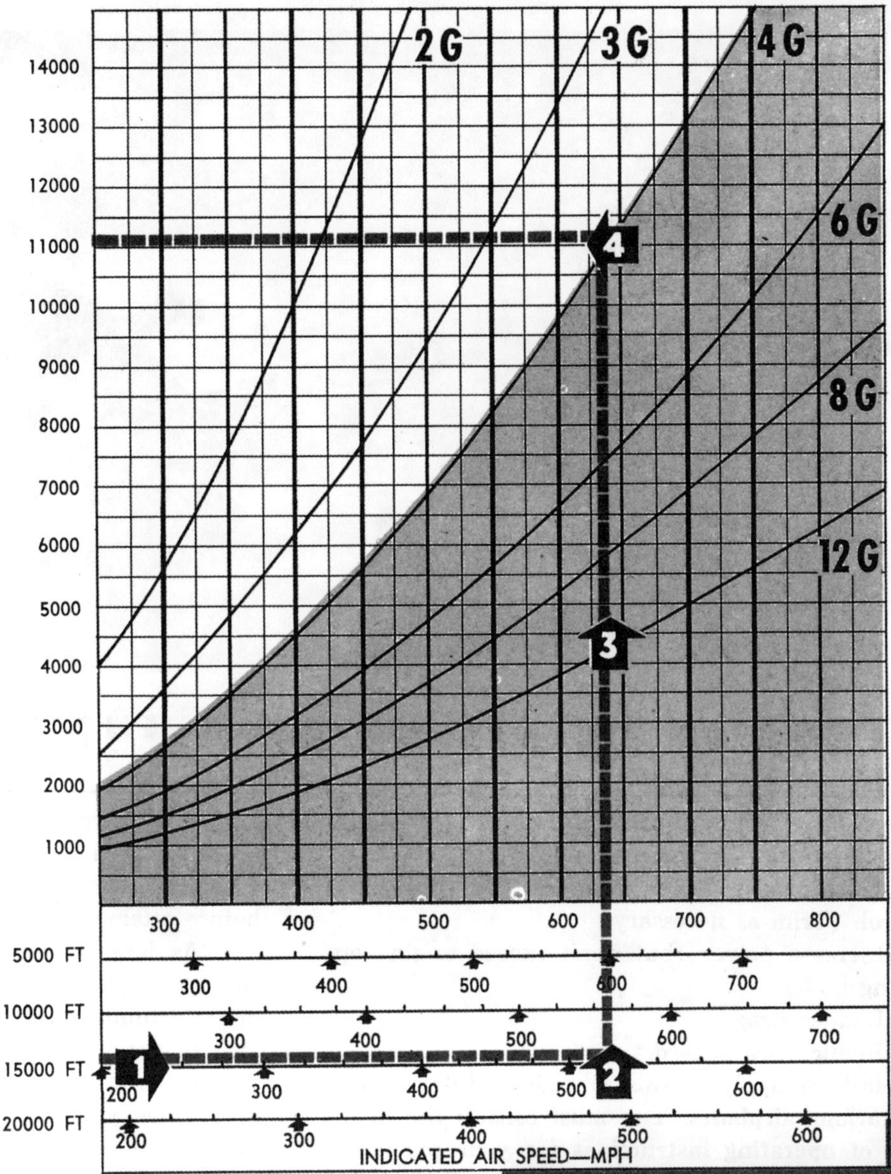


**PROBLEM** Find altitude remaining after making a pull-out from a vertical dive, starting the pull-out at 13,000 feet and 525 mph IAS and maintaining a constant acceleration of 4 G.

**SOLUTION** **1** 13,000 feet is nearest to the 15,000 feet altitude line. **2** Mark the 525 mph point on the 15,000 feet altitude line. **3** Sight up the vertical lines to point on 4 G line which is directly above point obtained in (2). **4** Sight back horizontally to scale on left to find minimum safe altitude required for pull-out (approximately 11,000 feet). **5** Subtract this figure from the 13,000 feet at which pull-out was started to find altitude remaining.

13,000 feet
11,000

altitude remaining = 2,000 feet

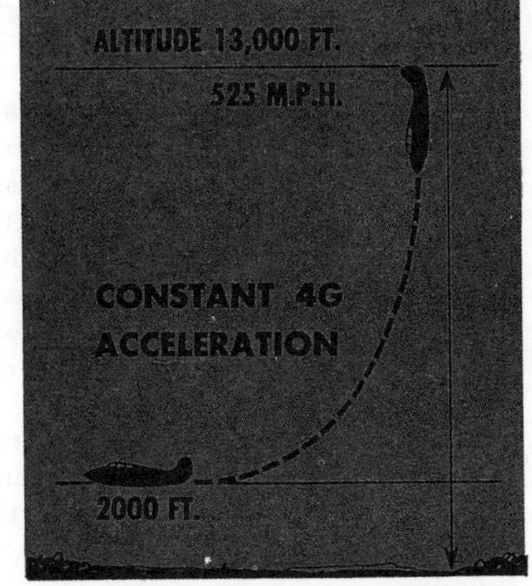

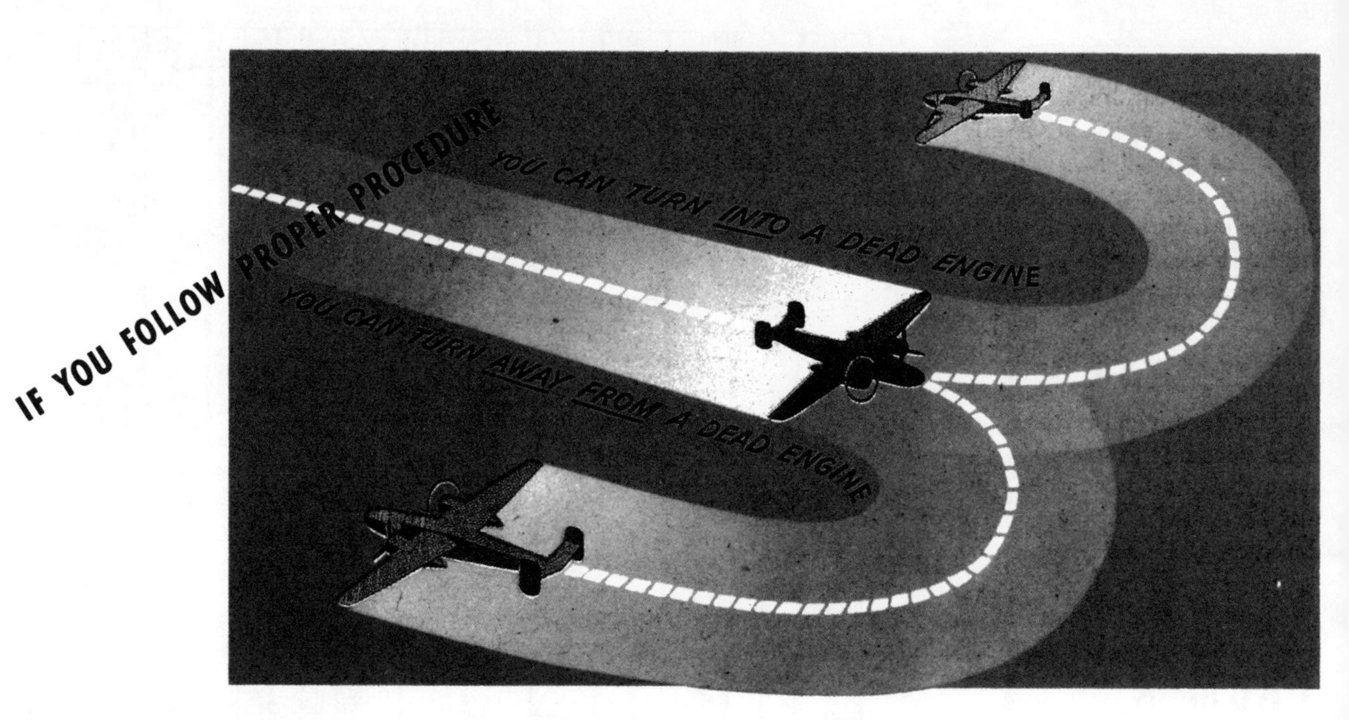


# ENGINE FAILURE ON MULTI-ENGINE AIRCRAFT

The loss of one engine in multi-engine aircraft is not serious if you will remember to do four things immediately. In sequence, they are:

1. Establish necessary airspeed and directional control. (Trim as necessary.)
2. Increase power (but don't exceed engine operating limits).
3. Reduce drag.
4. Reduce fire hazard.

**Details of applying this procedure differ widely for various airplanes. You must consult the handbook of operating instructions for your particular plane. See these instructions also regarding engine failure at takeoff.**

### Turns with One Dead Engine

Get this straight: turning into a dead engine is a simple and logical procedure, **after you have established necessary airspeed and directional control.** Failure to realize this has probably given rise to the old superstition that turns into a dead engine are disastrous. That belief is one of the commonest fallacies of hangar flying. The sooner you realize it is not true the safer you'll be.

**If you can make a smoothly coordinated turn into a live engine, or engines, you can turn into a dead engine just as easily if you observe the same rules exactly.** Remember:

1. Holding constant the airspeed prescribed (or greater) for your airplane with one engine not operating is the most important factor in making turns. You must trim the airplane for this prescribed speed before attempting the turn.
2. As long as airspeed is constant the thrust of the live engine, or engines, is balanced by the effect of the trimmed rudder.
3. In turns there is always increased wing loading. Without additional power from the good engine, or engines, you will have to sacrifice some altitude in order to maintain airspeed.

**Warning**

Avoid sudden and violent applications of power while making turns with one dead engine. Such action requires compensating change in trim, and results in an uncoordinated turn and possible loss of control over the airplane.

Don't attempt any turn without proper trim.

## Propellers

Feathering the propeller permits the stopping of a disabled and vibrating engine, decreases the drag of the propeller and increases the performance of the airplane.

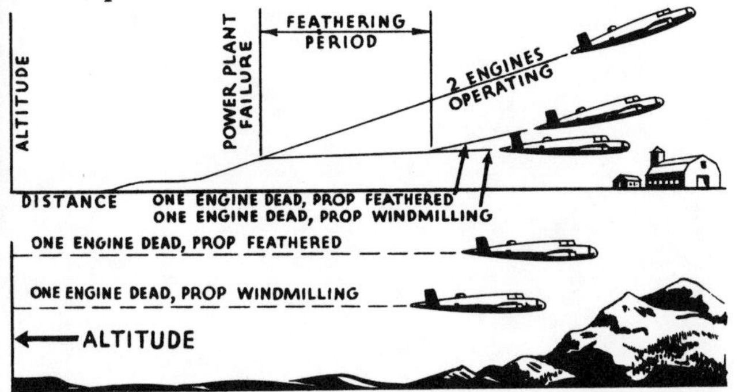


It is estimated that a feathered propeller on a twin-engined airplane increases the single engine ceiling 1,500 feet over that which can be maintained with the useless propeller "windmilling."

When a propeller is feathered, the momentary effect is the same as though the useless propeller were braked and, as the propeller fails to rotate, this effect exists until the propeller is unfeathered again. The single engine ceiling of the airplane during the unfeathering and until the engine is operating is approximately 3,000 feet lower than its ceiling would be with the useless propeller feathered.

See applicable Technical Orders for operation of propeller brakes on airplanes so equipped.

## Landing

When landing multi-engined airplanes with one or more engines inoperative:

1. Approach at an altitude of at least 1,000 feet above the field.
2. Lower landing gear.
3. Lower wing flaps approximately 20 degrees (or one-half flaps for indicators graduated in fractions).
4. Trim airplane for a lower power condition and a speed comfortably in excess of "flaps up stalling speed." (The margin of speed over "flaps up stalling speed" will vary from airplane to airplane, but normally will be 20 to 30 miles per hour.)
5. Maintain constant power and constant speed until a successful landing is reasonably assured, particularly until the danger of "undershooting" is eliminated.
6. Then lower the wing flaps to "full down," leaving power as previously set, and land in the normal manner.

If you have to go around again:

1. Raise landing gear promptly.
2. Apply power gradually, keeping the airplane trimmed directionally. Avoid excessive yaw caused by applying power too rapidly and failing to adjust directional trim.
3. Don't permit speed to fall below that maintained in the approach. If altitude is available, increase speed while setting power and trim.
4. Raise wing flaps **gradually** as soon as power and directional trim are set. Don't raise wing flaps until after checking to be sure air speed is more than "flaps up stalling speed."
5. If an engine has been shut off because of low oil pressure or some other fault not actually putting it out of commission, start it and operate at reduced power to make a landing with all engines operating.

## Failure On Take-Off

In case of engine failure on take-off:

Quickly throttle live engine (or engines) from take-off power to a lower value. Thus the airplane can be controlled directionally.

CAUTION: There is immediate danger from loss of directional control due to stalling of the vertical tail in attempting to compensate the uneven thrust of the engine (or engines).

Choose at once one of the following procedures:

1. Throttle down and land straight ahead if terrain permits. Less damage probably will be done with landing gear retracted unless there is a perfect field ahead.
2. Retract landing gear, feather propeller on dead engine and carefully build up flying speed and altitude until a safe landing can be made.

## Starting Engines In Flight

If an engine is stopped, when restarting **run at reduced rpm and power until the oil and cylinder temperatures indicate safe operation.**

## Practice Flights

Practice flights involving the operation of less than all engines may be made at any altitude provided that at no time will propellers be feathered, unfeathered, or braked at less than 5,000 feet above the surface over which the flight is being made. This restriction does not apply to two seater P-38 type airplanes when being used for training purposes provided the trainee pilot is accompanied by a qualified instructor.

REFERENCE: Technical Order 01-1-17, dated April 23, 1942.

# THE RECORD OF ONE STORM



# SECTION 3 WEATHER FLYING

# Weather Symbols

There are many symbols and combinations of symbols used in reporting weather by teletype. Being familiar with them and the standard procedure used in placing them in their proper sequence in the report will hasten your understanding of the weather picture. A typical teletype report appears below:

**WA N SPL 281624E E30⊕15⦶2VTRW-BD- 152/68/60↘22+/996/+⊕NW OCNL LTNG IN CLDS**

STATION — CLASSIFICATION OF REPORT — TYPE OF REPORT — DATE — TIME — CEILING — SKY — VISIBILITY — WEATHER — OBSTRUCTIONS TO VISION — BAROMETRIC PRESSURE — TEMPERATURE — DEW POINT — WIND — ALTIMETER SETTING — REMARKS

**ALL TELETYPE REPORTS FOLLOW THE ABOVE EXAMPLE**

It is deciphered as follows: Washington; observance of instrument flight rules required; special report at 1624, Eastern War Time; ceiling estimated at 3,000 feet; sky overcast, lower scattered clouds at 1,500 feet; visibility 2 miles, variable; thunderstorm; light rain shower; light blowing dust; barometric pressure 1015.2 millibars; temperature 68°F; dewpoint 60F°; wind, west northwest 22 miles per hour, strong gusts; altimeter setting, 29.96 inches; dark overcast to northwest, occasional lightning in clouds.

## SYMBOLS USED ON TELETYPE SEQUENCES

### SKY

○ Clear (less than 1/10 covered)

⦶ Scattered clouds (1/10 to 5/10 covered)

⦷ Broken clouds (6/10 to 9/10 covered)

⊕ Overcast (more than 9/10 covered)

No more than two sky symbols are grouped together. When there is no slant / all cloud bases are below 9,751 feet. When a sky symbol precedes a slant / those cloud bases are above 9,750 feet; when a sky symbol follows a slant / those cloud bases are below 9,751 feet.

⦶/, ⦷/, ⊕/ high scattered, etc.

⦶/⦶, ⦶/⦷, ⦷/⦶, ⦷/⦷, ⊕/⦶, ⊕/⦷ high scattered with lower scattered; high scattered with lower broken; etc.

⦶⦶, ⦶⦷, ⦷⦶, ⦷⦷, ⊕⦶, ⊕⦷ (no slant) scattered, with lower scattered; scattered with broken; etc.

The plus (+) or minus (—) sign preceding the cloudiness symbol indicates "dark" and "thin", respectively.

### WEATHER

A plus (+) sign following indicates HEAVY, a minus (—) sign following indicates LIGHT, no sign indicates MODERATE

- **R** Rain
- **S** Snow
- **L** Drizzle
- **ZR, ZL** Freezing rain, etc.
- **E** Sleet
- **A** Hail
- **AP** Small hail
- **SP** Snow pellets
- **SQ, RQ** Snow squall, etc.
- **T** Thunderstorm (no —)
- **SW, RW** Snow showers, etc.
- **TORNADO** (always written out in full)

### CLASSIFICATION

- **C** Satisfactory for contact flight
- **N** Requiring observance of IFR
- **X** Take-off and landing suspended
- (none) Station not at a controlled airport

### OBSTRUCTIONS TO VISION

A plus (+) sign following indicates HEAVY, a minus (—) sign following indicates LIGHT, no sign indicates MODERATE

- **F—** Damp haze (no +)
- **F** Fog
- **GF** Ground fog
- **IF** Ice fog
- **H** Haze (no + or —)
- **K** Smoke
- **D** Dust
- **BS, BD** Blowing snow, etc.
- **BN** Blowing sand
- **GS** Drifting snow
- (figures) Miles and/or fractions of miles
- (none) Visibility 10 miles or more
- **V** (Following visibility figures) variable visibility

### MISSING DATA

Indicated by the letter M entered in the place of the missing data.

### WIND

The velocity is indicated by figures representing its value in miles per hour, calm being indicated by the letter **C**. Signs shown below, following velocity figures, indicate:

- **+** Strong gusts
- **—** Fresh gusts
- **E** Estimated

### DIRECTION

Arrows flow with the wind as:

↓ North

↓↙NNE ↙ NE, etc.

### CEILINGS

Preceding ceiling figures

- **E** Estimated
- **M** Measured
- **W** Indefinite
- **A** Aircraft
- **P** Precipitation
- **V** (Following ceiling figures) variable ceiling
- **O** (Figure naught) ceiling is below 51 feet
- (no figures) Ceiling above 9,750 feet

CIRRUS
CIRROSTRATUS
25,000 FT.
CIRROCUMULUS
ALTOCUMULUS
ALTOSTRATUS
10,000 FT.
CUMULONIMBUS
STRATOCUMULUS
3,000 FT.
NIMBOSTRATUS
CUMULUS
STRATUS

# You and the Weather

There are three major weather hazards which every flyer is bound to encounter: Fog, thunderstorms, and icing.

The purpose of this information is to emphasize the conditions which lead to the formation of **Fog** so that you may anticipate it and take advance precautions against it, to show you **Thunderstorms** as you will encounter them in actual flight and what to do about them; to remind you of the nature and types of **Icing Conditions** and how to anticipate, recognize and analyze them.

The object of this discussion is not to make you a weatherman or a forecaster, but to offer you some "Cockpit Meteorology" based on Army, Navy, and Airline experience, to assist you in estimating weather during flight and to supplement your own good judgment, for which no formula or advice can be substituted.

### The Flight Plan

You are planning a flight. From take-off to landing, you will have control of practically everything **except** the weather. It makes little difference whether your flight is local or cross-country, you must fit your procedure into the pattern of winds and weather. If you don't, you may lose the opportunity to fly again.

First, check the weather map if one is available. It is a small scale picture of the weather in which are included the clues of weather trends. A weather map scales down the atmosphere to such an extent that symbols are necessary to represent measured quantities. It is to your advantage to know these symbols and to read them as you would a book.

Second, check current weather on latest teletype, radio and pilot reports. Correlation of the latest weather map and weather reports will give you a "motion picture" of flight conditions.

Third, be sure enroute weather is flyable. Icing, thunderstorms, strong head winds, extensive low ceilings and poor visibilities can keep you from reaching your destination especially if a few of these factors occur simultaneously.

Fourth, be sure both ceiling and visibility at your destination will be ample for landing procedures. Fogs, low clouds, drizzle, rain, snow, dust, smoke, blowing snow; all limit vision near the ground where you need it most.

Fifth, use every available means in estimating winds for your flight. It is much safer to have enough fuel for a 50 mph head wind which turns out to be only 30 mph, than to run out of gasoline because winds were 20 mph stronger than estimated.

Sixth, plan one or more alternate procedures against the possibility of flying into unforeseen or suddenly developing weather that makes your original plan impossible.

Seventh, work with a forecaster. Check your ideas against his and have him estimate enroute as well as terminal weather. He's there to help you make your flight a safe one. When your forecaster suggests you postpone your trip, you'd better double-check the weather before taking off against his advice because you're the one who will be up there battling the elements.

## Fogs

Fog may be defined as a cloud on the ground. It usually forms at night, as a result of the air being cooled by its contact with the ground, causing the air to become saturated. It also forms when surface winds carry air over terrain that permits slight super-saturation or when rain or snow falls into colder surface air.

### Ground Fog

This type of fog forms in nocturnally cooled surface air. It first appears in valleys and depressions as isolated patches, or, if terrain is level, where saturation of air is greatest. Patches of fog join to form a layer which deepens until an hour or two after sunrise.

If you are planning a night flight, particularly if arrival at your destination is planned near sunrise, a careful search for factors favorable to ground fog in the destination area is mandatory. When doubt exists as to presence of fog, arrival a few hours after sunrise should be the alternate plan. Do not exhaust fuel circling an airport, waiting for fog to lift.

Airports on hill tops are last to become foggy; airports in moist valleys tend to fog-in early.

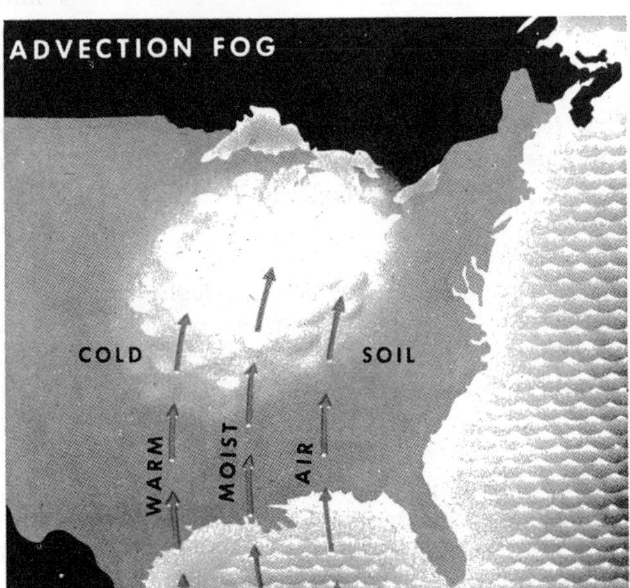


## Advection Fog

This fog develops in winter or early spring as a result of warm moist air drifting over cold ground or snow. Winds blowing northward off the Gulf of Mexico cause normal advection fog. This is likewise true in Spring over the cold Great Lakes.

Normal type advection fog is usually extensive. If you get caught over a region where it is forming or drifting you may not be able to reach an alternate airport, therefore exercise extreme caution in flying over or toward regions where there is any chance of advection fog.

## Sea Breeze Fog

Monsoon (sea breeze) fog is a feature of coastal areas where cool water lies close to sunheated land and is associated with a sea breeze. Inland from the shore the fog usually lifts into a layer of low clouds and then clears.

Monsoon fog often forms and spreads along coastal land in less than an hour. Because of the tricky nature of this fog, your flight plan to a coastal airport where a vigorous sea breeze is blowing, should include an alternate procedure to an inland field.

## Precipitation Fog

Precipitation (rain) fog is the result of relatively warm rain or snow falling into a layer of colder air. Precipitation fogs are usually associated with temperate zone cyclone fronts, particularly warm fronts. Precipitation fog frequently is preceded by broken low scud clouds or by stratus which thickens to the ground. Sometimes, fog forms rapidly at ground level and extends quickly over large areas. If there is any possibility of precipitation fog at your destination, an alternate plan is your safety factor.

**Upslope Fog**

Upslope fog develops in uphill winds. It is a cloud resting on a slope or hill top. Upslope fog, therefore, is confined to hilly terrain, particularly in the western half of the United States and the Appalachian Mountain region.

Upslope fog often forms rapidly and over large regions. An alternate airport on the lee side of a hill, preferably as far down the base of the hill as possible, is a good alternate plan if you are going to fly where upslope fog is a possibility. If you have no alternate, don't fly blind; landing on hilly fogged-in terrain is disastrous.

**Fog Warnings**

1. When temperature and dewpoint are only a few degrees apart.
2. A widespread precipitation area.
3. Within 200 miles of seacoast—if the wind is from the water.
4. In the Fall and Spring if the airport lies near a large river or lake.
5. If the flow of air is directed up a broad, fairly steep slope.

# *Estimating Weather in Flight*

### Cumulus Clouds & Thunderstorms

Cumulus clouds are billowed in appearance as the result of vertical currents. A pilot seeing cumuli is warned immediately that vertical currents are present in and below the clouds. For practical purposes, cumuli can be classed in three categories:

1. Flat cumulus usually having considerably more lateral development than vertical.
2. Towering cumulus frequently having more vertical development than lateral.
3. Cumulonimbus, or the shower and thunderstorm cloud.

### Flat Cumulus

Flat cumulus clouds are usually associated with favorable weather. They are readily identified by their lack of vertical development.

Air is mildly to moderately rough in and below the clouds but smooth above. Precipitation does not fall from them. Icing is light, occasionally moderate when their temperatures are below freezing. There is no particular need for staying out of flat cumulus unless light turbulence must be avoided or there is icing in the clouds.

### Towering Cumulus

Towering cumulus has considerable vertical development and a great tendency toward turret-type finger tops each of which contains a strong vertical current. Towering cumuli are forerunners of showers and thunderstorms.

Turbulence in the towering cumulus is normally moderate and occasionally severe. Turbulence below the cloud is mild to moderate. Rain or snow does not fall from this type of cloud. Its presence, however, indicates that rain or snow fall is imminent. Icing in towering cumulus is moderate to heavy where temperatures are 32°F or less, particularly in the towering fingers and domes.

Avoid towering cumulus. A flight around the cloud will take only slightly more time than one through, and in flying around it you can avoid exposing your plane and yourself to danger.

### Cumulonimbus

Cumulonimbus clouds are showers or thunderstorms. They assume many shapes and sizes but, in general, they appear as a massive towering cloud of great vertical and lateral development from which

rain or snow is falling. A thunderstorm is a typical cumulonimbus.

Turbulence is moderate to severe in the cumulonimbus. Updrafts and downdrafts are of such strength that aircraft, if caught, can be lifted thousands of feet upward or pushed downward against the ground. Icing in cumulonimbus where temperatures are 32°F and less is moderate to heavy. Hail, lightning, static in receivers, and St. Elmo's Fire in addition to icing, up and down drafts, and severe turbulence are characteristic dangers of the thunderstorm.

It is mandatory that cumulonimbus be avoided in all your flights; only when trapped with no alternative plan should you proceed through a massive cumulonimbus. Several procedures are possible upon encountering thunderstorms or very large showers. In order, they are:

1. Circumnavigate, preferably about that side from which the storm came.
2. Fly over the top of a saddleback between two towering clouds.
3. Fly underneath when (1) and (2) are impossible but only when the base is several thousand feet

above highest terrain. Turbulence is moderate to severe and downdrafts are dangerous below the storm base. Hail might also be present.

4. Land at an airport and await the storm passage on the ground if there is an airport or suitable field available.

If it is necessary to fly through the storm clouds, you have several choices.

1. Enter a thunderstorm preferably through the thin spots where you can see sunshine or blue sky. Storm clouds, however, shift continuously and sometimes imprison a plane before it can get through a thin spot.

2. In flying through black spots you are likely to find rough air, hail, up and down drafts, and lightning. Black spots certainly are not pleasant but also not extremely dangerous.

3. Avoid greenish and slightly off-color spots. They are usually regions of severe turbulence. Off-color is due in part to presence of electrical charges.

4. Flying at altitudes above the freezing level is sure to cause moderate to heavy icing.

5. Never land at an airport during a thunderstorm. Wind shifts may be disastrous.

**No flight through a thunderstorm is safe; some flight paths are only less dangerous than others.**

## HAZE LEVELS, OVERCAST BASES AND TOPS

Haze level is the top of a layer of air in which dust, smoke, haze and other debris are present. If impurities are numerous as they are in a dust storm, the haze level is as definite as the ground and horizon itself. Above the haze level, air is clear.

Fly above the haze level if you can. You have more visibility on top and there is no danger of getting dust into the engine. Air is smooth and carburetor icing is less likely.

### Flights on Top of Cloud Layers

**Advantages:**

1. If a temperature inversion caps the cloud layer, air is smooth on top and usually carburetor icing will not occur.

2. If the cloud top is flat little icing will occur.

**Disadvantages:**

1. You cannot estimate ceilings from on top a solid cloud layer. Rather than guess, call ground stations for weather report.

2. If the cloud top is undulating and has a tendency to tower, icing is certain at or below 32° F.

### Flights Under Cloud Layers

**Advantages:**

1. Contact procedure.

2. Icing in flat-based clouds is likely to be less than in undulating-based clouds.

**Disadvantages:**

1. You cannot estimate the top of a solid cloud when flying underneath. Pilot reports or sounding data must provide you with the necessary information.

2. Contact flights are dangerous in rough or hilly terrain; peaks sometimes merge with cloud bases causing zero-zero conditions.

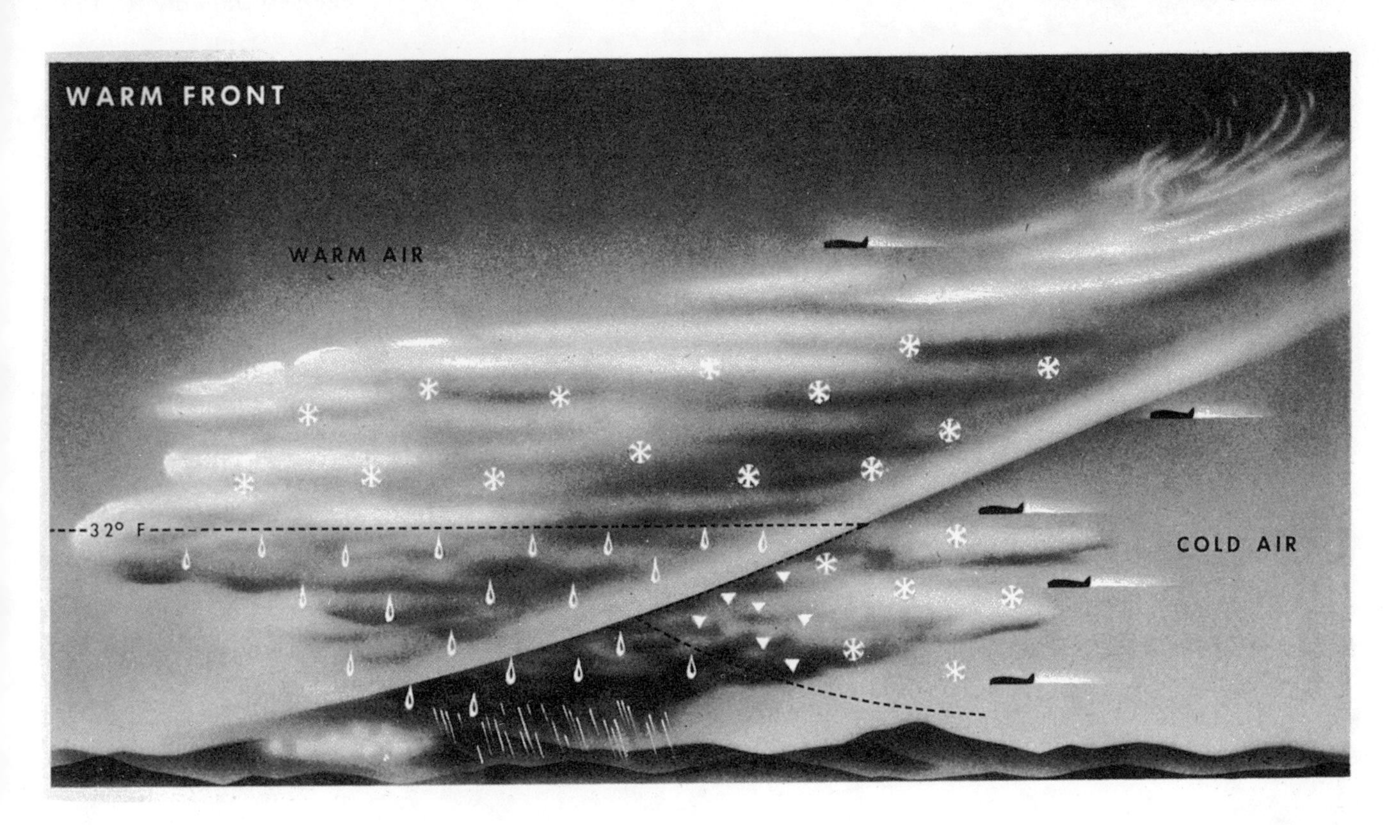


# Fronts

## Warm Fronts

Warm Fronts form when a warm air mass moves into a region occupied by a cold air mass, the warm air causing mass cooling as it rises. When you fly toward the front from the cold air side, warm front clouds will first appear as cirrus. Cirrus merge with cirrostratus which descend and merge with altostratus from which rain or snow falls. In the rain region, scud, stratus and stratocumulus lie below the altostratus, and higher clouds and fog frequently cover a considerable area near the surface front.

You will often find rough air in the cold air below the frontal surface; mild or moderate turbulence is frequent in the warm air clouds above the frontal surface. There is also moderate turbulence in the frontal surface itself.

Expect precipitation static in a frontal cloud. It sometimes is too harsh to permit voice communication and may smother reception from radio ranges.

Choice of flight paths through warm front clouds are numerous. Five are illustrated in the warm front sketch above.

1. Over the top is smooth and ice free but it is often too high to be practical or even possible.

2. You may encounter limited icing in clouds at sub-freezing temperatures in both the cold and warm air; you are certain to find considerable static and may find turbulence of moderate degree in warm-air clouds.

3. If you fly at an altitude just less than the freezing level in the warm air you eliminate serious icing possibilities, except for a relatively narrow band in the cold air before crossing the front. You must, however, expect rough air and static.

4. Altitudes somewhat below the freezing level of the warm air are free of icing in the warm air but present a very serious hazard in the cold air where sub-cooled rain droplets are present.

5. Contact flying in the frontal zone is dangerous. Clouds in nearly every well developed warm front include fog and scud at tree top heights.

The best choice is over-the-top. When you cannot get on top at some distance from the heavily clouded region your next best choice is a flight just below the freezing level of the warm air. Estimate the possibilities and plan your altitude accordingly. **Remember, it is usually easier to pick the right altitude approaching the frontal zone from the warm air side rather than the cold.**

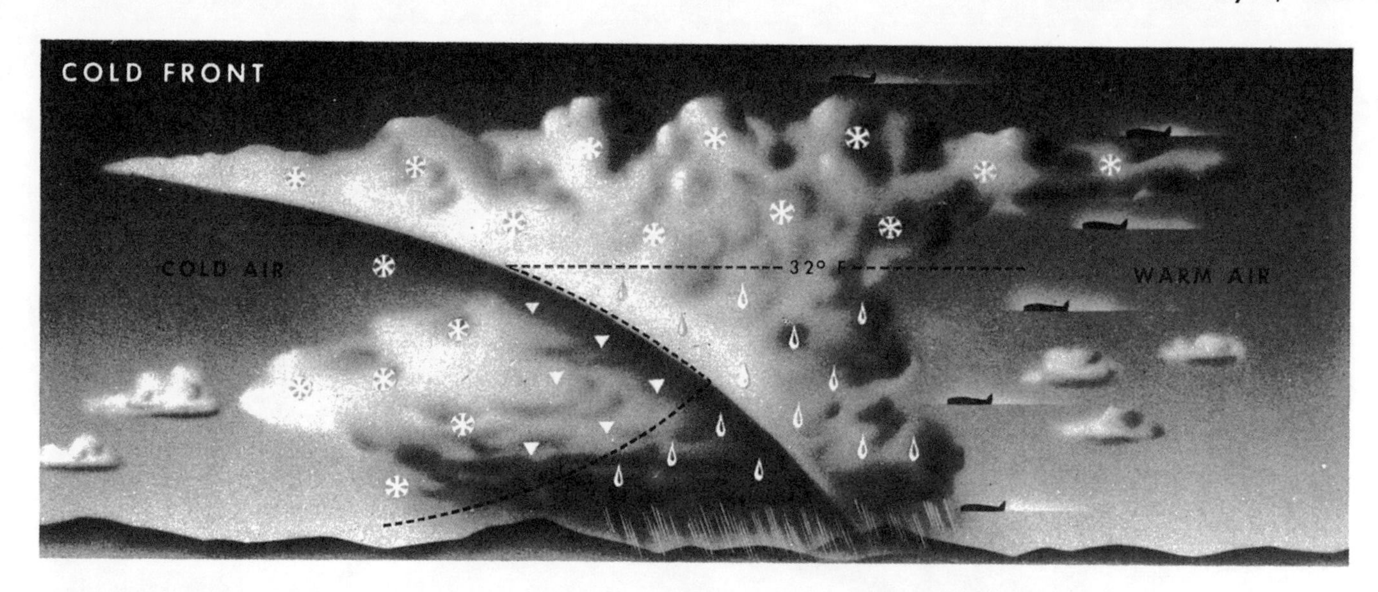
COLD FRONT
COLD AIR
32° F
WARM AIR

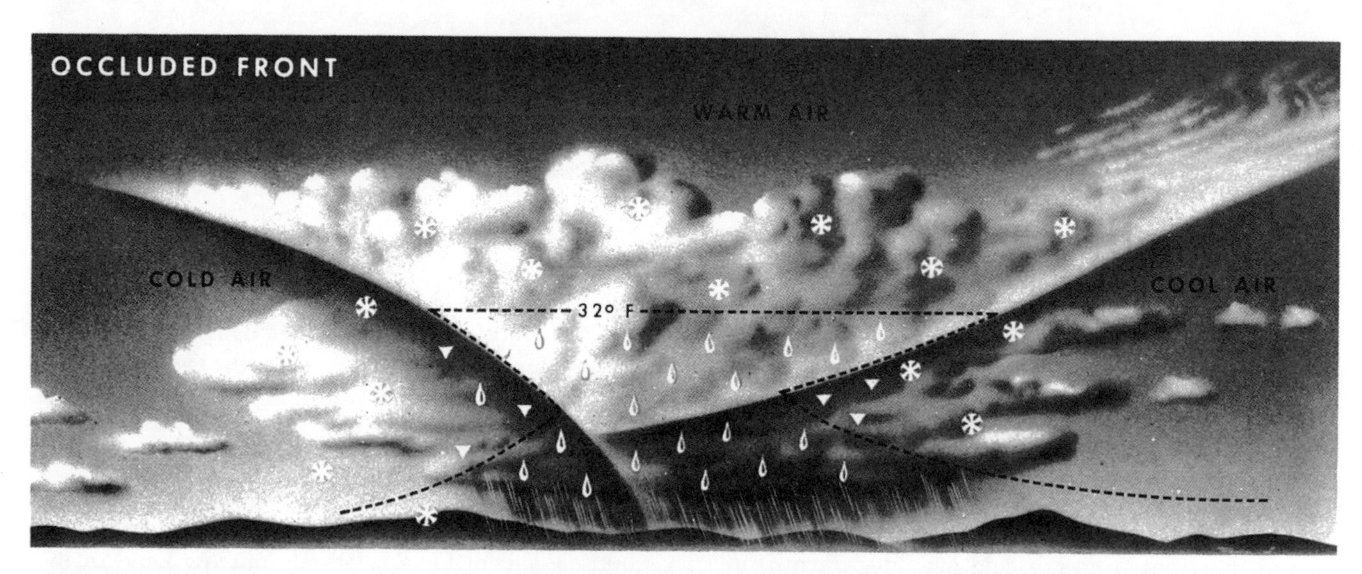
OCCLUDED FRONT
WARM AIR
COLD AIR
32° F
COOL AIR

COLD FRONT LINE SQUALL

### Cold Fronts

Cold Fronts form when a wedge of cold air moves into a region of warmer air. The warm air is forced upward and undergoes the process of cooling that may cause thunderstorms. These are generally concentrated in a zone 15 to 50 miles wide but may extend hundreds of miles in length.

Choose your flight path through Cold Fronts on the same basis as for Warm Fronts. Several flight paths with the plane approaching the Front from the warm side are indicated in the illustration above.

## *Line Squalls*

Line squalls are intense cold fronts along which unstable air and cumulonimbus occur in a line for hundreds of miles. Towering thunderstorms in Summer and snow showers in Winter are the usual features of a line squall. Summer time prefrontal line squalls are vicious and violent thunderstorm lines. They lie some distance ahead of a cold front and travel with greater speed than the front which is reduced to a partially clouded windshift line.

Avoid both cold front and prefrontal line squalls. Only when trapped with no other alternative should you attempt to go through the squall clouds. Choice of path through line squalls:

1. Best choice is find an airport nearby and wait on the ground for it to pass.
2. Fly through a saddle back between two towered clouds but be sure this is in the clear.
3. Pick the clearest spot between two storm centers and head directly into the squall at 90 degrees to the line. It will be rough but don't turn back when halfway through or you might get lost.
4. Fly through a thin spot where blue sky or sunshine is discernable through the break.
5. Enter through a dark spot, it will be rough but not as bad as other spots.
6. Fly underneath, turbulence will be moderate to heavy with possibility of hail, also up and down drafts.

Avoid off-color greenish or other off-hue spots, they are extremely turbulent and highly charged.

Remember that line squalls are violent by comparison with ordinary airmass thunderstorms. They often appear in an hour or two without warning. Be on the alert when flying in the warm sector of a temperate zone cyclone especially near the cold front. These are the places where line squalls usually first appear.

### Occluded Fronts

**Occluded front clouds** occur when a warm front collides with a cold front. The weather along an occluded front contains a combination of warm and cold front characteristics.

When an occluded front winds up into a low pressure center, that has been intensifying, be prepared to look for weather that will combine the worst features of both cold and warm fronts.

Plan flight paths through occluded front clouds with the characteristics of both fronts in mind.

### Cloud Features

Cloud features often help identify the cloud type and its probable origin, but you should never put full faith in what you can see from the ground or cockpit unless evidence is absolutely complete.

In general, frontal clouds will come across the sky in bands and in regular sequence from high types merging with and obscured by lower types.

Layers of stratus and stratocumulus not associated with fronts usually occur over somewhat irregular areas without smoothly cut edges. In contrast to frontal clouds they are not orderly. Stratocumulus clouds often occur in regions of thunderstorm activity. They may hide a thunderstorm if you are flying beneath them. Flying conditions are generally good over the top of stratocumulus, away from a front. When they occur alone, they do not indicate a front.

Cumulus clouds not associated with fronts occur at random. If you are unable to identify unexpected clouds properly in your flight path, proceed cautiously until you are sure they are harmless. Moisture in the air and clouds that result are responsible for practically all unfavorable and dangerous flying weather.

## *Getting the Weather*

1. Weather in flight is no longer confidential. In an emergency, you can get any weather information you ask for "in the clear" during one transmission.
2. You may request Terminal or Landing Weather from Range Stations or Control Towers. Reply will be limited to any two stations "in the clear" for any one request.
3. You may transmit such weather information during flight as is requested.

# Rules for Flying Weather

**1.** When approaching a thunderstorm, analyze it before you encounter the surrounding clouds. They may obscure important characteristics of the storm after you get into them.

**2.** Before attempting to fly any thunderstorm, study the situation thoroughly.

**3.** Whenever possible, circumnavigate a storm. Always fly around isolated air mass thunderstorms.

**4.** In coastal regions, where thunderstorms prevail along the mountains, fly a few miles to seaward and avoid them.

**5.** Thunderstorms over islands may be thousands of feet higher than those over the open sea. Fly around them.

**6.** Cold front thunderstorms generally stretch too far to fly around. Remember, the storm front is a series of individual storms linked by intervening clouds. If you must go through, fly between the storm centers or over the saddlebacks.

**7.** If you can't see blue sky beyond the storm and must go through, determine the direction the storm is taking and head in at a right angle.

**8.** Once you have headed into a storm, don't turn around on account of turbulence, rain, or hail. If you do, you'll have to fly through the same condition twice, and you may get lost. Hold your original course.

**9.** In entering the front of a thunderstorm, you will encounter updrafts. Go in low, and if conditions permit, fly underneath the base of the storm.

**10.** Entering a storm from the rear, you will experience downdrafts first. Go in high.

**11.** In flying under a storm, the higher the flight level, the rougher the trip. Fly about one-third of the distance from the ground to the base of the clouds if you can; but don't go underneath unless you can maintain contact flight.

**12.** Don't try to fly underneath a storm along mountain ranges unless there is a good ceiling and you can see peaks and ridges clearly. At sea you can usually count on being able to fly under any thunderstorm in daylight.

**13.** Never land at an airport when a thunderstorm is advancing toward the field. Shifting surface winds make it too hazardous. Wait until the storm center has passed, and the winds have stopped shifting, before you land.

**14.** When you expect to try level flight in flying a storm, get altitude before approaching it, so that you can inspect the storm line before selecting your course.

**15.** The altitude necessary to fly around the tops and over the saddlebacks of a thunderstorm will vary with the seasons and the latitude in which you encounter the storm. In high latitudes, 12,000 to 15,000 feet is generally sufficient. In the tropics, the tops of the saddlebacks may be above the ceiling of your aircraft and you may have to fly through the saddlebacks on instruments, a procedure recommended only for high performance aircraft. Over the open sea, 15,000 feet of altitude usually will clear the saddlebacks.

**16.** Lightning is of little consequence when you are flying an all metal, closed-cockpit plane, which acts as a perfect conductor. Don't worry about it. Switch on the cockpit light and keep your eyes on the instrument panel, so that the bright flashes won't blind you. If you are flying an open cockpit plane, or a plane with a plywood or plastic fuselage, better keep away from the lightning.

KEEP YOUR HEAD, KEEP YOUR COURSE—YOU'LL COME THROUGH

# *Remember This*

1. Don't fly cross country over a low cloud layer which extends for hundreds of miles even though ceilings may be 1000 feet and the top is only a few thousand feet. Get your cross country instrument practice when the cloud layer is less extensive.

2. When an active warm front is approaching, delay your take-off until the front has passed, and even then, check to be sure you will not cross the warm front.

3. In winter be on the lookout for snow. If the temperatures are near, or below freezing, the precipitation is likely to be in the form of snow. If it is, remember that snow can cause a low effective ceiling in just a few minutes where before there was only an intermediate cloud layer with unlimited visibility. Snow static often makes your radio useless, and it is much easier for heavy snow to fall than heavy rain.

4. Carefully consider the intensity of each cold front before you attempt to fly through it. If there is any doubt in your mind, sit down and wait until it passes.

5. Avoid flight where icing conditions are likely to be found.

**Special conditions you should watch for in Spring:**

1. Formation of secondary storms on cold fronts.
2. Widespread low cloud area is still a hazard.
3. Cold fronts may be at their most violent stage.

During Summer, there is one paramount rule: **Thunderstorms are dangerous.**

If your flight plan and forecast work out as expected, and they will most of the time, you know what you are encountering. If, on the other hand, your forecast of winds and weather and your flight plan do not turn out as planned, you have no foolproof method of knowing actually what is ahead from what you can observe from the cockpit. Contact ground stations often for weather reports, ask for advice or revised forecast, call other pilots in the same area if possible, **consult Pilots' Advisory Service.** If you don't like the looks of what you are seemingly encountering, turn around while you still have time and proceed to your point of departure or some convenient field. Remember, you must play along with the weather and it's no disgrace to admit you are out-maneuvered.

**No formula has ever been devised to fly bad weather nor has any substitute for experience and good judgment been discovered. The information offered here is intended merely as a guide; for detailed information study Technique in Weather T. O. No. 30-100 D-1.**

# SUMMER FLYING

### Before Starting

Check status of airplane on Form 1A.

Be familiar with check list for the airplane you are flying.

Don't over-prime the engine.

### Warm-Up

If no oil pressure shows on the oil pressure gage within 30 seconds after starting, stop the engine and have it checked.

Run the engine at 600 to 800 rpm until oil pressure is steady. Don't exceed 1,000 rpm until oil temperature reaches 40°C (104°F).

**Keep the warm-up time to a minimum,** and nose the airplane into the wind for better cooling, as cylinder head temperatures will rise very rapidly during warm weather.

### Prior to Take-Off

Keep full-throttle engine and instrument test as short as possible during the summer.

Remember that take-off distances will be longer because of the thinner air encountered in warm weather.

### Care During Flight

Do not climb the airplane at less than the specified climbing speed for that particular airplane. Low flying will result in higher temperatures than normal.

### Landing

In hot weather the air is not as dense as in cold weather, other things being equal; so your true stalling speed is greater in summer than in winter—but your indicated stalling speed is the same.

**Therefore, don't expect to land in the same distance that you would during the winter, as the airplane will come in hotter than usual.**

#### Caution

Remember that although the air-speed indicator reads the same the ground speed will be increased because of the thinner air.

### Care of the Airplane

**Keep canvas covers on the windshield and windows when they are exposed to the sun.** If this is not done the heat of the sun will soften and bend the Plexiglas and will also cause malfunctioning of radio equipment, gyro equipment, and instruments.

**Keep doors and hatches open** so that air can circulate through the airplane, except when it is necessary to keep the airplane locked.

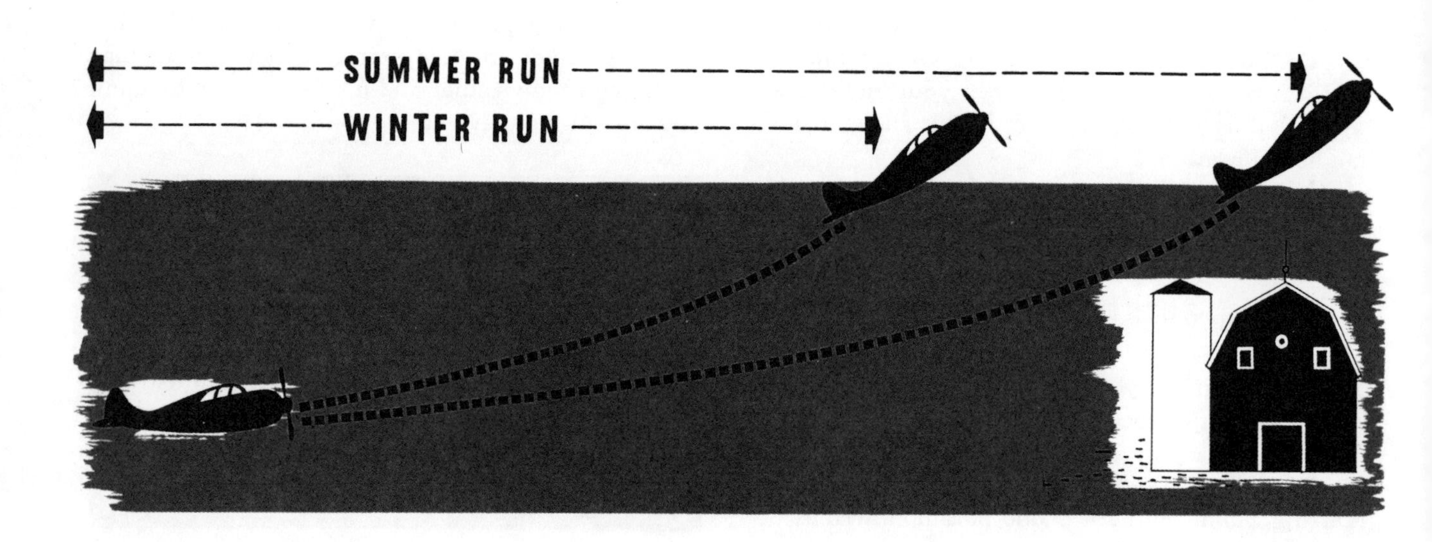

# WINTER FLYING

### To Start Engines

1. Before starting the engine, check the "Y" drain and oil tank sump drain to be sure that fluid oil is present at these points. If ice or congealed oil prevents flow at either point it will have to be thawed out.

2. In winter, prime the engine more than in warm weather—before the starter is engaged and **after** the engine fires. Turn the electric primers "ON" just prior to engaging the starter and hold "ON" until the engine fires steadily.

3. If the engine fails to start in three or four attempts, moisture may have accumulated on the plugs. To remove this moisture, take at least one plug from each cylinder, heat the plugs until warm, and replace.

4. If the engine fails to start after a reasonable time, inspect for probable causes and consult the Handbook of Service Instructions for that engine.

### Warm-Up

If no pressure shows on the oil pressure gage within 30 seconds after starting, stop the engine and have it checked.

Run at 900 to 1,100 rpm until oil pressure is normal. Do not exceed 1,200 rpm until oil temperature reaches 20°C (68°F). However, emergency take-offs may be made as soon as the oil pressure is normal, oil temperature begins to rise, and the engine is smooth. Refer to Handbook of Operating Instructions for your engine.

With tightly baffled air-cooled engines, nose into the wind for the warm-up as the cooling of cylinders will be poor on the ground. Keep cowl flaps or engine shutters full open for all ground operation.

Use necessary carburetor heat. See "Carburetor Icing" in "Weather Flying" section.

### Take-Off

1. Do not take off with frost or snow on the wings. **Ice must be removed completely. The thinnest film of ice forms a base upon which more ice can rapidly form.** Pay special attention to control hinges and surfaces. Use alcohol for cleaning frost from windows and windshields. Steel wool may be used on plate glass windshields.

2. Moisture condensation may cause ice accumulation **inside** the wings. **It must be removed with heat before take-off.**

3. If frosting is severe, taxi to the take-off position with frost covers in place.

4. Use carburetor heat in extreme cold for take-off because fuel does not vaporize sufficiently for good combustion at low temperatures. **Use only enough heat to insure smooth engine operation.** Do not exceed +40°C (104°F) carburetor air temperature at any time.

5. Don't take off on soft snow. Taxi along the runway a few times to pack the snow.

### Care During Flight

1. Following a take-off from snow, wet, or slush-covered fields, operate the landing gear, flaps, and bomb bay doors through several complete cycles to prevent freezing in the UP position.

2. **You will not see, hear or feel carburetor ice until it is too late. Watch your instruments so you will detect carburetor icing conditions. For complete instructions see ICING—CARBURETOR, PIF No. 3-6.**

3. **Icing of surfaces, wings, and propellers may occur under conditions of high humidity or visible moisture near freezing temperatures. See ICING—WINGS—PROPS, PIF No. 3-7 section for complete instructions.**

4. Increase propeller speed by about 200 rpm every half hour to assure continued governing. Return at once to the desired cruising rpm.

5. If carburetor icing is expected, move the throttle frequently to prevent freezing in one position (throttle ice).

6. Stay on a prearranged flight course so searchers will be able to find you if you are forced down. Except in extreme emergency, it is better to land or crash-land than to bail out.

7. Ice formation can cause engine failure by clogging the air filters. Remove or by-pass filters before long flights or before flights over water even though take-off must be made from a dusty field. **See ICING—CARBURETOR, PIF No. 3-6, for complete instructions.**

8. While letting down for a landing, watch engine temperatures closely. Temperature inversions are likely in winter and the ground air may be 15 to 30°C (50-86°F) colder than at altitude.

9. If oil temperatures continue to rise with the oil cooler shutters open, close shutters fully. The oil in the cooler may have congealed.

### When to Dilute Engine Oil

During cold weather, dilute (thin) the engine oil with gasoline after each engine run to aid starting and reduce starting wear on the engine. Engine oil should be diluted before stopping the engines if the expected temperatures are 5°C (41°F) or below. The amount of dilution will depend on the number of minutes the dilution switches are held "ON."

### How to Dilute Engine Oil

1. Run engine at 1,000 to 1,200 rpm. **Oil cannot be diluted unless engine is running.**
2. Maintain an oil temperature less than 50°C (122°F) and an oil pressure above 15 lb. sq. in. **If oil temperature rises or oil pressure falls beyond these limits, shut down and allow engine to cool.**
3. If the airplane has an automatic dilution switch installed follow dilution instructions on the placard.
4. If the airplane has the manual dilution switch hold the switch on for the period given in the instructions for that particular airplane. At the end of that period keep the switch on until the engine is stopped.
5. Proper operation of the dilution system is indicated by considerable drop in fuel pressure. If fuel pressure does not drop off investigate the dilution system.
6. A complete redilution of the engine is required only after one half hour or more operation at normal oil temperature, as this is the time required to boil off the gasoline.
7. If it is necessary to service the oil tank split the dilution period in half and service between the two periods.

### General Rules

If specific dilution instructions are not available, follow these general rules:

| Anticipated Lowest Outside Air Temperature | Dilution Time Minutes (One Period) |
|---|---|
| +4° to −12°C | 3 |
| −12° to −29°C | 6 |
| −29° to −46°C | 9 |

For each 5°C below −46°C add one minute to the time given.

### How to Dilute Propeller Oil

If the engine is equipped with Hamilton Standard Hydromatic propellers, near the end of the dilution period depress the propeller feathering button long enough to give a maximum of 400 rpm drop. Then pull out the button. Repeat three times to insure proper feathering and governing after the next start.

### Turbosuperchargers

If the airplane is equipped with turbosuperchargers regulated by engine oil, move the turbo contact handles back and forth during the last two minutes of the dilution period.

**NOTE: After dilution, drain all moisture from the "Y" drain and oil tank sump drain.**

### Winter Care of the Airplane

**Winterization.**—Check with your crew chief to see that all necessary winter changes have been made on your airplane.

**Preheat the Engines.**—Portable ground heaters should be used, if available, to preheat engines and compartments prior to starting.

**Preheat the Oil.**—Under extreme conditions Oil Immersion Heaters may be used. But they should not be necessary if the oil is properly diluted. For details on the use of Immersion Heaters see the Handbook of Cold Weather Operations for your particular airplane.

**NOTE: The oil tank must not be filled to capacity when the oil system is cold.**

**Conserve Batteries.**—Use an auxiliary power source for starting or for any power required when engines are not running. Batteries must be filled in a warm place and the charge maintained at 1.275 to 1.300 specific gravity. If auxiliary power is not available put the airplane batteries in a warm place until just before the start.

**Use Protective Covers.**—Use wing and engine covers to prevent the accumulation of snow, ice, and frost.

**Protect Your Tires.**—Place insulation under the tires to prevent them from freezing to the ground.

**Parking Brakes.**—May be set after they have cooled. Wait until after the oil has been diluted and the brakes have had plenty of time to cool before setting them. Otherwise they may freeze in locked position.

**Prevent Frosting.**—When parked, leave a hatch or panel open to prevent windows and windshield from frosting. Do not open a panel that will permit snow to enter and accumulate in the airplane.

# Icing- CARBURETOR ICING

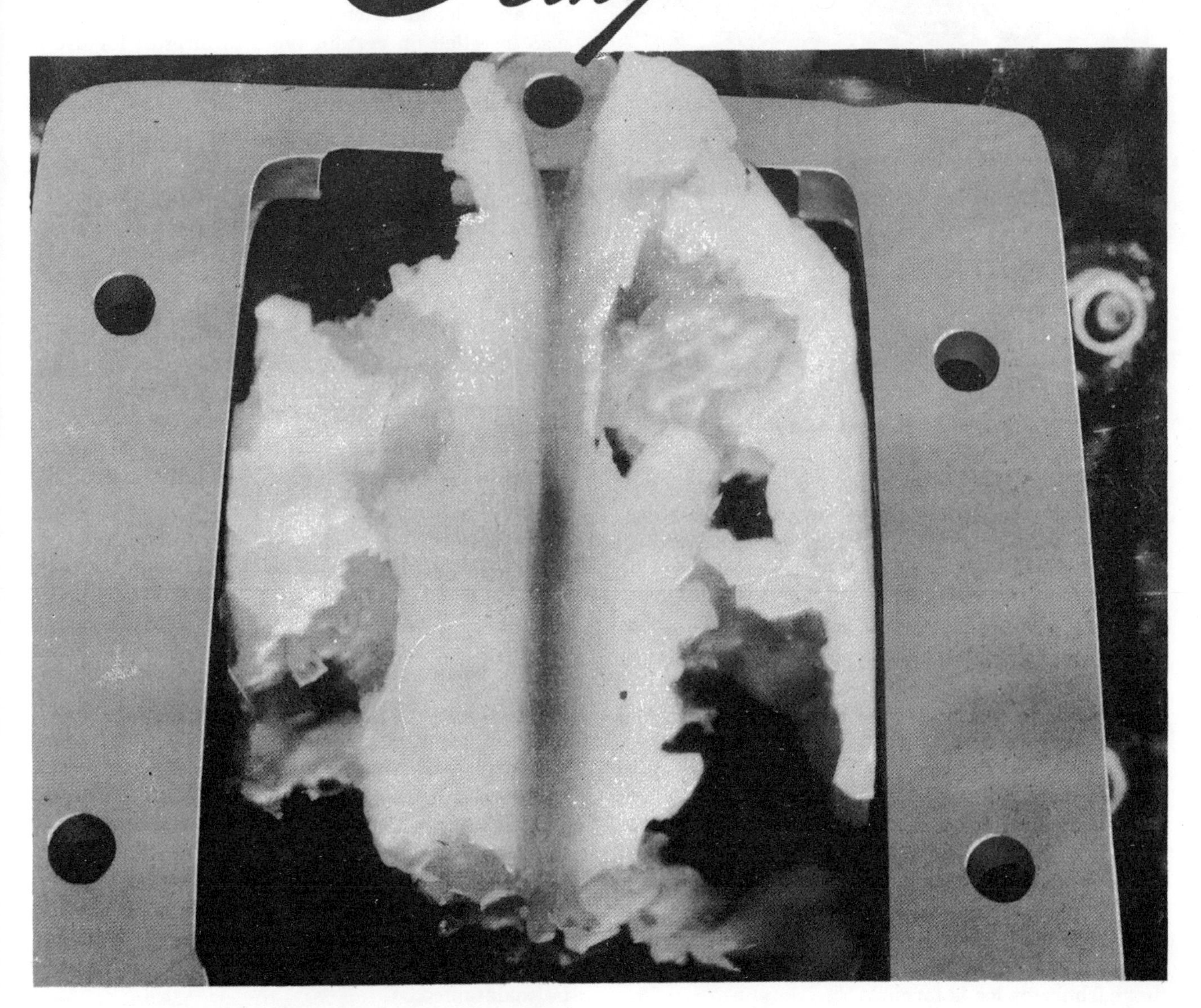

## There is no such thing as a NON-ICING CARBURETOR

Basically, a carburetor functions a great deal like the expansion valve in a mechanical refrigerator with the result that a temperature difference as great as 60°F can exist between the free outside air temperature and the carburetor-mixture temperature. A carburetor literally can manufacture its own ice . . . and at any season of the year—winter or summer.

**Your carburetor can ice when the free-air temperature is as high as 95°F (+35°C).**

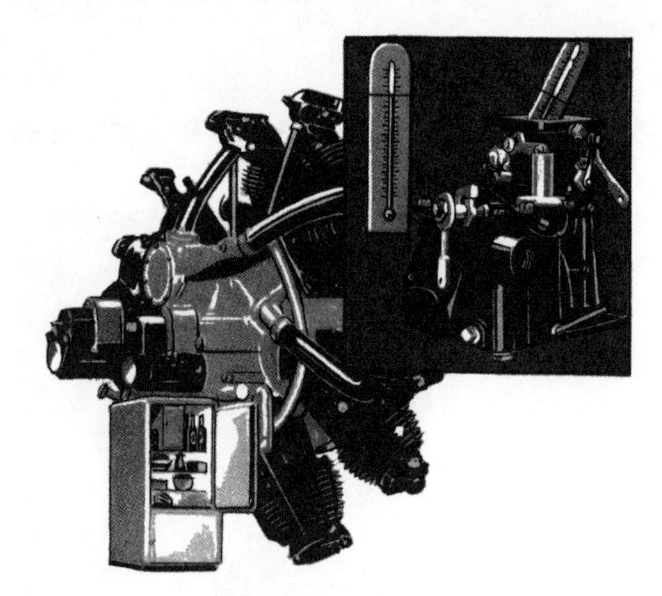

## Induction-System Icing

### Cause

Induction-system icing is likely to occur when flying in clouds, rain, fog, sleet, wet snow, or when relative humidity is high. Ice is formed by a cooling effect on the evaporated fuel after it has been introduced into the carburetor air stream. Under humid conditions ice has been known to form when the free air temperature is as high as 95°F.

### Symptoms

Normally, if the airplane has a fixed pitch propeller, the formation of ice in the induction system can be detected by a gradual loss of rpm and/or a gradual loss of manifold pressure. This may be noted without any change having been made in the throttle position or in the attitude of flight.

Under most conditions, the formation of ice is a slow process. Therefore, a pilot not on his guard might advance the throttle gradually to maintain constant rpm or manifold pressure without realizing that carburetor ice is forming.

**Rare, local conditions may change these indications somewhat, so be on the alert.** Under extreme conditions, ice may form so rapidly that power loss will be abrupt.

When an airplane has automatic boost control as well as a governing propeller, the pilot will have no warning of ice formation until it is so bad that the regulator can no longer maintain the boost. In such aircraft, **use carburetor heat when any possibility of icing exists.**

### Cures

**Carburetor Heaters.** The function of the carburetor air heater is to prevent or eliminate the formation of ice within the carburetor, around the throttle valve adapter, or elsewhere in the passages of the induction system.

**De-Icing Equipment.** In addition to carburetor air heaters, alcohol carburetor de-icing equipment is also installed on certain types of aircraft for use in emergencies and when descending through an overcast with reduced power.

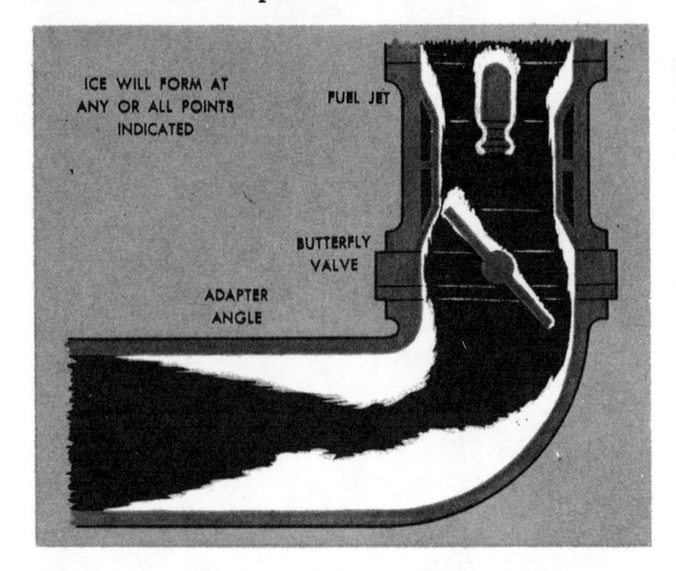


Carburetor Ice

## Types of Carburetor Air Heaters

Carburetor air heaters may be divided roughly into the following classes:

**Exhaust**—utilize heat from the exhaust piping and have an adjustable control.

**Alternate air inlet**—provide two-position selection of air inlets.

**Hot spot**—found only on early Pratt & Whitney engines. Located between the carburetor and the engine. Mixture-temperature reading is affected by such heaters, so temperatures of 50°F higher must be maintained.

Warmed air from behind the cooling radiators may also be utilized. Intensifier tubes employing this principle may be operated the same as exhaust heaters.

Turbo supercharged engines may use intercooler shutters for carburetor heating. Some later models use hot exhaust gas introduced directly into the carburetor.

## When Ice Forms

With a fixed pitch propeller, if rpm or manifold pressure begins to drop for no apparent reason **put your carburetor heat full on.**

There will be a slight loss of power when heat is applied. If the manifold pressure, or rpm, continues to drop:

Adjust the mixture until you have obtained smooth operation. However, be sure to avoid getting too lean a mixture, as this may cause detonation.

**Always act quickly under these conditions. Don't delay. The accumulation of ice is progressive and often you may have only a minute to do something about it.**

**As a Last Resort.** Turn on the alcohol de-icing pump if the plane is equipped with one. Or put the carburetor heat on full cold and lean the mixture until backfiring occurs. The backfire may loosen the ice and shake it clear of the intake passages.

**This is a dangerous procedure, however, and should be used only as a last resort** and then only with the carburetor heat on COLD.

Don't cause a violent backfire by extreme over-leaning. Lean the mixture just enough to cause normal backfire.

### Prevention

If icing conditions prevail, don't wait for engine troubles to warn you of carburetor icing. **It is easier to prevent ice formation than to remove it.**

Prevent ice by using whatever carburetor heat is necessary. Remember, the control may have to be **nearly full on** to add much heat. Also bear in mind that an automatic boost control will give no warning

### Manifold Pressure Drop

When the carburetor heat is turned on, a slight loss of power will be indicated by a drop in manifold pressure due to the reduced air-intake pressure and the lower density of the warm air. This loss of power is no sacrifice in engine efficiency when operating with the carburetor mixture control in the automatic position, since an increase in carburetor air temperature does not enrich the mixture. **However, when operating with manual mixture control** any increase in carburetor air temperature will enrich the mixture so **it will be necessary to readjust the manual control whenever the carburetor heater control setting is changed.**

### CARBURETOR AIR THERMOMETERS

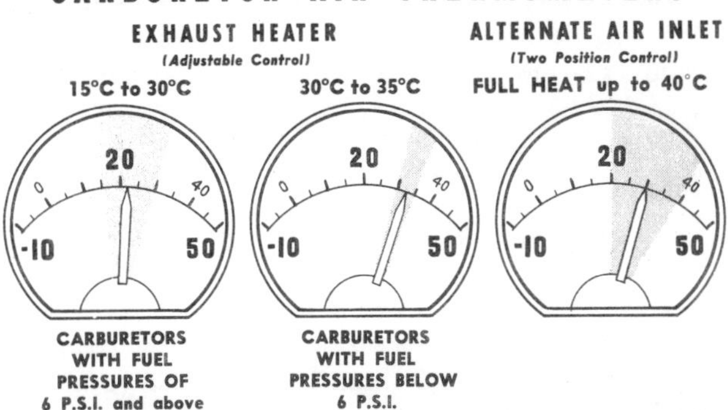


### CARBURETOR MIXTURE THERMOMETERS

(Installed only with Carburetors having Fuel Pressures below 10 p.s.i.)

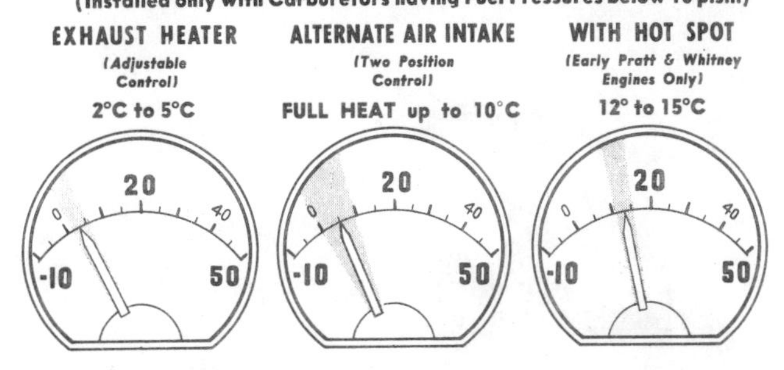

of ice until nearly too late. So turn on the heat as soon as any icing conditions are indicated.

**Be conscious of moisture in the air.** Check for moisture by applying heat. Note the effect on the engine's operation. With precipitation near the freezing range, under icing conditions, use full carburetor heat. Make sure the ice is eliminated. Then maintain the following temperatures (at the proper range for the model involved and conditions encountered). If these temperatures cannot be maintained, place the control in the **full cold** position.

**Exhaust Turbine-Driven Superchargers.** With inter-cooler shutters and/or carburetor air heater control, adjust to above temperatures. Without inter-cooler shutters, adjust to the above temperatures using carburetor air heat control.

**When There Are No Carburetor Thermometers.** Play safe. Place carburetor air heater control in one-half to two-thirds "HOT" position under all icing conditions. Do not exceed engine manifold pressure of 2″ Hg less than normally allowed for take-off. (If no manifold pressure gage is installed, limit maximum rpm to 200 less than rated take-off rpm.)

**When There Is No Manual Heat Control.** Certain types of primary training airplanes have no cockpit control for the carburetor air heater. The air-heater valve will be wired in the "ON" position unless excess loss of rpm is experienced in hot weather. During such periods the heater valve may be wired in the "OFF" position. WARNING—check the plane before take-off.

**Starting.** Carburetor air heaters should always be in full "COLD" position when starting. This will eliminate heater valve damage due to backfire (except in airplanes having no manual heat control).

**Ground Running.** Ordinarily, the carburetor air heater control should be in the "COLD" position during ground running. However, during extended ground operation under icing conditions and before take-off preventive measures should be used to stop ice formation and eliminate all ice from the induction system. At outside air temperatures below —20°C (—4°F) sufficient carburetor heat should be used to vaporize fuel properly and insure smooth operation.

**Take-off.** When outside air temperature is below —20°C (—4°F) use enough carburetor heat to prevent rough running. Some carburetor heat may be used at take-off under severe icing conditions, but carburetor air temperature should never be allowed to exceed 40°C (104°F).

**Cruising.** When cruising under severe icing conditions at least 75 percent engine power should be used and the mixture control set on the rich side for best power. Seek an altitude where precipitation can be avoided and where the temperature is farther from the freezing range.

**Landing.** In cold climates it is especially important that carburetor heat be FULL ON during the approach glide and landing. Otherwise, if power is applied, the cool engine will sputter. It may die. Many crashes have been caused when a cold engine refused to pick up because the fuel was too liquid to burn properly. Do not allow the carburetor air temperature to get into the icing range or to exceed 40°C (104°F). Remember that the carburetor heat control setting for a given carburetor air temperature will vary with the power output.

**Carburetor-Mixture Thermometers.** Stratification of hot air and intake air ramming conditions may make it necessary to move the exhaust heater control two-thirds of its travel before the hot air strata affects the thermometer bulb. However, once the bulb temperature is affected, a large range of temperature change will occur with small movements of the heater control.

**Changing Flight Level.** If free air temperature is below 20°F (—7°C), danger of ice formation is lessened because the quantity of water vapor in the air is slight except in extreme northern latitudes, such as Alaska, Canada, and the North Atlantic where moisture has been found in the atmosphere during a super-cooled condition of —40°C. Since ice formation depends partly upon humidity, the mere act of flying into a less humid region (such as leaving a

cloud bank or climbing above a fog) may arrest the ice formation.

### Alcohol De-Icing Equipment

On airplanes equipped with alcohol carburetor de-icing systems, the following additional operational instructions must be followed:

**Take-off.** When icing conditions prevail the alcohol pump should be turned on immediately before the take-off and the carburetor heat control placed in the **full cold position.** Due to the richer carburetor mixture caused by the alcohol, a slight loss in engine power may be expected. When a safe altitude has been reached, the carburetor air heater should be adjusted and the alcohol pump turned off.

**During Flight.** Except in emergencies, the alcohol system should not be used in flight.

**Emergencies.** When the alcohol system is used, remember—the alcohol supply is limited to about one hour of continuous use. Attempts should be made periodically to adjust carburetor temperature.

**WARNING—The use of the alcohol de-icing system at very low power and low engine idling speeds produces a rich mixture which may choke the engine.**

### Air Filters

Under icing conditions, airplanes equipped with air filters other than those installed in the auxiliary cold-air intake system or on the pressure side of the turbo-supercharger should be operated without filters unless they are operating regularly from dusty fields. This precaution is particularly important on long ferry flights and flights over water. The filters should be removed and stowed.

The following is a list of the various filters used on Army Air Forces airplanes together with instructions in case of icing difficulties:

#### Temporary Expédient Filters

**Hood Type.** This filter consists of a sheet-metal hood containing a filter screen. It is mounted by means of rapid-acting fasteners directly in front of the existing carburetor air scoop. The joint between the hood and the air scoop permits sufficient air to enter the engine in the event that the filter does become clogged. If icing is encountered, no difficulty is experienced other than some loss of power. Use carburetor heat as outlined.

**Insert Type, Automatic.** This filter consists of a screen inserted in the existing cold-air intake duct on the airplane. An automatic spring door installed in the air duct on the engine side of the filter opens automatically if the filter becomes clogged. The operation of this type of filter under icing conditions is identical with that described for the hood type.

**Insert Type, Non-automatic.** This filter consists of a screen inserted in the existing cold air intake scoop. The only means of making air available to the engine without passing through the filter is to open the hot-air door (carburetor heat **full on).** If air-filter icing is indicated, adjust the carburetor air-heater control valve to obtain necessary engine air and reduce engine power. **Avoid icing conditions with this type of filter.**

**Insert Type, Hot-Air Replacement.** This filter consists of a screen installed over the hot-air door in the air intake system of the airplane, thereby eliminating the original hot-air system. Air is drawn from the accessory compartment through the filter and into the engine. This air is at a slightly higher temperature than the air obtained from the ramming cold-air intake, but is not as hot as the air previously obtained with the original hot-air system. Use filtered air only during ground operations under severe dust conditions. **Use the cold ramming air normally for all flights.** If icing occurs in flight when the ramming cold air is being used, a small amount of carburetor heat can be obtained by using filtered air.

#### Permanent Filters

**Filters Installed in Ramming Cold-Air Intake.** This filter installation consists of a standard type flat-panel filter in the cold-air intake system, the only other source of air to the engine being the hot-air system. If filter icing occurs, adjust the carburetor air-heat control valve to obtain necessary engine air and reduce the engine power.

**Filters Installed in Auxiliary Cold-Air Intake System or on Pressure Side of Turbo-Supercharger with Cockpit Control.** This installation consists of a standard flat panel-type filter installed in a third air intake system entirely independent of the ramming cold-air intake or hot-air intake system. Use filtered air normally only for landing, take-off, and ground operations under dusty conditions, or when dust is encountered in flight and the mission is such that reduced power is permissible. Filters installed in the auxiliary cold-air intake system with a cockpit control will eventually be provided on all production airplanes.

**Filters Installed in Auxiliary Cold-Air Intake System and Controlled by Landing Gear.** This installation consists of a standard flat panel-type filter installed in an auxiliary cold-air intake system. The valves for the selection of filtered air or ramming air are controlled by raising or lowering the landing gear. If an icing condition may exist, do not conduct extensive operation with the landing gear in the "DOWN" position.

The pilot should be thoroughly familiar with the type of air filter installed on his plane and the possible ways of eliminating engine stoppage due to ice on the air filters. Report any engine stoppage which may be attributed to icing of the air filters immediately through channels by radio to the Commanding General, Air Service Command, Patterson Field, Fairfield, Ohio; Attention: Maintenance Division.

REFERENCE: T. O. 02-1-5, dated June 16, 1942; T. O. 01-1-13, dated Sept. 2, 1942.

**Warning**

Pilots should consult the Technical Orders and Technical Order Operation and Flight Instruction Handbooks for information on specific airplane models.

## CONVERSION TABLE CENTIGRADE TO FAHRENHEIT

(Formula: °C × 9/5 + 32 = °F)

| Centigrade | Fahrenheit | Centigrade | Fahrenheit | Centigrade | Fahrenheit |
|---|---|---|---|---|---|
| —60° | —76° | 30° | 86° | 120° | 248° |
| —55° | —67° | 35° | 95° | 125° | 257° |
| —50° | —58° | 40° | 104° | 130° | 266° |
| —45° | —49° | 45° | 113° | 135° | 275° |
| —40° | —40° | 50° | 122° | 140° | 284° |
| —35° | —31° | 55° | 131° | 145° | 293° |
| —30° | —22° | 60° | 140° | 150° | 302° |
| —25° | —13° | 65° | 149° | 155° | 311° |
| —20° | — 4° | 70° | 158° | 160° | 320° |
| —15° | 5° | 75° | 167° | 165° | 329° |
| —10° | 14° | 80° | 176° | 170° | 338° |
| — 5° | 23° | 85° | 185° | 175° | 347° |
| 0° | 32° | 90° | 194° | 180° | 356° |
| 5° | 41° | 95° | 203° | 185° | 365° |
| 10° | 50° | 100° | 212° | 190° | 374° |
| 15° | 59° | 105° | 221° | 195° | 383° |
| 20° | 68° | 110° | 230° | 200° | 392° |
| 25° | 77° | 115° | 239° | 205° | 401° |

# ICING ON AIRCRAFT

Icing can be a hazard to flying. You can fly safely during icing conditions if you know your plane's limitations under these conditions, if you know icing formations and how to remove ice when you do encounter it.

Avoid flying through icing zones when possible. Know the correct procedures for removing ice.

Ice accretion may occur whenever there is visible moisture in the air at temperatures from near-freezing down to more than 40° below zero. A small accumulation, concentrated on critical surfaces of aircraft with high wingloading, may alter the normal airflow so that their flight characteristics are greatly affected.

## EFFECTS OF ICE

1. Reduces the efficiency of the airfoil, adds drag, and raises the stalling speed.
2. Makes the airplane difficult to control and maneuver.
3. Increases the drag of struts, fuselage, radio masts, fixed landing gear, etc.
4. Increases the load.
5. Causes certain flight instruments to give false indications or to fail completely.

## TYPES OF ICE

Ice which forms on an airplane in flight is of two general types, rime and glaze or clear ice. Often, it is difficult to classify as one or the other, but usually one predominates.

### Rime Ice

Rime is a granular, translucent crust, generally encountered in stable air, such as that marked by stratus cloud formations. It is formed when droplets of supercooled moisture hit the airplane.

Rime builds out from the leading edges of wing and

tail surfaces, spinners, props, and fuselage. Even a light coating on the leading edges of the wings, by altering the airfoil, may raise the stalling speed.

### Glaze or Clear Ice

Glaze is a crystalline coating of hard ice which tends to conform to the airfoil of wing and tail surfaces. It occurs generally in regions of turbulence, in freezing rains, and frontal regions where the temperatures range from 20°F (−7°C) to freezing. It is formed when drops of supercooled moisture splash and spread out on the airplane, then freeze.

It forms a continuous shell, sometimes hard to detect, building back from the leading edges. Often it completely covers the wing and tail surfaces. Occasionally, glaze builds forward from each rivet head or protuberance, forming cones of ice that stick out into the airstream. Glaze ice, if allowed to build up, often is difficult to remove from the airplane, even when de-icing facilities are provided.

Use caution when flying at a high angle of attack; large quantities of ice can form on the lower surfaces of the wings. Such ice can be hazardous because you cannot see it and because of its weight.

## OTHER FORMS

**Frost** is dangerous for takeoff, but not on an airplane in flight. Remove it from the wings and tail surfaces just before any takeoff. Use wing covers on the ground when conditions are favorable to frost; remove them just before takeoff.

**Fog, Clouds, Etc.** Minute particles of moisture in the air rarely cause icing of the wings, fuselage or tail. They are too small to break through the surface friction but follow the airstream without contact with the skin. **They may cause propeller icing.** It is possible to have propeller icing without icing on other parts of the airplane.

## WHERE ICING OCCURS

**Prepare for icing whenever there is visible moisture in the air at temperatures approaching or below the freezing level:**

1. In freezing rains, in all frontal zones.
2. Sleet itself may not be hazardous, but if there is sleet on the ground, somewhere aloft there is usually a layer of freezing rain and above that, a layer above freezing. Such temperature inversion is a condition favorable to icing.
3. In cumulus clouds and others with vertical development, wherever they occur.
4. Along fronts, in stratus and strato-cumulus cloud formations.
5. In mountain (orographic) clouds, formed when moisture laden air is forced upward over hills and mountain ranges.

## TEMPERATURES

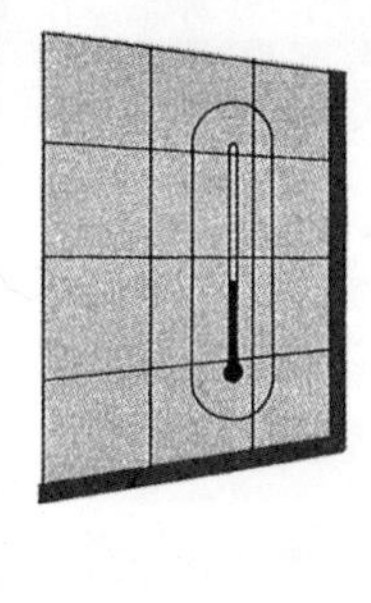

Look for most severe icing when the temperature is between freezing and 18°F (–8°C). Icing may occur down to –40° or colder. Low pressure areas on the airfoil may cause mild icing at temperatures a few degrees above freezing when other conditions are favorable to icing.

Avoid icing where possible. Plan your flight to go around or over frontal formations when conditions are favorable to icing. **Consult your forecaster before taking off.** Analyze reports of icing carefully before following a flight plan similar to one where icing was encountered. And plan your flight accordingly.

**Icing conditions never remain constant.** The pilot ahead of you may have reported light icing; a few minutes later, in the same spot, you may find the icing severe.

**If you encounter severe ice at a given altitude, try at once to get out of it:**

1. Climb above it, if icing levels do not extend too high. If it is impossible to get on top, remember that most ice occurs at temperatures between freezing and 18°F (–8°C). Figure a free-air temperature drop of 3° for each thousand feet you climb. There will be less icing where the temperature is 18°F (–8°C) or colder. If the icing zone is shallow, get above it as rapidly as possible.
2. Descend below the icing level if you will still have plenty of altitude left for safe instrument flight when you get below the icing.
3. Change your course and fly out of it. If you enter a zone of severe icing from one comparatively free from icing, make a 180° turn before making your climb or descent. Then resume your original course when you reach the desired altitude. **Always get out of severe icing zones as soon as possible.**
4. **Go around or over mountain clouds, never under them; you'll need all the altitude you can get.** Icing can be dangerous if you don't know what to do to avoid it or can't remove ice from critical surfaces when it occurs. **Know the limitations of your airplane in icing conditions.**

## STALLS

A stall caused by ice does not feel the same as a normal stall. It occurs at a higher speed and is preceded by noticeable sloppiness of the controls. Also, the stall will not break cleanly; there is a gradual transition from normal flight, through heavy, mushy flight, to the full stall. The airplane will fall off on one wing or the other, according to the individual characteristics of the plane. As your stalling speed in straight and level flight increases because of ice accretion, it increases even more in banks.

## PREVENTING AND REMOVING ICE

If your airplane has de-icing equipment, use it when needed. But don't fly into icing zones when you can avoid them, simply because you have this equipment. De-icing equipment is for safety when icing is unavoidable. Know how to use it.

### Wings and Tail

**De-Icing Boots** are installed over the leading edges of the wing and tail surfaces of many airplanes. They are flexible rubber sheets containing inflatable fabric tubes. The tubes inflate and deflate from air pressure supplied to them by pumps. Distorting and stretching of the surface of the boots cracks and loosens the ice. The ice then is blown away by the airstream.

The de-icing system operates through a 40-second cycle. When you encounter icing, wait until a coating approximately ¼″ thick builds up on the leading edge of the wing, turn your de-icers "ON" to break it loose, then turn "OFF." Every time a thin coating of ice builds up, repeat the process. **Test on the ground before takeoff when you anticipate icing in flight.**

**Never operate boot de-icers during takeoff or landing. They act as spoilers.**

**Don't use de-icers when heavy ice forms behind the boots back on the wings.**

**Anti-Icing Systems,** heated wings and tails, are built into some planes for preventing ice formation. Air is heated and then passed through ducts to the leading edges of the wings and empennage; to the cabin and windshield. The surfaces are heated enough to prevent the formation of ice. When not in use, the air passes on out through a dump valve.

If you anticipate icing immediately, turn the system "ON" just before takeoff. Leave it on as long as you encounter icing conditions. In flight turn "ON" whenever you notice or anticipate icing. Allow about 30 seconds for surfaces to heat up; ice will seem to explode off the wings and tail surfaces. If your airplane is not equipped with either system and you

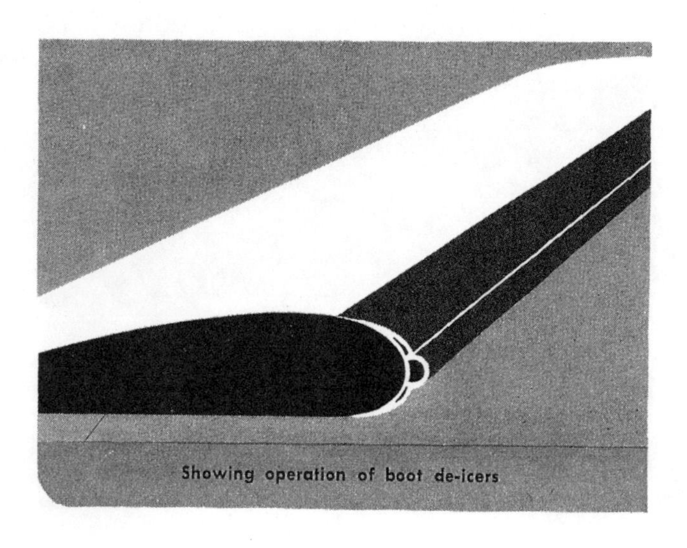

Showing operation of boot de-icers

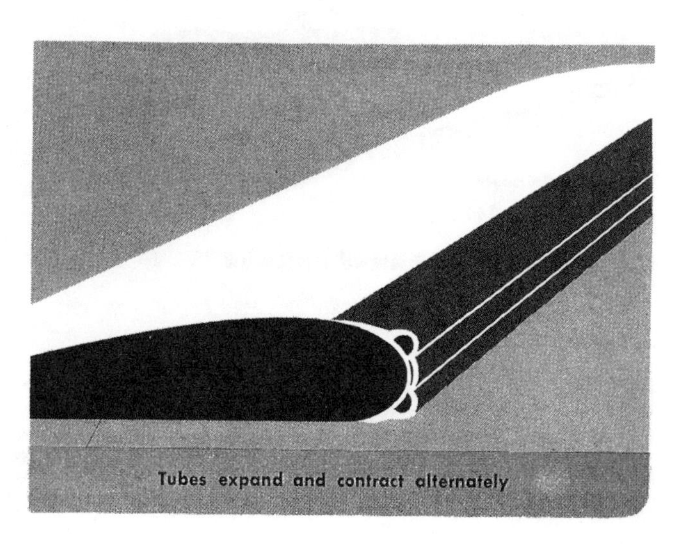

Tubes expand and contract alternately

encounter icing, change your course and altitude at once. You may be able to dislodge rime ice from leading edges by increasing speed, but leave the icing zone before trying this method of removing ice.

Airplanes with high wingloading or critical control characteristics may be seriously affected by even a small amount of wing ice. **Know the flight characteristics of your airplane in icing conditions.**

## PROPELLER

Icing on propellers reduces propeller efficiency and causes vibration. It may occur in any condition favorable to icing on wings and in below-freezing zones where there is fog or very stable clouds, consisting of extremely small moisture particles.

Ice on propeller spinner

Remove ice from propellers by one or more of the following methods:

### Increase RPM of Propeller

The added centrifugal force often throws off the ice accumulation. This method is especially helpful when used in conjunction with de-icing fluid and anti-icing spinners.

### De-Icing Fluid

De-icing fluid is effective in removing ice from propellers. The fluid is pumped from a tank in the airplane to a slinger ring, which distributes it to the blades. To operate:

Turn the rheostat full clockwise to drench propellers with fluid. Then turn counterclockwise to "NORMAL." If rheostat is not marked for normal setting, turn it counterclockwise approximately half way to deliver normal output to blades.

Approximately 4 quarts of fluid per hour, for each propeller, are required for normal operation of the propeller de-icing system. Fill the tank before take-off on any flight which may encounter icing. Turn on propeller de-icer just before you enter a zone of anticipated icing. **Test the system on the ground before a flight when conditions are favorable to icing.**

### Anti-Icing Spinners

Some pitch-control mechanisms are provided with anti-icing spinners, usually coated with rubber. When covered with an oil before takeoff on flights where icing is anticipated, they will help keep the propeller hubs free from ice.

## PITOT TUBES

Ice on the pitot-static head affects readings of the altimeter, rate-of-climb, and airspeed indicator. There is a heater in the pitot head to keep it free of ice. Whenever the free-air temperature is near or below freezing and there is visible moisture in the air, turn the pitot heater "ON."

An alternate source of pressure is provided so that the altimeter and rate-of-climb indicator will continue to function should the pitot static source fail. However the readings will differ. So when you start out on a flight, check the difference in readings at normal cruising speed, by switching from normal pitot static source to "ALTERNATE SOURCE." Then, should it be necessary to use the alternate source, you will be able to figure satisfactory readings. **The air speed indicator will not operate at all if the pitot-static tube fails.** (See PIF 3-8.)

Flush-static vents are installed on many tactical airplanes. They are designed for use where the pitot source and static source are separate; there is no alternate source with this installation.

The flush-static vent is a circular plate, 2″ in diameter, mounted on the side of the fuselage. The holes in this plate must be kept free from paint and dirt to function properly.

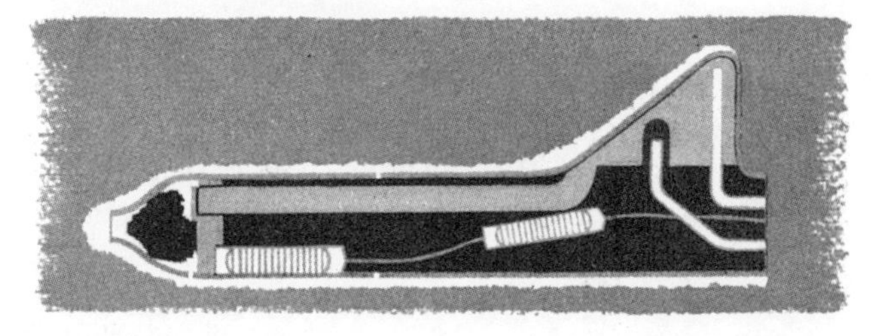

# WINDSHIELDS

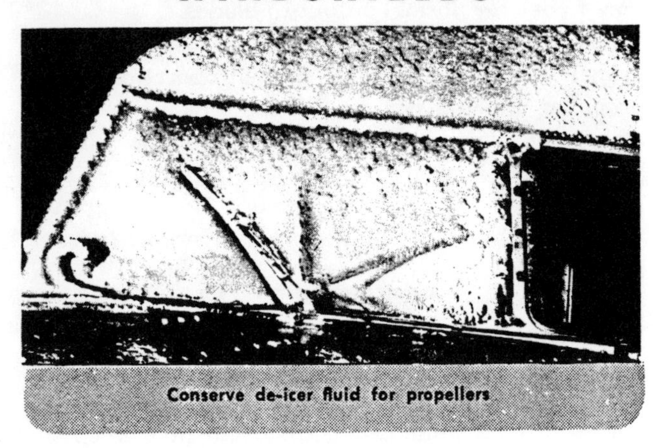
Conserve de-icer fluid for propellers

### De-Icing Fluid Systems

Fluid de-icers are installed on the windshields of some airplanes. When windshields so equipped begin to ice up, turn the rheostat full clockwise for maximum output, then adjust to the minimum output at which it will operate effectively. Don't waste de-icing fluid. You may need it for the propellers if icing gets more severe.

### Heated Windshields

Heated windshields, with double panes, are installed on many airplanes. Heat is supplied from the cabin heater or, when planes are equipped with heated wings, through special ducts connected to the heat exchangers. If normal vision is possible with the inner pane in place, leave it in position at all times. If inner pane is stowed normally, put it in place before taking off when icing is anticipated. Leave windshield heaters "ON" at all times in weather favorable to icing. The system also is satisfactory for defrosting.

# PILOT REPORTS

When you encounter icing unexpectedly, report it at once by radio. The information may help other pilots to avoid that area. Give location, upper and lower limits of the icing levels if you know them, and the amount of ice encountered.

# ICING SAFETY NOTES

### On the Ground

1. Check the weather. See your forecaster. Don't plan a flight into known hazardous icing conditions.
2. Check de-icing equipment before takeoff when conditions are favorable to icing.
3. **Remove all frost before takeoff.**
4. Before takeoff into known icing conditions cover props with de-icing fluid.
5. Avoid slushy and icy spots when taxiing. Be easy with your brakes; you may skid.
6. During a winter fog or rain, watch for icing on wings from propeller blast during warmup.
7. Don't take off in wet snow, in borderline temperatures if you can avoid it. It may freeze before you can gain altitude.
8. Carburetor Heat—Read PIF 3-6.

### In the Air

1. Following takeoff from a wet or slushy field, operate landing gear, flaps, etc., through several cycles, when possible, to prevent freezing.
2. Turn pitot heater "ON" if you anticipate ice.
3. Make your climb or descent through icing zones as fast as possible.
4. Exercise controls occasionally when flying in icing conditions. Otherwise, they may freeze.
5. Avoid turbulent areas. Fly above or below clouds when possible. Look for temperatures below 0°F (–18°C) or well above freezing. You'll find less ice there.
6. Radius of action drops when ice forms on your plane. Plan your flight accordingly.
7. **Avoid ice where possible.**

### In Landing

1. Fly in with plenty of power for a wheel landing. Remember that stalling speed increases with ice. Approach from a safe altitude.
2. Make wide turns. Increase airspeed in all turns over normal. Avoid steep turns with icy wings.
3. Turn wing de-icers "OFF" on base leg. They act as spoilers if left on.
4. Use flaps with care. They may aggravate the condition if there is ice on the airplane.

**Know the Handling Characteristics of Your Airplane Under Icing Conditions. Don't Take Chances With Ice!**

REFERENCE: Technical Order 30-100D-1

# FLIGHT INSTRUMENTS

## PITOT-STATIC SYSTEM

The readings of the airspeed indicator, altimeter and rate-of-climb indicator depend upon two different pressures. One is called pitot (impact or dynamic) and the other is called static (atmospheric). Actually the only instrument affected by both types of pressure is the airspeed indicator. The altimeter and rate-of-climb indicator depend solely upon static pressure.

### Pitot-Static Head

A pitot-static head supplies both pitot and static pressures to the airspeed indicator and static pressure alone to the altimeter and rate-of-climb indicator.

The pitot pressure opening is on the forward end of the head, while the static pressure openings are located a short distance behind the forward head.

The head is so designed that any moisture or dirt entering it is trapped and kept from reaching the instruments themselves. A heating element is provided inside the head to prevent icing.

To deliver true pressure to the instruments the pitot-static head must be streamlined with the airflow. This can be accomplished usually only at cruising speed. Thus, inaccuracies may develop at other than cruising speed, but they are negligible except during takeoff, sharp pullouts, level-offs, and landings.

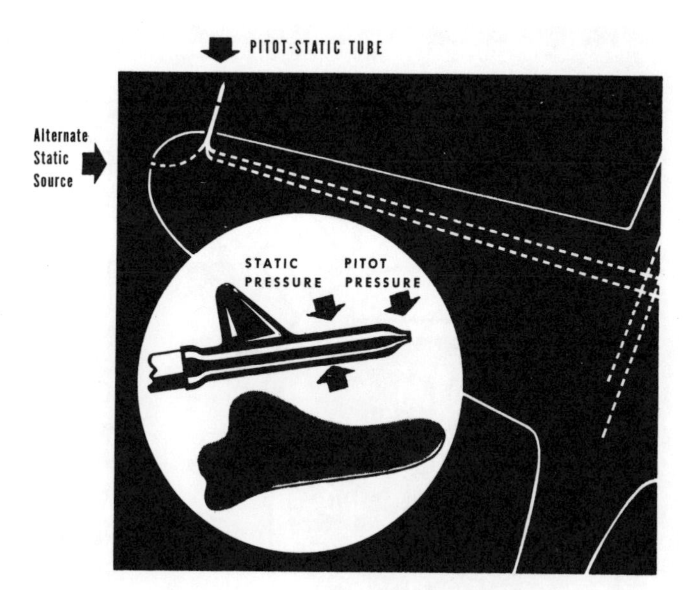


**REMEMBER—Always make sure the pitot-static head cover is removed before takeoff. Don't leave the pitot heater turned on for extended periods on the ground. Except for testing purposes, turn it on only in flight when icing conditions are anticipated.**

### Flush Static Source

Static pressure is sometimes derived from a flush static pressure source, by means of a perforated flush-mounted plate, located usually well back on the airplane's fuselage. Pitot pressure in this system, however, is still obtained from a pitot head.

### Alternate Static Source

Some airplanes have an alternate source of static pressure controlled by a switch or valve in the cockpit provided in case the usual source fails. For example, if the pitot-static head were broken off or the openings became iced, all three instruments in this group would become unreliable. By switching to the alternate you regain use of the rate-of-climb indicator and the altimeter but not the airspeed indicator, depending as it does on both pitot and static pressure.

The static pressure provided by the alternate source is usually lower than true or actual pressure. Thus, upon first switching to it, you find the rate-of-climb indicator ordinarily indicating a climb, while the altimeter ordinarily indicates an increase in altitude, though none has occurred. The rate-of-climb indicator returns to a normal indication shortly. The altimeter, however, continues to indicate an altitude other than actual.

Try the effect of the alternate source of your own airplane if so equipped, so you can be prepared for an emergency.

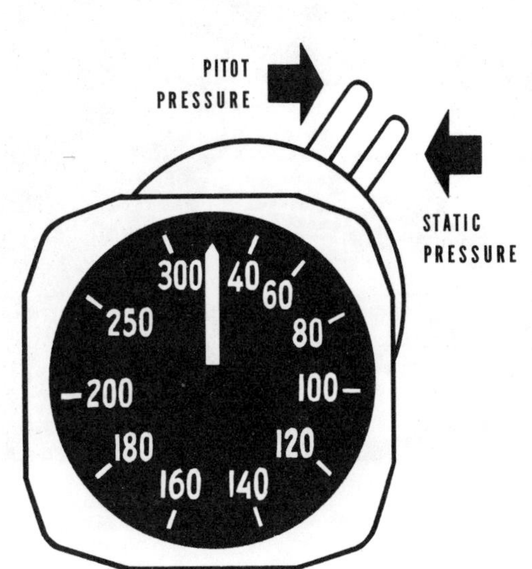


### AIRSPEED INDICATOR

The actuating mechanism of the airspeed indicator is a metal diaphragm housed in an airtight case. Pitot pressure is admitted to the inside of the diaphragm. Static pressure is admitted to the case around the outside of the diaphragm. The difference between these two pressures causes the indications of this instrument and gives indicated airspeed (IAS). Any change in either pitot or static pressure causes a change in the IAS.

You can estimate true airspeed (TAS) roughly by adding 2% of the IAS for each 1000 feet of altitude. For example, if the reading is 140 IAS at 20,000 feet of altitude, add 40% of 140 to 140, which gives an approximately correct reading of 196 TAS. For more accurate calculations use a computer.

**REMEMBER—With the exception of your having wing ice or an increased load, the IAS at which your plane will stall is the same, no matter what the altitude or temperature. The airspeed indicator tells you accurately your safe airspeed limits, maximum or minimum. Make corrections for stalling speed if you have picked up wing ice or increased your normal load. See PIF 3-4-1 and High Speed Stalls, PIF 2-20-1.**

## RATE-OF-CLIMB INDICATOR

The rate-of-climb indicator has the same actuating mechanism as the airspeed indicator. A diffuser valve simply restricts the flow of the air from the case to the inside of the diaphragm. Only static pressure is admitted to the inside of the case and the outside of the diaphragm.

You get a reading when there is a difference in pressure between the inside and the outside of the diaphragm. The difference occurs chiefly in climbing or descending, when pressure inside and outside the diaphragm tend to equalize, but this takes some time because of the restriction in the flow of air between the diaphragm and case. Because of this restriction the instrument has a lag in its indications of several seconds.

Depend on the instrument only when establishing a constant rate of climb or descent in relatively smooth air. Use it in conjunction with the altimeter at all times as it indicates the **rate of change** and not the **amount of change** in altitude. The instrument is relatively useless in turbulence.

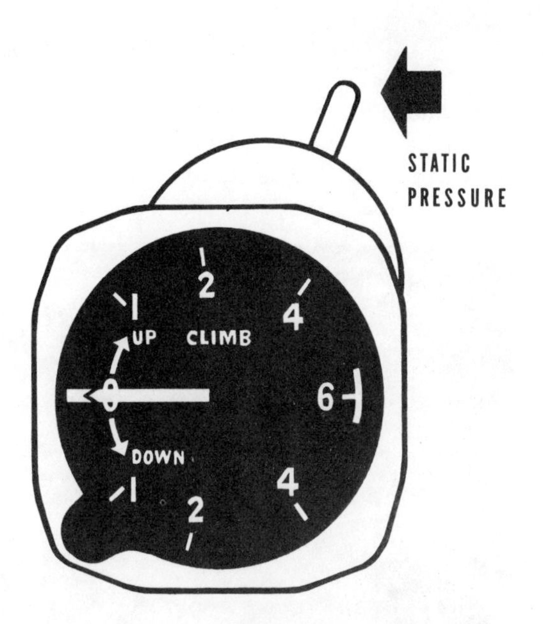


## ALTIMETER

The actuating mechanism of the altimeter is called an aneroid, which is a device for measuring absolute pressure. It is housed in an airtight case. Static pressure is admitted to the inside of the case, and acts on the aneroid so as either to compress it or allow it to expand. This compression or expansion causes movement of the hands on the face of the instrument.

The altimeter actually measures the weight of the air column (barometric pressure) above it and is so calibrated that a decrease in pressure causes an increase in its indication of altitude. The weight of the air column measured by the altimeter will vary because of changes in the air's density, which in turn is affected by variations in temperature. These factors often cause incorrect altimeter readings.

The altimeter is calibrated according to standard conditions of temperature and pressure, but these conditions seldom prevail. In order to correct for non-standard conditions modern altimeters are provided with a barometric scale.

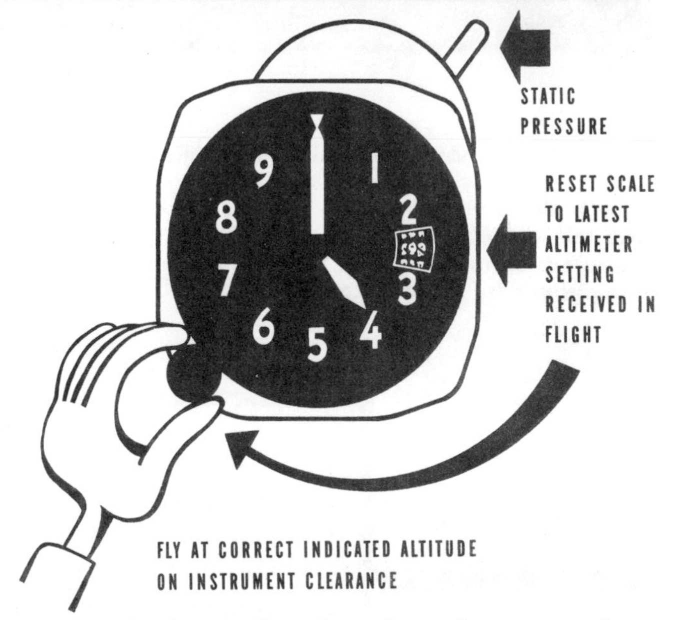


### Standard Altimeter Setting

**Remember this system will always give you your height above sea-level, not above the terrain.** Use these steps:

1. Before takeoff obtain the **latest altimeter setting** for the field from the tower. Then set the barometric scale on the instrument to this setting. The hands of the altimeter should then indicate the surveyed altitude of the field above sea-level. (If they do not, within the limits allowed on the scale-error card on your instrument panel, ask a mechanic to find the trouble. Often a small adjustment is enough to correct the error).

2. During flight you must continually correct the altimeter by resetting the barometric scale according to the latest altimeter setting of the area in which you are flying. Request it from the nearest radio facility.

3. Before landing again request the latest altimeter setting from the tower, and reset the instrument accordingly, so that you can depend on its accuracy.

## Effects of Pressure and Temperature on Altimeter

Standard atmospheric conditions at sea-level exist when the barometric pressure is 29.92 and the temperature is 15°C. The standard temperature for any altitude above sea-level decreases at the rate of 2°C for each 1000 feet of altitude. Remember these two rules:

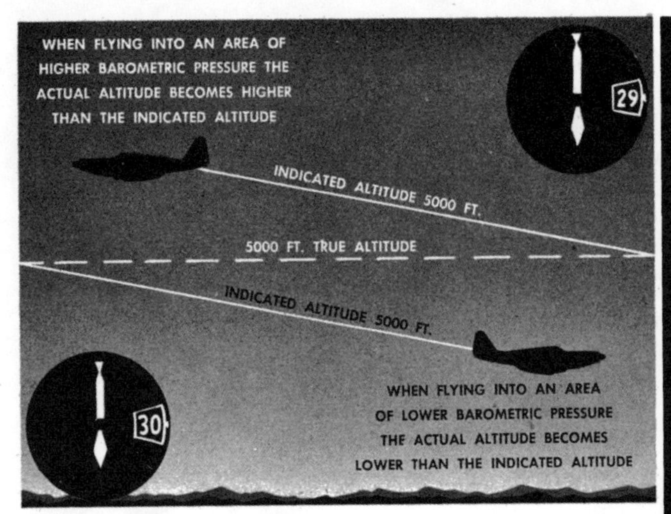


### ALTITUDE AND PRESSURE

**Rule 1—In flying from an area of relatively high pressure into an area of lower pressure the actual altitude will become lower than the indicated altitude. This is a dangerous situation.**

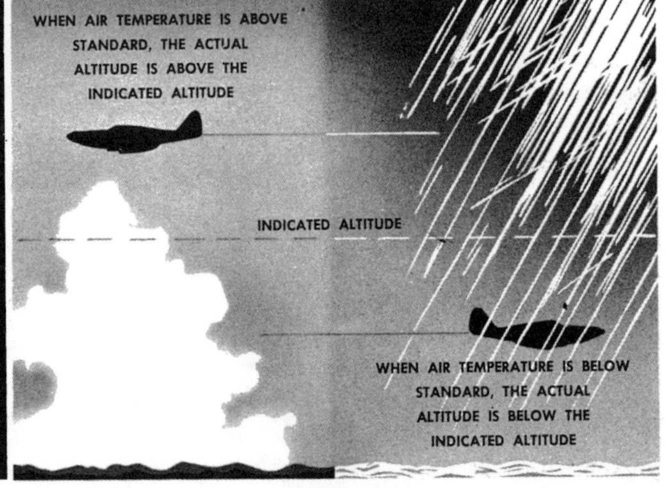


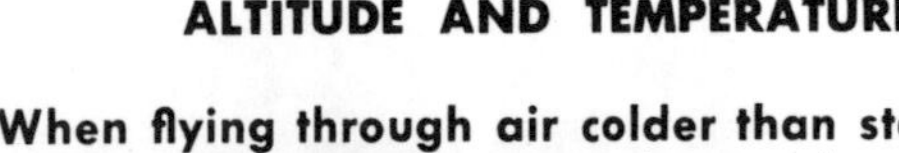

### ALTITUDE AND TEMPERATURE

**Rule 2—When flying through air colder than standard the actual altitude becomes lower than the indicated altitude. WHAT TO DO: Calculate your true altitude by use of a computer.**

### WARNING

Before clearing for an instrument flight, be sure the altitude you have requested gives you sufficient terrain clearance over your entire route. Then keep your altimeter reading at this altitude. Don't make mental corrections and try to figure your true altitude. Keep your indicated altitude corrected by resetting your altimeter from settings given you over your radio check points.

# MAGNETIC COMPASS

The magnetic compass is the basic navigational instrument. It consists of an airtight case filled with a special fluid in which a compass card assembly is pivoted. It is subject to many errors; mainly deviation and the errors produced by turning, accelerating or decelerating the airplane.

A compass correction card mounted on the instrument panel shows the amount of deviation.

Note the following errors in turning:

When turning toward North the indication of the compass lags considerably. In most sections of the United States this lag amounts to approximately 30°. When turning toward South the compass indication leads the airplane's direction on the heading by the same amount. These errors decrease as you make a turn toward East or West, becoming approximately zero on those headings.

The rule becomes, then; **Overturn the heading when turning toward directions on the Southern half of the card** (points between East and West

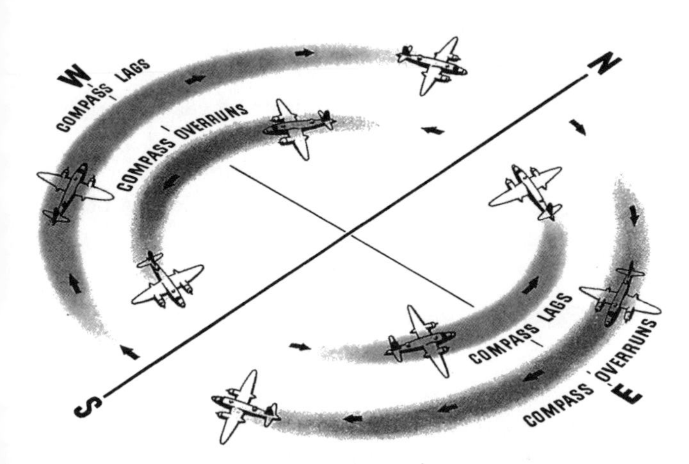


through South). **Underturn the heading when turning toward directions on the Northern half of the card** (points between East and West through North).

When you accelerate or decelerate the airplane in an Easterly or Westerly direction you will notice an error as the compass turns off the direction in which you are flying.

When heading East or West acceleration causes the compass to indicate a turn toward the North. You get the same effect by going abruptly into a dive on these headings.

Deceleration on a heading of East or West causes the compass to indicate a turn toward the South. You get the same effect by going abruptly into a

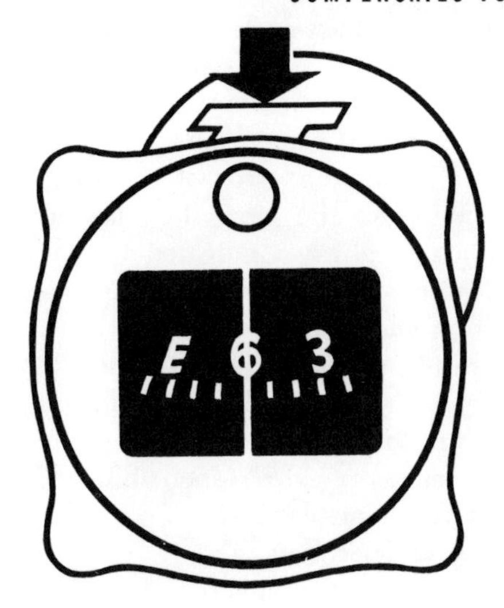


climb on these headings. These rules apply only in the Northern hemisphere. They are exactly the opposite in the Southern hemisphere.

The magnetic compass is difficult to use in turbulence.

The compass should be swung frequently in order to keep the compass correction card up to date. If you notice serious compass errors report it on your Form 1-A. Keep metal objects and electrical equipment (such as headphones) away from the compass.

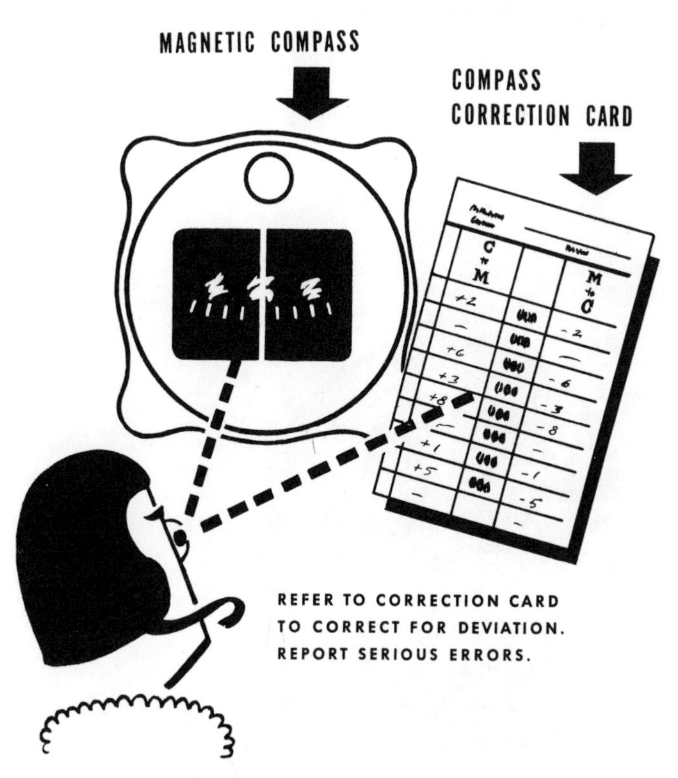

# GYRO FLUX GATE COMPASS SYSTEM

The Gyro Flux Gate compass system is a remote indicating earth-inductor compass consisting of gyro-stabilized flux gate transmitter, an amplifier, a master indicator and from 1 to 6 remote indicating repeaters. It is not subject to the error of the magnetic compass except when the gyro flux gate is caged. A true heading is always indicated by the master and the repeater indicators, because of compensation for variation and deviation.

Its upset limits are 65° climb, glide and bank.

Although the responsibility for the operation of this compass rests on the navigator, you should know how to use it, and it is your responsibility if there is no navigator in the airplane.

Follow these steps in operation:

1. Be sure the amplifier switch is always left "ON."
2. Turn on the inverter after starting engines.
3. Cage the gyro after the compass has been in operation 5 minutes.
4. Leave it caged for 45 seconds.
5. Uncage, and it is ready for use.

If you exceed the upset limits of 65°, cage and uncage the instrument just as you did before.

**Leave the gyro uncaged at all times except those indicated above.**

On the newer models the gyro is designed so it cannot be left in the caged position. Follow this procedure for erecting the gyro: After gyro motor has been operating 5 minutes, hold push button in until red light is on, then release push button. When red light goes out an erection cycle has been completed.

# THE SUCTION SYSTEM

Gyro instruments are operated normally by suction, supplied by a vacuum pump driven by the airplane engine. Actually each gyro instrument works on the same principle as a turbo. A gage is provided to indicate amount of suction.

On modern single-engine airplanes there is no alternate source of suction. On multi-engine airplanes there are usually two or more suction pumps. A selector valve in the cockpit allows you to switch from one suction pump to another in case of engine or suction pump failure.

**If your suction system fails completely you have approximately three minutes dependable use of the gyro instruments.**

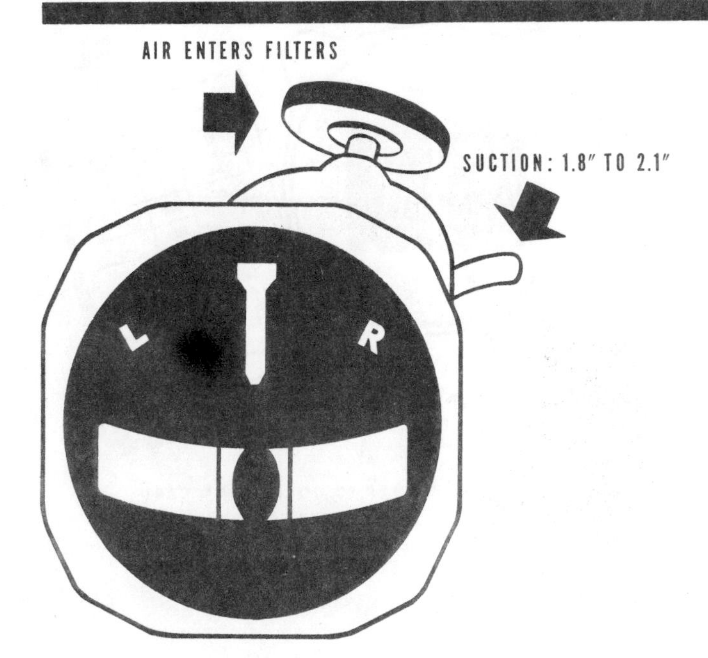


## BANK-AND-TURN INDICATOR

The bank-and-turn indicator is a combination of two instruments; the bank indicator and the turn indicator. Since you use them together, however, they are housed in one case. The bank indicator consists of a ball free to move back and forth in a tube filled with a special liquid. It shows the relation between angle of bank and rate of turn. When a turn is properly executed the ball will remain in the center of the tube between the two reference lines. If the ball is on the low side you are making a slipping turn, if on the high side a skidding turn. You may use the ball in level flight to determine whether you are skidding or slipping, or if the airplane is properly trimmed.

**The standard bank-and-turn indicator shows rate of turn and does not show the amount of turn.** A needle width's deflection means that with a properly functioning instrument you are turning at the rate of 180° per minute. This is a standard rate turn used in instrument procedure.

The standard bank-and-turn indicator indicates rates of turn up to 1080° per minute.

The rate of turn is the result of two things—airspeed and angle of bank. The greater the airspeed the greater must be the angle of bank to produce a given rate of turn.

**For a one needle-width standard turn (180° in one minute) the following speeds and angles are required:**

| MPH | Angle of Bank |
|---|---|
| 100 | 13.5° |
| 150 | 19.8° |
| 200 | 25.6° |
| 250 | 31.0° |
| 300 | 35.75° |

It is obviously impossible to make angle-of-bank readings with sufficient accuracy in instrument procedure turns. But any of the above can be read on the bank-and-turn indicator as a single needle-width turn.

At speeds above 200 mph make only one-half needle widths turns in order to avoid a dangerous angle of bank, providing the instrument has the standard calibration. To read the rate of turn in turbulence average the oscillations of the needle.

## DIRECTIONAL GYRO

Use the directional gyro in connection with the magnetic compass. It is not a direction seeking instrument, and because of precession, must be checked every 15 to 20 minutes against the magnetic compass, and reset.

Set the desired heading by the caging knob underneath the dial.

Its approximate upset limits, are 55° bank, climb, and glide. When these limits are exceeded the card generally begins to spin; level the airplane, cage the gyro and reset by reference to the magnetic compass.

The instrument is reliable in turbulence, in contrast to the magnetic compass.

In ground checking the instrument turn the caging knob gently and pull out in one smooth operation. The card should stop and hold its position. This is a check on the rigidity of the gyro. **Do not twist the knob sharply and pull it out at the same time.**

If the instrument drifts off cardinal headings more than 3° in 15 minutes report it on your Form 1-A.

Leave the instrument uncaged at all times except in maneuvers which exceed its upset limits.

## GYRO-HORIZON INDICATOR

This instrument indicates realistically the attitude of the airplane.

The upset limits of the instrument are 100° bank and 70° climb and glide.

After starting engines allow at least 5 minutes at sufficient rpm before takeoff for the instrument to obtain proper speed. Be sure before takeoff the horizon bar has settled into its proper 3-point position. This may be speeded up by caging and uncaging the instrument after engines are started. **This is the only time the instrument should be caged while on the ground.** Leave it uncaged at all times except in maneuvers which exceed its upset limits.

**When caging and uncaging the instrument in flight always make sure the airplane is level both longitudinally and laterally at the moment of uncaging.** Always cage the instrument gently to avoid damaging it.

When using the gyro-horizon be sure it is locked in the full "ON" position (uncaged).

Whenever you make a turn the gyro-horizon develops a small error in indication, known as turn error. After straight and level flight is resumed the error will disappear shortly. Therefore, during turns

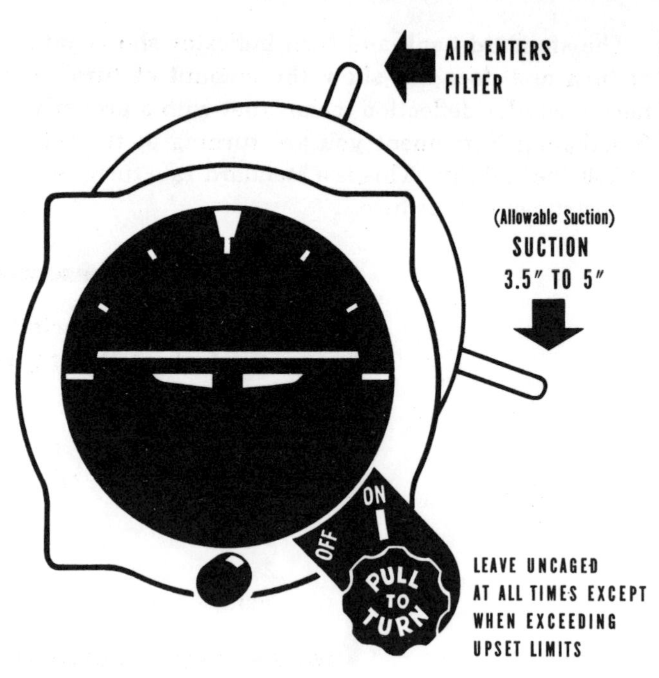


and immediately thereafter refer to the altimeter and airspeed indicator to maintain altitude and compensate for the error in the gyro-horizon.

During flight do not depend solely on the gyro-horizon. Cross-check it with the other instruments.

## AUTOMATIC PILOT

Two types of automatic pilots are in general use: suction driven or electrically operated. Operating instructions for each vary widely. Consult instructions for the equipment installed on your airplane, and don't try to operate both types if you are familiar with only one.

**In operating either type remember to trim the airplane for hands-off flight before engaging.**

Make sure all indices or control indices are in alignment or neutral before engaging the instrument.

Some slight adjustments of the automatic pilot to correct for small changes of load balance are permissible. However, any large change of load balance (such as transfer of fuel, dropping bomb loads, or dropping auxiliary tanks) necessitates your disengaging the pilot and trimming the airplane for hands-off flight before you attempt to use the instrument again.

### WARNING

Never use the automatic pilot for locking the controls while the airplane is parked, or while you are warming up.

# SECTION 4 MAN GOES ALOFT

# EFFECT OF OXYGEN-WANT ON HANDWRITING DURING ASCENT

| ASCENT TO 25,000 FEET WITHOUT OXYGEN | EXPLANATORY REMARKS |
|---|---|
| A sample of normal hand-writing in flight at 2000 ft | Control specimen of normal handwriting. |
| 10000 ft - breathless | No apparent effect. |
| 15000 ft - feel uneasy generally [illegible] feeling some numbness in legs and hands | Beginning muscular incoordination. |
| 1800 ft very bad | Definite physical and mental inefficiency. |
| 2000 ft: faint - numbness in legs: vision fading | Last zero off both 18,000 and 20,000—marked incoordination. |
| 22000 ft: [illegible] pale to me feel better | Feeling better? Evidence of false feeling of well-being. |
| 23000 ft: feel good except for [illegible] legs numbness | Feel good. Insight, judgment and coordination very faulty. |
| 24000 feet - [illegible] | Mental and physical helplessness. |
| 2500 ft oxygen turned on | Improvement with few breaths of oxygen. |
| [illegible] ft - things look brighter - Hearing returning feel O.K. now - | Last zero left off—general improvement, but not completely normal. |

# EFFECTS OF HIGH ALTITUDE

## General

The effects of high altitude flying on the human body include:

1. Oxygen want (Anoxia).
2. Expansion of gases in the body.
3. Need for clearing (equalizing pressure in) the middle ear and nasal sinuses.
4. Escape of dissolved blood gases in the form of bubbles ("Bends", "Chokes", "Creeps").
5. Increasing cold with altitude. (See PIF No. 4-7.)

All but the last of these effects are due to a decrease in the total barometric pressure with higher altitudes. The atmospheric pressure at 18,000 feet is one-half that at sea level; at 27,000 feet, one-third, and at 40,000 feet, one-fifth.

## Oxygen Want (Anoxia)

Oxygen want means lack of oxygen in the body tissues due to a decrease in the quantity of oxygen in the air breathed. Oxygen want increases with altitude, since as the air becomes less dense, a given lungful of air contains less oxygen (and, of course, less of all other gases as well).

In flight, oxygen want begins at about 5,000 feet, but, except for less ability to see in dim light, it is not noticeable to the flyer until an altitude of about 10,000 feet is reached. Healthy persons may occasionally become unconscious while breathing air at 18,000 feet after short exposure, and unfit persons may suffer collapse at lower altitudes.

**The chief symptoms of oxygen want are:**

1. Loss of insight and lack of realization of danger.
2. An early false sense of well-being.
3. Lessening of judgment, inability to think clearly, and tendency to make errors.
4. Smaller field of vision and decreased hearing.
5. Sluggishness and clumsiness.
6. Lack of emotional balance.
7. Greatly reduced ability to see in dim light, or at night.

## Know What Happens

**All practical rules for the use of oxygen equipment and the probability of your having "bends" or oxygen want depend on the reading of the altimeter in the airplane (indicated altitude above sea level).** That is, regulations call for your use of oxygen at any time you fly above an indicated altitude of 10,000 feet **above sea level,** not above a zero setting for a given field.

**Become thoroughly familiar with the various stages of oxygen want.** In the first stages the senses are dulled and you have a false feeling of well-being. This may start at 10,000 feet or less. At any altitude above 10,000 feet the effects of anoxia may develop without warning. Always judge your need for oxygen by the altimeter. Never wait for symptoms.

**At 10,000 feet,** the effects of oxygen want are definitely present, but they may be subtle and insidious. You may not be aware that anything is wrong. Use oxygen, however, at this altitude if you are going higher, and under all circumstances, **start it on the ground for night flying, except for low level training flights such as those routinely encountered in primary, basic, and advanced training.**

**At 15,000 to 18,000 feet** the first effects on the brain become marked, and you may experience loss of judgment, dulling of the mind, loss of emotional balance, and the development of fixed irrational ideas. Muscular control is impaired, memory fades temporarily, and the ability to solve navigational and other problems is impaired seriously.

**At 20,000 feet** there may be fits of laughing or crying, impatience, rage, or other emotional disturbances, and great muscular weakness or paralysis. Vision is affected at this altitude. Depth perception may become faulty, and double vision occur. With some there is a feeling of high efficiency, even though unconsciousness is approaching. Others get sleepy and pass into a stupor.

**Above 20,000 feet** most people lose consciousness within a short time and death follows.

These symptoms may, of course, develop in different sequences and in various forms, depending upon:

1. Rate and duration of ascent to altitude.
2. Activity or excitement, cold, and the presence of dangerous gases. (See PIF No. 4-11.)
3. Differences in the individual's reaction.

If oxygen has been supplied up to **30,000 feet,** sudden removal of the supply produces great mental and physical inefficiency in from 30 to 60 seconds, and unconsciousness in 30 to 90 seconds.

**After having suffered from oxygen want, see your flight surgeon.**

**Maintain physical fitness.** This has a great deal to do with your ability to withstand high altitudes. (See PIF No. 4-4-1.)

If you must fly at altitudes in excess of 40,000 feet, be absolutely sure that your oxygen equipment is functioning perfectly, that you know how to use it, and that you know what to do in an emergency.

### Body Gases

Expansion of body gases in the stomach and intestines occurs as the atmospheric pressure decreases. At about 18,000 feet the stomach might contain more than a half-pint of air; normally it contains one-quarter pint. The expansion of gases in the intestinal tract may cause some discomfort, but only rarely serious difficulty in a normal person. Relief usually can be obtained by belching and the passing of flatus. In severe cases descent to a lower altitude may be necessary.

When you are flying frequently, avoid foods that tend to produce gas. Beans, cabbage, beer and other carbonated beverages are likely to make trouble. There probably are others that affect you. Notice which they are and save yourself distress.

Practice clearing your ears rapidly by swallowing, yawning, or closing your mouth and nose and gently attempting to exhale while swallowing. This will considerably increase the rate of descent you can stand comfortably.

**Do no flying when you have a head cold** unless it is absolutely necessary. Congestion of the lining of the nose and throat causes difficulty in the equalization of pressure, and pain in the ears or sinuses often results. Infection sometimes spreads to the sinuses and ears.

### Aeroembolism ("Bends", "Chokes", and "Creeps")

Decreasing the pressure on the body not only affects the gas already present, but allows the gases dissolved in the body fluids (carbon dioxide, oxygen, and especially nitrogen) to escape and form bubbles. Trouble seldom develops below 30,000 feet, but the likelihood of it increases the higher you go, the longer you stay, the colder you get, the more you exercise, and the faster you ascend, particularly if you have not taken oxygen from the ground up.

Bubble formation usually is indicated by pain in or near the joints (bends). A feeling of oppression in the chest or tightening and pain in the throat (chokes), and itching and irritation of various parts of the skin (creeps) sometimes are felt. Any of these symptoms may gradually or suddenly become worse and force you to descend. The symptoms then become less noticeable, and descent to 20,000 feet usually brings complete recovery.

If you have had any physical difficulties, especially with your ears or sinuses, during a flight, see your flight surgeon as soon as possible.

**Four Golden Rules For Oxygen:**

**1. Use oxygen for all flights above 10,000 feet.**
**2. Use oxygen from the ground up for all night flying, except for low altitude training flights.**
**3. Judge your need for oxygen by your altimeter, not by your sensations.**
**4. Check all oxygen matters with your oxygen officer.**

REFERENCE: "Your Body in Flight," T. O. 00-25-13, dated July 20, 1943; T. O. 03-50-1, dated July 1, 1943; "Physiology of Flight."

# OXYGEN EQUIPMENT

## General—Oxygen Systems

Aircraft oxygen systems in use by the Army Air Forces consist of two general types:

1. **Demand system.**
2. **Continuous flow system.**

**The demand system** is automatic and furnishes the oxygen on demand in just the right quantities for all altitudes. Every time the user inhales, a shot of oxygen with the proper mixture of air is supplied. The demand system is now being installed in all Army Air Force combat aircraft.

**The continuous flow system,** on the other hand, supplies the oxygen in a continuous flow. In use, an altitude dial on the regulator must be manually adjusted to correspond with the reading of the altimeter to insure the delivery of the proper amount of oxygen for that altitude. Until recently, Army Air Forces aircraft were provided with this type of equipment.

The A-14 demand oxygen mask is the latest type of demand mask available.

## Demand System

The demand oxygen system includes a demand type mask, A-12 type regulator, pressure gage, pressure indicator lamp, and ball or blinker type flow indicator. In addition, a portable recharger hose is supplied at each crew position in heavy bombardment aircraft for recharging portable (walk-around) oxygen equipment from the oxygen system of the airplane.

The A-10 Revised demand oxygen mask. Fit is important in all demand masks. Let your Oxygen Officer check the fit.

Three types of demand oxygen masks are available—type A-10, type A-10 Revised, and type A-14. They are used with the demand type regulators, the A-12 regulator for permanent installations within the airplane and the A-13 regulator for portable use.

The demand regulator (A-12) is essentially a diaphragm-operated flow valve which is opened by suction when the user inhales and closed when he exhales.

The demand regulator (A-12) is provided with two manual controls for use under special conditions—the AUTO-MIX lever and the EMERGENCY VALVE.

**With the AUTO-MIX lever in the normal ("ON") position, the A-12 demand regulator automatically mixes just the right amount of oxygen with the air for the altitude at which the plane is flying.** This is accomplished by an evacuated metal bellows like the aneroid control used in the altimeter. At sea level, the bellows is fully contracted and the air intake port is wide open while the oxygen port is closed. However, as the altitude increases and the pressure decreases the metal bellows expands and as it does so it gradually closes the air intake port. Finally, at an altitude of approximately 30,000 feet, the

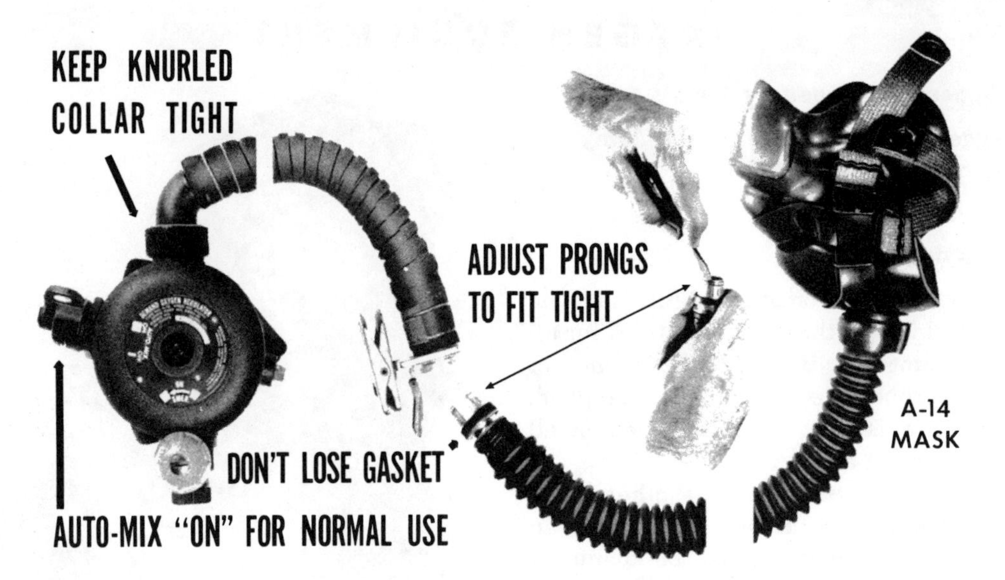


air intake port is entirely closed and the oxygen port is wide open to deliver pure oxygen to the mask.

**Remember, the normal position for the AUTO-MIX lever is "ON."** When it is in that position the regulator automatically furnishes the proper amount of oxygen for all altitudes. **When the AUTO-MIX lever is in the "OFF" position, the air intake port is closed and pure oxygen is supplied at all altitudes.**

**Note: It is not necessary to turn the AUTO-MIX lever to the "OFF" position above 30,000 feet.**

When the red colored EMERGENCY VALVE knob on the regulator is turned on, the oxygen by-passes the demand mechanism in the regulator and enters the mask in a steady flow, regardless of breathing and of altitude. The valve, therefore, should not be opened except in case of emergency, as the oxygen supply will be quickly exhausted.

**Use the AUTO-MIX lever in the "OFF" position ONLY in the following cases:**

1. Below 30,000 feet to give a wounded man pure oxygen as treatment for shock due to wounds.
2. In special cases where the flight surgeon may advise breathing pure oxygen on the ground before flight and from the ground up as a protection against bends.
3. As protection against poison fumes, like carbon monoxide, or poison gas used by the enemy.

**Turn on the EMERGENCY VALVE only:**

1. To revive an unconscious crew member.
2. In the case of failure of the regulator. (Watch the flow indicator!)
3. In case of obvious failure of oxygen. In this case, turn on the emergency until the cause of the failure is discovered and corrected.

## Precautions

**Before Take-Off:**

1. **Make sure mask fits properly.** Check for leaks by holding thumb over end of hose and breathing in gently. **Have Oxygen Officer check size and fit, with special Oxygen Officer's Test Set whenever possible.**
2. Check the pressure of the oxygen system. It should not be less than 400 pounds per square inch.
3. Crack the emergency valve on the regulator and see that you get a flow. (Caution: When the emergency valve is open do not pinch the hose or block the outlet or the regulator diaphragm will blow out.) Then be sure to close the valve tightly.
4. Check knurled collar at outlet end of regulator. It should be tight.

5. Check rapid-disconnect fitting on mask hose. Be sure rubber gasket is in place. Make sure male end of fitting fits snugly into the regulator hose. If it is too loose, pry open the prongs to get a tight fit.
6. Clip oxygen-supply hose to clothing or parachute harness close enough to your face to allow movement of the head without kinking or pulling the hose.
7. Be sure the AUTO-MIX lever is in the "ON" position.

**In the Air:**

1. When the mask is first put on, check for leaks by holding your thumb over the end of the mask hose and breathing in gently.
2. Manipulate the mask at frequent intervals when the temperature is low, to free it of any ice that may form.
3. Check the oxygen pressure gage frequently. Note: the regulator does not function properly with a pressure of less than 50 pounds per square inch.

**After A Flight:**

1. Wipe the mask dry. Wash it frequently with soap and water, rinse well, and dry thoroughly. Note: Masks with microphones should not be immersed in water; they should be washed with a cloth.
2. Inspect mask for cracks and possible punctures.
3. Don't lend your mask to anyone except in an emergency.
4. Take care of your mask and it will take care of you.

## Continuous Flow System

Three types of masks—the A-7A, A-8A, and A-8B —are used with the continuous flow oxygen system.

These are rebreather types of masks. The function of the bag is simply to conserve useful oxygen from the exhaled air. These masks must be used with the oxygen regulators A-9 and A-9A. The A-9 and A-9A regulators consist of an altitude flow indicator and a pressure gage. The flow indicator must be set at the altitude at which the plane is flying.

The A-7A mask, used with the A-11 regulator, is procured for use in cargo aircraft only. Since it leaves the mouth uncovered, it would be dangerous for general use above 20,000 ft.

## Precautions

**Maintenance:**

1. Have rate of flow checked every ten days with a flow meter.

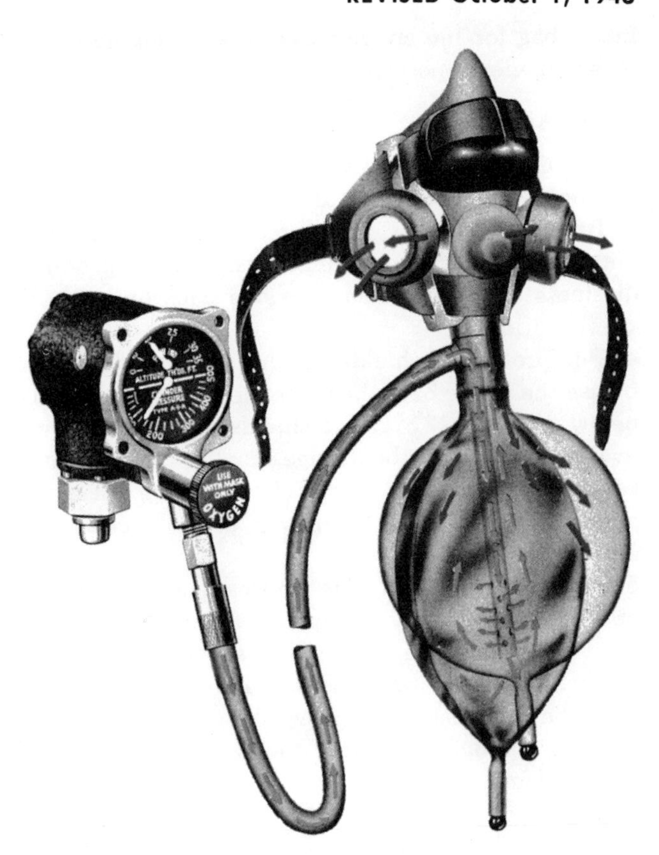


2. Never apply oil to any part of your oxygen equipment.
3. See that all parts are free of dirt.
4. Check entire system for leaks. Pressure should be maintained overnight with all regulators in the "OFF" position, if there has been no appreciable change in temperature.
5. Make sure that the valve adjustment knob of the regulator has fair resistance against turning to prevent any possibility of its being accidentally moved during flight. If it is loose, tighten the valve gland packing nut.

**Before Take-Off:**

1. Check the cylinder pressure. It should show 400 to 450 pounds per square inch for A-9 or A-11 regulators.
2. Check connections between mask, bag, and plastic connecting tube.
3. Check the rebreather bag for holes. Be sure plug is in bottom of bag.
4. Check to see that the exhalation disks are in proper position.
5. Know where your regulator is located.
6. Carry extra sponge rubber discs and protective shields for the exhalation turrets, or a protective

fabric bag for the entire mask. Take along an extra mask whenever possible.

**In the Air:**

1. Be sure regulator is set at proper altitude.
2. Check the cylinder pressure occasionally.
3. Breathe normally; overbreathing accomplishes nothing. In fact, it is dangerous; it may produce dizziness and other more serious consequences, if persisted in.
4. Put protective shields on the exhalation turrets or use the fabric bag whenever the temperature falls below 10°F (−12.3°C). If shields and bag are not available, examine the sponge discs at intervals and remove any ice that forms by squeezing them, or change the sponges. Whenever possible carry an extra mask.
5. Above 30,000 feet, the rebreather bag should never be completely collapsed when breathing in. If it does collapse, the valve should be opened further no matter what the flow indicator reads.
6. When exercise is necessary, be sure the valve is open far enough to prevent the bag from collapsing at 25,000 feet.
7. When you change your station at altitude be sure the new cylinder valve is **FULL ON and that the bayonet fitting is locked.**

**After a Flight:**

1. Shut all flow valves and make sure they are tightly closed.
2. Wash mask frequently with soap and water, rinse well, and hang up to dry.
3. Don't lend your mask to anyone except in an emergency.
4. Keep your mask in a safe place and away from sunlight.
5. Check the bayonet connection to see that the rubber seat is in place.

**Warning**

Extreme caution must be exercised in the use of oxygen equipment to insure that none of it becomes contaminated with oil or grease. **FIRE OR EXPLOSION** may result if slight traces of oil or grease come in contact with oxygen under pressure. Be sure that all lines, fittings, instruments and other parts are free of oil, grease, and other foreign matter. **NEVER USE LUBRICANTS ON ANY PART OF THE OXYGEN SYSTEM.**

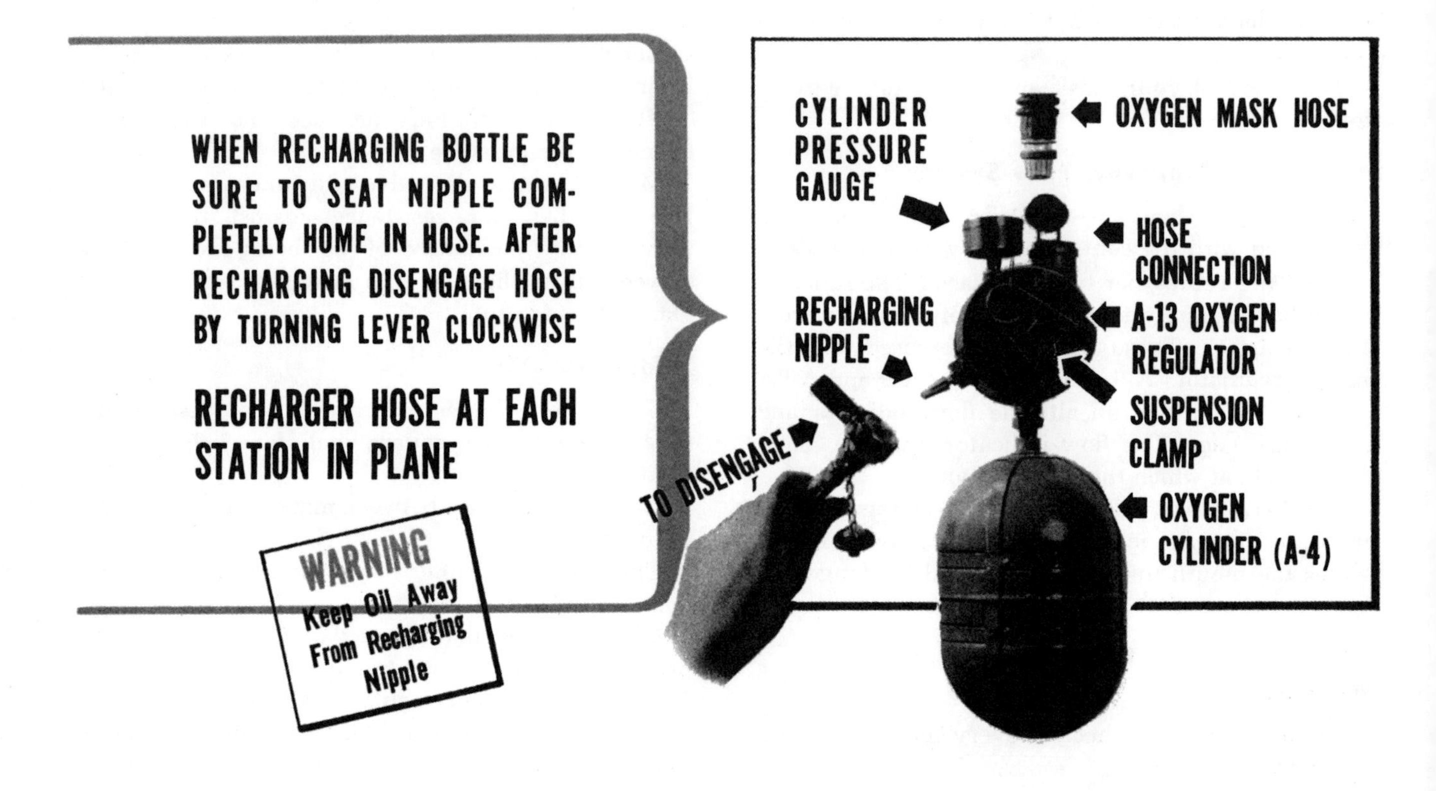

# *Portable* OXYGEN EQUIPMENT

### Walk-Around Bottle

In large airplanes, portable oxygen equipment in the form of a walk-around bottle fitted with an A-13 pure demand regulator is provided for use by the crew when changes of station are necessary. When movement from one part of the airplane to another is necessary, the mask hose is simply disconnected from the regular oxygen system hose and connected to the outlet on the A-13 regulator.

The walk-around bottle holds about a six-to-eight minute oxygen supply. The supply may last a shorter or longer time, depending on the altitude, depth of breathing, and the pressure in the bottle. The bottle can be replenished directly from the airplane oxygen system, but obviously only up to the remaining pressure in the system. The bottle is replenished by means of the portable recharging hoses at each station in the plane.

### To Use the Portable Unit:

1. Before using, check the pressure gage. If pressure gage shows less pressure than that of airplane oxygen system, the cylinder should be refilled.
2. Inhale deeply, hold your breath, then disconnect mask from regular oxygen system hose.
3. Quickly open the spring cover of the regulator connection on the walk-around bottle and snap in the male fitting on the end of the mask hose.
4. Clip the portable unit to your clothing or parachute harness by means of the spring suspension clamp attached to the regulator.
5. During use, watch the pressure gage and refill cylinder from oxygen system whenever pressure falls below 100 pounds per square inch.

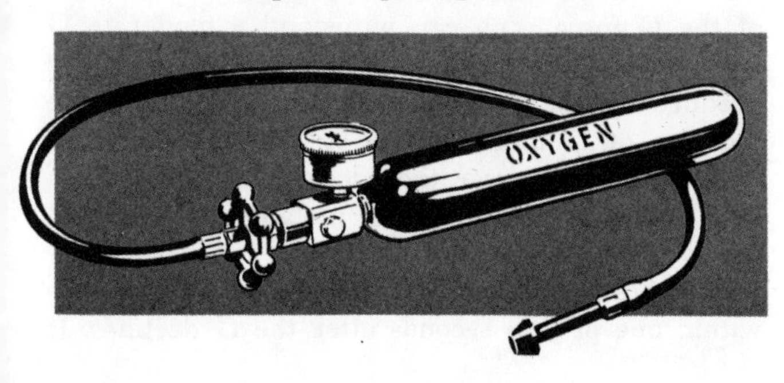


### Bail-Out Oxygen Cylinders

Two bail-out oxygen cylinder assemblies are used for parachute descents from high altitudes. Both are completely self-contained units with pressure gage and release valve.

**Either cylinder must be tightly fitted and securely tied in a pocket sewn to the flying suit or harness.**

Before takeoff, check the cylinder's pressure gage. It should read at least 1800 pounds per square inch. Either cylinder assembly can be used in parachute descents above 30,000 feet, and as an emergency oxygen supply in fighter aircraft if the regular oxygen supply suddenly fails.

Type H-1: Before jumping, grip pipe stem between your teeth and completely open flow valve.

Type H-2: Before jumping, pull the release to open flow valve. Then, disconnect the main oxygen tube and jump.

The symbol "G" is used to compare any given force with the pull of gravity. Under normal conditions your body experiences a force of 1 G, the pull of gravity. A force of 4 G is a force four times the pull of gravity. If you weigh 200 pounds under ordinary conditions, you weigh 800 pounds during exposure to 4 G.

In level flight you can endure any speed if you are protected from the windstream. During changes of direction in high speed maneuvers, however, centrifugal force imposes increased G forces. The increase in G forces results in proportionately severe stress on you and your airplane.

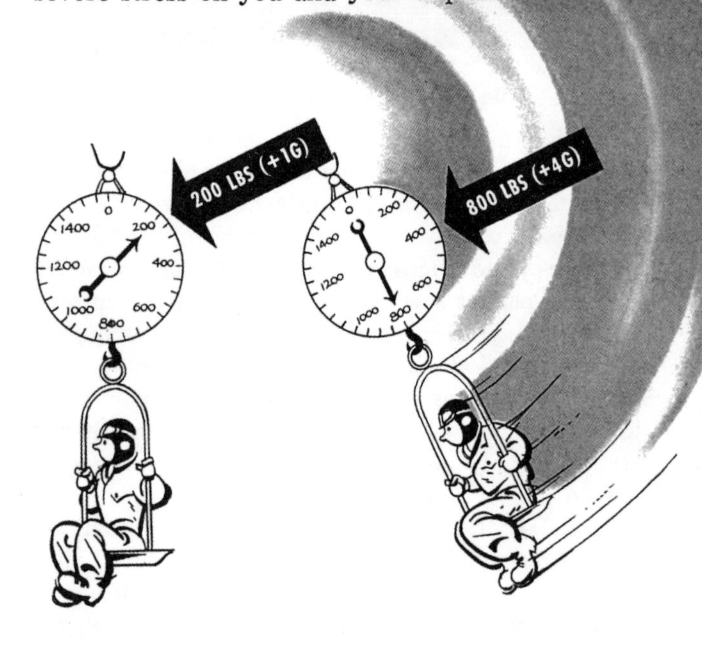


**Positive G ( + G )**

When you pull out from a dive or make a spiral turn, your head is toward the center of the circle which your airplane describes. Centrifugal force acts from your head to your seat. Force acting in this direction is called positive G; it pushes you down into the seat and seems to pull you downward.

The physiological effects of positive G result from a decrease in the heart's efficiency.

1. The pumping action of your heart may not provide sufficient pressure to raise the blood to your brain, since the blood is heavier under increased G.

2. The blood tends to be forced to the lower parts of your body and must return to your heart against the G force. As a result less blood gets back to your heart, and like any pump it loses efficiency with this decreased intake.

The decrease of the blood supply to the eyes and brain brings on serious effects: dimming or loss of vision, and unconsciousness. With less blood supply the eyes and brain do not receive their normal supply of oxygen and cannot function properly.

These effects vary with the intensity and duration of the G force. You can withstand a moderate G force with clear vision. A slightly greater force may produce visual dimming, called graying. With the increase of G in successive steps you experience temporary blindness at the sides of your visual field, then blackout, or complete inability to see, and finally, unconsciousness.

If you remain conscious, recovery of vision occurs within one or two seconds after the G declines. If you lose consciousness, you may suffer loss of memory and of muscular coordination lasting up to 20 seconds, and may remain confused for a minute or more. The danger in this situation is obvious.

Besides the magnitude of the G force, its duration is also a critical factor. If G is applied suddenly, it takes approximately 3 to 5 seconds for visual symptoms to develop. Therefore, you may withstand successfully for 2 or 3 seconds a force which, if prolonged, would result in unconsciousness.

On the other hand, at any particular G level, all your difficulties develop within the first 10 seconds of exposure. In fact, some improvement in your first visual impairment occurs after about 15 seconds. **This means that if you ease off on the stick as visual graying occurs, your blood vessel system adjusts to the situation and you can withstand more G in the latter part of a long spiral turn than at the beginning.**

Men differ in their reaction to G forces. The average man blacks out at 5 G, if maximum G is reached in 2 or 3 seconds and maintained for 10 seconds. Some men black out at 3½ G, while a few retain some vision until they reach 7 G. Most men experience some visual dimming between 3½ and 4 Gs, held for 10 seconds.

### Tolerance to "G" Force

While standing, your tolerance to G force is greatly reduced. **If possible, warn your crew before you perform any sudden maneuver, so that they won't be caught standing up.** Otherwise, even a relatively low G force may cause a man to lose consciousness and to injure himself by falling.

Your own G tolerance is characteristic for you and remains constant. Practice by frequent exposure to G neither raises your tolerance nor lowers it. In some flight maneuvers you are able to tolerate higher G levels than you could otherwise because the G force is of short duration.

There are ways in which you can increase your tolerance for G forces. If you can manage to crouch slightly, you decrease the height your heart must pump blood against the G force up to your brain. You can also employ the muscle straining technique described in this section.

A man with a high G tolerance who is an expert at using the muscle straining technique can withstand forces which exceed the safety limits of his airplane. You should know the limits of your plane and never exceed them. If you are going to experiment with G limits, install an accelerometer in the cockpit of your plane during the tests. Watch the accelerometer and learn how your own G tolerance compares with the stressing on your airplane.

### Negative G (—G)

Negative G is the force you experience during inverted flight, outside turns, and push downs. In these maneuvers, centrifugal force operates from your seat to your head, driving the blood upward and increasing the pressure within the blood vessels of the head.

At the relatively low force of —2 or —3 G, you feel fullness or throbbing pain in your head, a gritty sensation in your eyes. Prolonged stresses of —3 G may cause hemorrhage in the eyes and, in certain cases, fatal hemorrhage into the brain tissue.

Unlike the situation with positive G, the disorders caused by negative G may outlast the stress by several minutes, or even by several hours.

Neither you nor your airplane are as well equipped to withstand negative G forces as you are to withstand positive G forces.

### Play Safe

1. Stay within the safe load limits of your plane's structure and safe limits of your capabilities. Consider graying as a warning signal of excessive +G.
2. Learn to perform the muscle straining maneuver properly so that you will be able to use it when you need it.

### Raise Your +G Tolerance by Proper Muscular Straining

1. Pull your head and neck down between your shoulders like a turtle.
2. Tense the muscles of your abdomen, arms, and legs. Strain against the safety belt.
3. **You must keep breathing!** To insure this, yell "Hey" over and over again.
4. Once you begin this maneuver, **keep it up until the +G is past!** You may have reached a higher level than you can tolerate when relaxed.

# Physical Fitness

You are responsible not only for your own physical fitness but also for that of your crew. **This means daily exercise for all of you.** Exercise can take the form of sports and games, or of conditioning calisthenics, but it is essential, whether or not an exercise period is provided during duty time.

Haphazard exercise is of little value. It must be regular, intense, and of proper duration to do you any good. You should be really tired out when you are finished.

Of course physical exercise alone will not keep you fit. **It must be supplemented with good food, and with sufficient sleep, rest, and relaxation.** Otherwise the fatigue and strain will ultimately wear you down.

The importance of physical fitness is obvious. You wouldn't take out a plane in poor condition. Neither should you fly with a crew in anything less than perfect physical shape. **Flying requires perfection.** You cannot attain perfection nor maintain it without daily exercise. If you can't get in a game, get a half hour of conditioning exercises each morning, and some running during the day. Join in at least one or more of the following activities as the occasion and facilities permit. Above all make exercise a **daily routine.**

Some of the sports listed here can always be used as part of your personal physical conditioning program:

**Team Activities:** touch football, basketball, soccer, speedball, volley ball, softball, baseball.

**Dual Activities:** badminton, tennis, handball, squash, medicine ball.

**Individual Activities:** Swimming, cross-country running, weight-lifting, rope skipping, setting-up exercises.

Mix these up as practical needs require. The main thing is to get a maximum of fun and hard physical exercise out of them. In this way they become literally recreational.

**Remember your physical condition is, in the last analysis, strictly up to you.** You can and should check up on your physical fitness from time to time by giving yourself the work-out prescribed in the AAF Physical Fitness Test. The test is simple, practical, and specific, and you and your crew can find out how you compare with men in top physical shape by trying your own ability against the standards. Use the sit-ups, chinning, and 300-yard shuttle run as you would use a scale to keep yourself informed of your physical state.

A man takes justifiable pride in knowing that he is trim and fit; when he is fighting a war as a member of a team like the AAF more than pride is involved. He has a direct responsibility to himself and to other members of his fighting team to keep himself ready for high precision physical performance at all times.

# VISION AT NIGHT

### To Develop Night Vision

1. Insure adaptation to dark preceding any night operation by staying in a dark room or by wearing red lensed goggles, (Goggles, Assembly, Polaroid Type D. A., Class 13, Stock No. 8300-343575), for 30 minutes.

2. Protect this adaptation by not exposing your eyes to any bright light, either inside or outside the aircraft.

3. Keep all nonessential lights within the aircraft turned out, and all essential lights dimmed.

4. Use red light within the airplane whenever possible.

5. Read instruments, maps, and charts rapidly; then look away. Or, use only one eye; the other eye will retain its dark adaptation.

6. Use oxygen from the ground up on all night flights, except in low altitude training.

7. Keep all windows scrupulously clean.

8. Avoid casual paper or unpainted spots on the instrument panel—they reflect light.

**Be sure to adapt your eyes to night vision before taking off.** The sense organs for night vision become insensitive under strong light and require time for "adaptation to dark." During this period nothing brighter than candlelight should be used. If a dark room is not available, red goggles will do almost as well. Then when you take off a 10,000 fold increase in sensitivity will have been achieved.

9. Experience brings much improvement in night vision. Practice off-center seeing on dark nights whenever possible.

**Keep goggles, enclosure windows, and windscreens clean and free from scratches.** Scattered light reduces the contrast between faint lights and their backgrounds; reflected light from windscreens does the same.

**Be sure you have an adequate supply of oxygen from the ground up.** Without it, night vision is impaired at an altitude of only 5,000 feet, and is only one-half as efficient at 12,000.

**Eat food rich in vitamin A.** This is a chemical factor essential to good night vision. Eggs, butter, cheese, liver, apricots, peaches, carrots, squash, peas, and especially cod liver oil and all types of greens are rich in vitamin A. Too much vitamin A will neither help nor harm you.

**Remember that in dim light, or at night, you cannot see what you look at directly.** A night blind spot, unrecognized by most people, lies at the center of the eye. **Therefore, in the dark, look to one side of the thing you want to see.**

For reading maps and instruments, or when caught in searchlight beams, close one eye and avoid looking at the source of light. The eye which is kept closed will retain its power to see in the dark independently.

Use roving vision when searching the night sky. Keep your line of sight fixed in one direction for about one second, then move on a short distance to another position, thus covering your entire field in a series of eye movements and pauses. Small adjustments of the line of sight are sometimes made more easily by moving the head than by moving the eyes alone.

REFERENCE: "Physiology of Flight" and AAF Memorandum No. 25-5.

### In Flight

The air you breathe while in flight should be free of all other gases except the oxygen from your mask. Exhaust gases, fumes from hydraulic fluid or coolant fluid, smoke, and poison gas may contaminate the cockpit air. In sufficient concentration, any of these gases is dangerous. You can protect yourself from them by knowing when to suspect their presence and by observing the necessary precautions.

Exhaust gas is a mixture of several substances. Among these are carbon monoxide and oxides of nitrogen, both of which are poisonous. Carbon monoxide acts by combining with the red cells of the blood and making them useless for carrying oxygen to the body tissues. This results in oxygen want or anoxia. As you ascend from sea level the dangers resulting from carbon monoxide increase, even below 10,000 feet, unless you use your oxygen mask.

The results of carbon monoxide poisoning are similar to those caused by oxygen lack—shortness of breath, headache, nausea, dizziness, dimming of vision, poor judgment, weakness, unconsciousness, and death. The higher the concentration of the gas, the longer you breathe it, and the higher the altitude (unless you use your oxygen mask), the more severe are the symptoms.

Like anoxia, carbon monoxide poisoning may give no warning. The gas itself has no odor, but you can be pretty sure of its presence if you smell exhaust gases. **The only safe rule for protecting yourself against carbon monoxide in flight is to wear your oxygen mask with the Auto-Mix "OFF" whenever you smell exhaust gas. In this way you get pure oxygen to breathe and are completely protected from any gases in the cockpit. As soon as you safely can, land your plane and report the trouble to your engineering officer.** If you had any unpleasant sensations while in flight see your Flight Surgeon as soon as possible.

Don't use your exhaust heater during combat. Enemy gunfire may cause dangerous leaks of the exhaust gas into the cockpit.

Be sure to wear your oxygen mask with the Auto-Mix "OFF" whenever you notice smoke or any unusual odor, such as that from gasoline, oil, hydraulic fluid, or coolant fluid, in the cockpit. If you are exposed to war gas and do not have your gas mask with you, wear your goggles and oxygen mask— **but be sure to have the Auto-Mix "OFF".**

### On the Ground

Carbon monoxide is always present in exhaust gases. It occurs in poorly ventilated hangars and garages when motors are running. It is important, therefore, not to keep any kind of engine running any longer than necessary inside closed hangars and garages, and to ventilate such buildings as much as possible. If the hangar or garage must be kept closed do not stay inside any longer than necessary.

In cold regions, where gasoline or oil stoves and lamps are used in closed building or tents, carbon monoxide is formed whenever fuel is burned. Cross-ventilation is your only reliable protection.

### First-Aid

If a person is overcome by carbon monoxide or any other gas, begin first-aid at once:

1. Remove him from the source of the gas.
2. Give him pure oxygen to breathe, if available (turn on EMERGENCY flow).
3. Begin artificial respiration immediately (see PIF 8-14-4).
4. Keep him warm.
5. Send for a medical officer.
6. Never exercise a person who has been poisoned by gas. This only makes him worse.

In flying operations extremes of temperature ranging from more than +130°F to less than -50°F may be encountered. It is important to know the precautions necessary to protect yourself against these wide thermal variations. The specific measures which must be adopted vary in different regions. Consult your Flight Surgeon for special information pertaining to health safeguards in unfamiliar regions. Knowledge of certain generalizations, however, is useful wherever you may be.

## HEAT

### Clothing

In the desert you must protect yourself from the severe burns which may result from exposure to the intense rays of the sun. This protection can be accomplished by wearing a sun helmet and lightweight clothing which leaves as little of the body surface as possible exposed to the sun. You can protect exposed surfaces of the body from sunburn by the use of sunburn protective ointments. Clothing should be loose-fitting and porous, to permit the evaporation of sweat. Tinted goggles or sunglasses will protect your eyes from the glare of the sun as well as from dust and sand.

Despite intense heat on the ground, it will be much cooler while flying and extreme cold will be encountered at high altitudes. Be prepared to increase or decrease the amount of clothing worn, according to the temperature met with in flight.

Even on the ground, desert nights are frequently cold. Be prepared for rather rapid changes in temperature by having available warm clothing for use at night. Woolen socks are most suitable for general use; they provide the greatest comfort because of their ability to absorb moisture and their good insulating properties. You will be troubled less by sand in your shoes if you wear the high-top variety. Oxfords have no place in the desert.

In the jungle or in regions where mosquitoes are prevalent, keep your body covered as much as possible and wear a head net for protection of your face and neck.

### Water

In most warm-weather operations, water supplies are limited. If water is rationed, use it sparingly for purposes other than drinking. You will be surprised at how little water you can get along on in relative comfort if you are careful. Cut down on your smoking; better still, don't smoke at all—for smoking increases your thirst. When your water rations are limited chewing gum will help decrease your thirst. Water from streams or sources of questionable purity should always be boiled for at least 15 minutes before drinking. This will kill any dangerous organisms which may be present. Water also can be purified by chemicals such as Halazone tablets, which are included in the Kit, First-Aid, Aeronautic. Add 1 Halazone tablet to a pint of water; wait a half-hour before drinking.

### Food

Food spoils quickly in warm climates. If canned rations are used they should be eaten soon after opening the cans or stored in a cool place. Food obtained from natives should be thoroughly cooked before eating. Thick-skinned fruits which can be peeled need not be cooked. Never use milk or milk products, such as butter, cheese, and ice cream, which are sold by the natives.

### Exercise

In warm regions, physical exertion should be kept at a minimum. Exertion increases your sweating; sweating results in increased thirst.

### Warm Climate Health

Sunstroke usually is caused by prolonged exposure to the direct rays of the sun, although it may occur even in cloudy weather. This is a serious condition. The symptoms may include headache, dizziness, red spots before the eyes, and vomiting. Even unconsciousness may occur. Heatstroke is a similar condition resulting from exposure to excessive heat from any source. Heat cramps in the various muscles of the body may develop following profuse sweating and the resulting loss of salt.

All of these conditions can be prevented by avoiding sun, sweat, and toil. Since this is usually impractical, do the next best thing: whenever possible perform your heavy work in the late afternoon, evening, or early morning when the temperature is lower. Provide all possible ventilation. Salt your food liberally and take salt tablets daily.

If your water rations are extremely limited, however, salt tablets may do more harm than good. Consult your Flight Surgeon before using them. He will instruct you as to the necessity for taking salt tablets and advise you concerning the dosage.

If sunstroke or heatstroke occurs, prompt treatment must be given. This should consist of cooling the body—immersion in cold water, covering with wet sheets, or simply pouring cold water over the patient's clothes, and fanning him.

In certain regions where malaria is prevalent it may be desirable to take atabrine or quinine to suppress the symptoms of this disease.

Take these drugs, however, only upon the advice of your Flight Surgeon. He alone can tell you when you need them and in what amounts.

## COLD

The principal hazard in cold weather operations is frostbite. Protect yourself against this condition by observing a few simple principles.

### Clothing

All clothing should fit loosely. This applies particularly to your socks, boots, and gloves. Tight-fitting clothing interferes with the blood circulation and makes you more susceptible to frostbite. Individual garments should be light and porous. It is desirable to wear several layers. Two light garments afford better insulation than a single heavy one. Long woolen underwear should be worn on all flights over cold regions. Wear several layers of woolen socks in sub-zero climates, but be sure that your boots fit loosely over them.

Leather shoes are of little value in Arctic regions. They not only afford poor protection, but may cause harm if they fit tightly over your socks.

Don't wear ordinary GI shoes inside your winter flying boots. Use woolen socks and felt liners or electrically heated shoes instead.

If possible wear two pairs of gloves or mittens—a rayon or other light pair inside heavier ones (either A-9's or electric gloves). Mittens are better than gloves, for they allow your fingers to come in contact with each other. This helps keep them warm.

Keep your socks and underwear clean. After they have become soiled by body oils and secretions, they lose a great deal of their insulating properties.

### Keep Your Clothing Dry

Wet clothing is almost worthless in protecting you from the cold. If any part of your clothing should become moist, either by accident or through perspiration, take it off and dry it over a fire or change to dry clothing immediately. Wet feet and hands are particularly dangerous in cold climates for they will fall easy prey to frostbite.

Exercise to keep warm but guard against overexertion in extreme cold. Overexertion makes you perspire and the perspiration may turn to ice inside your clothing. This is dangerous. If necessary to perform much physical work, open or remove some of your clothing in order to prevent perspiration. Don't put on a heavy suit until just before takeoff. Wipe your body dry; then dress slowly. Once dressed, exercise no more than necessary.

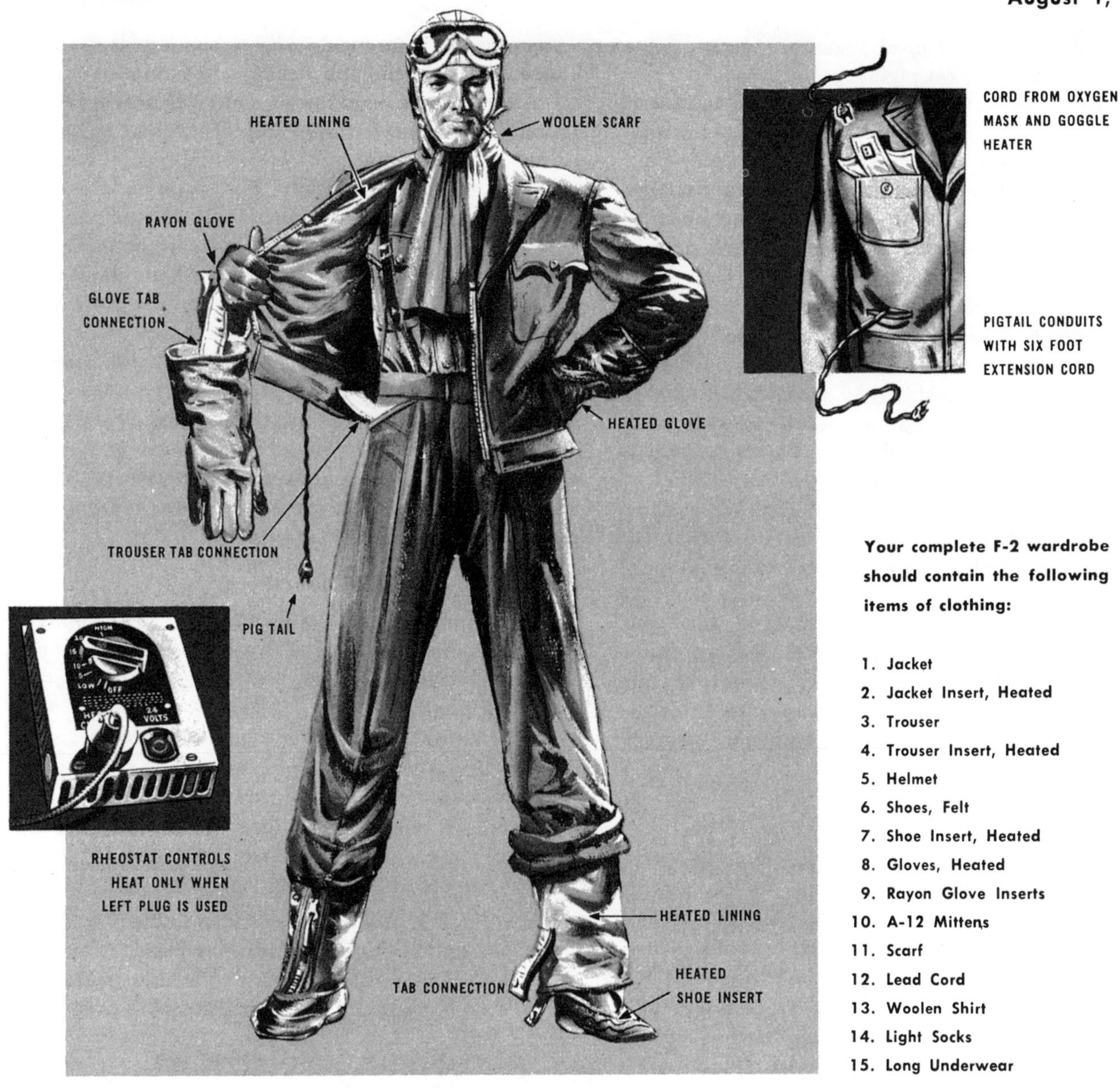


F-2 ELECTRICALLY HEATED FLYING SUIT

**Your complete F-2 wardrobe should contain the following items of clothing:**

1. Jacket
2. Jacket Insert, Heated
3. Trouser
4. Trouser Insert, Heated
5. Helmet
6. Shoes, Felt
7. Shoe Insert, Heated
8. Gloves, Heated
9. Rayon Glove Inserts
10. A-12 Mittens
11. Scarf
12. Lead Cord
13. Woolen Shirt
14. Light Socks
15. Long Underwear

### Electrically Heated Flying Suits

Electrically heated flying suits have the advantage of eliminating bulkiness and permitting greater ease in manipulating the controls of the airplane. However, there is a great disadvantage in wearing them on flights over cold regions. If your electrical system fails, if you are forced down, or if you have to bail out, you are left without protection.

Know how to use your electrically heated suit and treat it carefully. The electric heating elements are fragile. Dry your suit between flights, if possible, and have it tested by your Personal Equipment Officer.

1. Wear your F-2 electrical suit over long woolen underwear. (If your suit is the F-1 type, you should wear additional clothing over it as well.) The F-2 suit will afford adequate protection down to –40°F. If lower temperatures are encountered, add other flying clothing.

2. Put on the shoes with inserts over lightweight woolen socks. Then connect the snap fastener tabs on the trouser leg to the corresponding snaps on the shoe insert. Be sure that **both pairs** of snaps are properly connected.

3. Connect the tab at the top of the trouser to the corresponding snap fasteners on the inside of the jacket at the right. Make certain both pairs of

snaps are securely snapped together. Connect six-foot lead cord to jacket pigtail.

4. Put on regulation flying helmet and auxiliary equipment. Protect your neck from the cold by wearing a wool or silk scarf.

5. Put on lightweight rayon gloves. Snap the tabs on the jacket sleeves to the corresponding snaps on the heated gloves. Then put on the electrically heated gloves.

Connect your extension plug in the left receptacle of the built-in rheostat before takeoff and be sure that the suit is working properly. The plug can be locked into position by a simple clockwise twist. When in flight, keep the rheostat at the lowest comfortable heat. Don't ride hot; it will make you sweat.

Never rely on electrically heated suits alone when flying over cold regions. They are safe for use over temperate or tropical zones where cold is experienced only at high altitudes.

In extremely cold weather never touch cold metal with your bare hands, even for a moment. Your skin may freeze to it. If by accident this should happen, thaw your skin loose from the metal by warming the latter or by urinating upon it—don't pull your fingers loose.

### Frostbite

Frostbite occurs most commonly in the fingers, toes, nose, ears, chin, and cheeks. It may set in gradually and painlessly and without your being aware of it. Numbness, stiffness, and a whitish discoloration of the affected part are among the first signs. Wrinkle your face frequently when exposed to cold air. If you experience a sensation of numbness, warm the affected part with your ungloved hand until sensation returns.

Crew members should watch each other's faces and be on the alert for areas of blanching. In this way serious trouble can be prevented. Frostbitten tissues may later become painful. Such tissues should never be rubbed. Never apply snow or ice to a frostbitten part. If your hand becomes cold or numb, warm it by placing it inside your clothing or under your armpit. Frostbitten tissues should always be thawed gradually. When possible, thaw them at ordinary room temperature. They should never be placed near a heater or immersed in warm water or in kerosene. Any of these procedures may cause irreparable damage. If frostbite occurs, cover the frostbitten area with a loose sterile bandage; keep the patient comfortably warm with blankets or a sleeping bag and by giving him hot drinks; and give him 100 per cent oxygen to breathe (Auto-Mix "OFF"). Report to your Flight Surgeon immediately after returning from your flight.

### Snowblindness

This condition may result from exposure of your eyes, even for brief periods, to the glare which exists in snow-covered regions. The resultant damage to your eyes may cause intense pain and seriously interfere with your vision for several days—sometimes even longer. The hazard of snowblindness is particularly great on sunny days, but the glare which results from a bright overcast is almost as dangerous. Always protect your eyes by wearing colored goggles or sunglasses. In the Arctic, snowblindness may be brought on by merely lifting your goggles a half dozen times.

### Body Heat

Eat an abundance of fatty foods while in cold regions. Fats are rich in calories, which help you maintain body heat. Take hot drinks such as coffee, tea, cocoa, or soup along with you in thermos jugs on all flights. These also add to your body warmth. Never drink whiskey or other alcoholic beverages to keep warm. They give you a false sensation of warmth but may do great harm by actually robbing you of your body heat, since they cause flushing of the skin. This loss of heat, together with the false sense of security which alcohol produces, makes whiskey, brandy, and the like dangerous for anyone who is exposed to the hazards of cold. Don't eat snow. If it is necessary to use snow for drinking water, melt the snow first.

### Examination for Flying

Your Flight Surgeon can help you maintain your physical health, as well as your mental well-being. Consult him whenever you feel sick, tired, or rundown. Be sure to see him if you are injured. He is responsible in a large part for your flying status, as governed by AR 40-110. This regulation requires that you have a new physical examination for flying (Form 64) after:

1. Hospitalization for serious illness, or injury related to flying.
2. Grounding or removal from flight status, for cause other than minor illness or injury.
3. Return from sick leave.
4. Head injury with actual or suspected damage to the brain.

### Protection from Venereal Disease

Sexual relations outside of marriage are likely to result in infection with venereal disease. Abstinence is the best method of prevention. The next best way to prevent venereal disease is to use proper protective methods **each time** you have sexual relations. It is impossible to tell in advance whether you are exposing yourself to venereal disease. If you fail to protect yourself, you are apt to become infected.

Adequate protection consists of wearing a condom and taking a prophylactic as soon as possible (within an hour) afterwards. Prophylactic kits can be obtained from your Flight Surgeon or PX.

If you have been exposed and failed to protect yourself, don't wait for symptoms to develop. Consult your Flight Surgeon as soon as possible. Remember that he is neither a moralist nor a judge. He is interested only in preventing and curing disease and in keeping you in the best possible physical condition.

Disciplinary action will not be directed against military personnel who have acquired venereal disease or who have failed to take prophylaxis. **Appropriate action is authorized, however, if you try to conceal infection.** Treatment with quack medicine is dangerous, so don't take chances—**report to your Flight Surgeon.**

### Benzedrine

Benzedrine is a drug which may temporarily postpone sleep. The responsibility for the use of benzedrine on tactical missions rests with your Commanding Officer, who must decide what situations require it. Distribution and administration of benzedrine tablets, however, is the responsibility of your medical officer. Never take benzedrine on your own.

Benzedrine will not make you a superman. It merely postpones for a relatively short time your body's demand for sleep. Benzedrine is not a substitute for sleep, so be sure that you have a proper amount of rest after using it.

### Sulfa Drugs

The sulfonamides are potent drugs for the prevention and treatment of many infections. These drugs, however, may sometimes cause temporary mental confusion, difficulty in seeing and hearing, clumsiness, and other difficulties.

No one will be permitted to participate in aerial flights while receiving sulfa drugs or for six days following the last dose. Temporary grounding need not apply where sulfa drugs have been used for **external application only.**

The instructions and provisions of this regulation do not apply to passengers or patients being transported or evacuated by air, nor to the emergency use of sulfa drugs as supplied in the individual first-aid kits (AAF Regulation 25-13).

# SENSE OF POSITION IN FLIGHT

**Your sense of position, both on the ground and in the air, is governed by:**

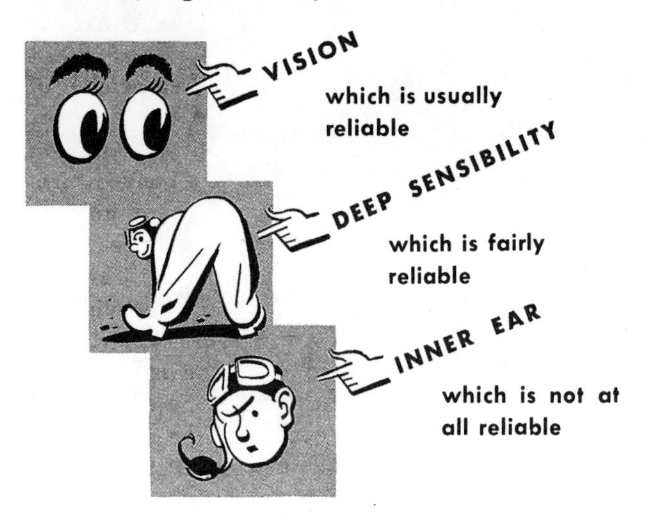


1. Vision is the all-important sense in contact flying. Vision furnishes your natural horizon. It is loss of vision which makes the transition from contact to instrument flight difficult, unless you learn to trust your gyro-horizon and other flight instruments with as much confidence as you do the natural horizon. In instrument flight, rely only on what you see in the cockpit.

2. The combination of sensations of pressure and tension on the skin, muscles, tendons and internal organs is known as deep sensibility.

When your feet are on the ground, sensations from your soles and the muscles of your legs tell you what position your feet are in. This deep sensibility is an important factor in helping you maintain your bodily balance.

In flight you get the effect of deep sensibility in the seat of your pants, since gravity and centrifugal force exert their action there. Changes in the position of your plane make you feel heavier or lighter, or as if you were being forced sideways in your seat.

3. When you can't check your position by visual reference, you rely mainly on your inner ear. The inner ear has two parts. The semicircular canals, which are placed in three different planes, are responsible for the sensation of rotation or turning, pitching and rolling. The static organ tells you the position of your head in relation to the ground.

## INSTRUMENT FLIGHT

The sensations provided by deep sensibility and the inner ear lead to confusion when you are flying on instruments. Never trust your feel or sense of position when you are flying on instruments or at night; if you do, you'll get into trouble since the sensations from the seat of your pants and your inner ear do not indicate the true position of your airplane.

You often get false impressions when on instruments even during straight and level flights and especially in turns.

### Sensory Illusions in Straight Flight

1. If your airplane tilts or tips suddenly in rough air and recovers slowly, you may not know when recovery is complete. Your inner ear retains the impression that your aircraft is still tipped or tilted. This impression may be so strong that you may lean to one side in an attempt to remain upright. **The leans** is one of the most common sensations experienced during instrument flight.

2. If you don't check your instrument readings every few minutes, your airplane may turn from the desired heading so slowly that you won't notice it. Then when you check with your instruments and correct your heading and position, you will feel that you are continuing your turn in the opposite direction.

3. The greatest single danger to an inexperienced pilot flying at night or on instruments is the **Graveyard Spiral.** This is a diving spiral which gets progressively tighter and steeper, and which is accompanied by a rapid loss of altitude.

Such a spiral results directly from relying too much on your sensations and too little on the instruments which indicate bank and turn. The spiral starts when the airplane gradually enters a turn without your realizing it. In such an involuntary turn the angle of bank and the rate of turn both increase so slowly that you have no sensation to warn you that the position of your airplane has changed. You retain the positive impression that you are still flying straight and level. Your first indication that anything is wrong is a change of noise, an increase in airspeed or a loss of altitude.

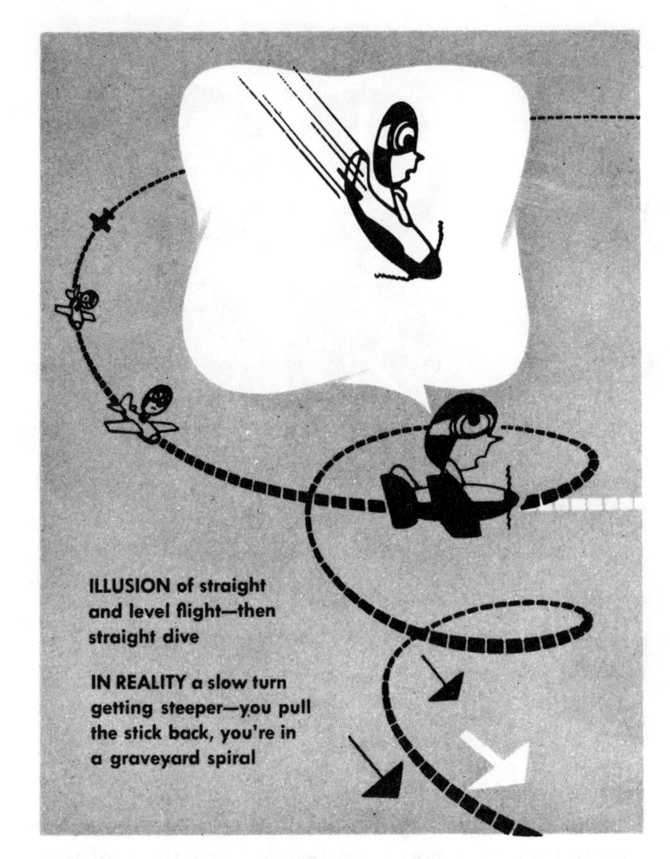


Under these circumstances, unless you look at and believe the instruments which indicate turn and bank, you may merely pull back on the stick, under the impression that by so doing you are recovering from a straight dive. If you do, this impression of being in a straight dive increases, since pulling back on the stick gives you the same feeling you have in a normal straight pullout from a dive.

Once you start to pull back on the stick, the turn gets tighter, the nose drops lower, and there is a great increase in airspeed and the rate of descent. After a few seconds the airplane may have more than doubled its original speed, and the rate of descent may have increased to several thousand feet per minute.

You can recover easily and quickly from such a spiral. **Remember: when anything seems to be going wrong, when the airplane is starting to lose altitude and gain speed, first look at the instruments which indicate bank and turn.** In all probability the airplane is banked and is turning, although your sensations make you feel it is in straight and level flight. Don't act according to your sensations. **Check and cross-check your instruments.**

4. The feel on the seat of your pants (deep sensibility) may make you think you are climbing when your airplane enters an updraft, or that you are diving when it enters a downdraft. You feel heavier in an updraft and lighter in a downdraft.

### Sensory Illusions in Turns or Spins

1. The standard rate turn is a slow turn. In it you experience much the same sensations you do in straight flight.

2. While performing steep turns, especially at high speed, your deep sensibility and the static organ of your middle ear may give you the sensation that you are in a loop rather than in a banked turn. When you come out of a steep turn your impulse may be to climb, because you begin to feel light. The force that has been pulling you hard against your seat lets up, so you think you are diving. If you trust your senses, you will stall the airplane.

3. If you hold your head steady in a turn, you have little sensation of turning. If, however, you move your head to look down or to the side in the cockpit, you will experience the sensation of doing a rapid roll, or of snapping around in the turn. You get this feeling because the new position of your head places another semicircular canal in the plane of rotation of the airplane.

4. During a spin you are subjected to high rotational speeds. If the spin is continued for more than an extremely short time, you will have, upon recovery, a marked sensation of falling off into a turn

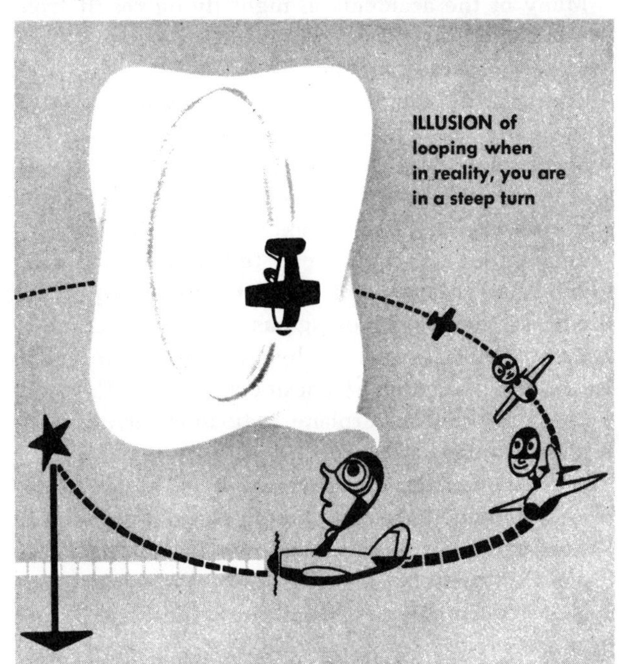

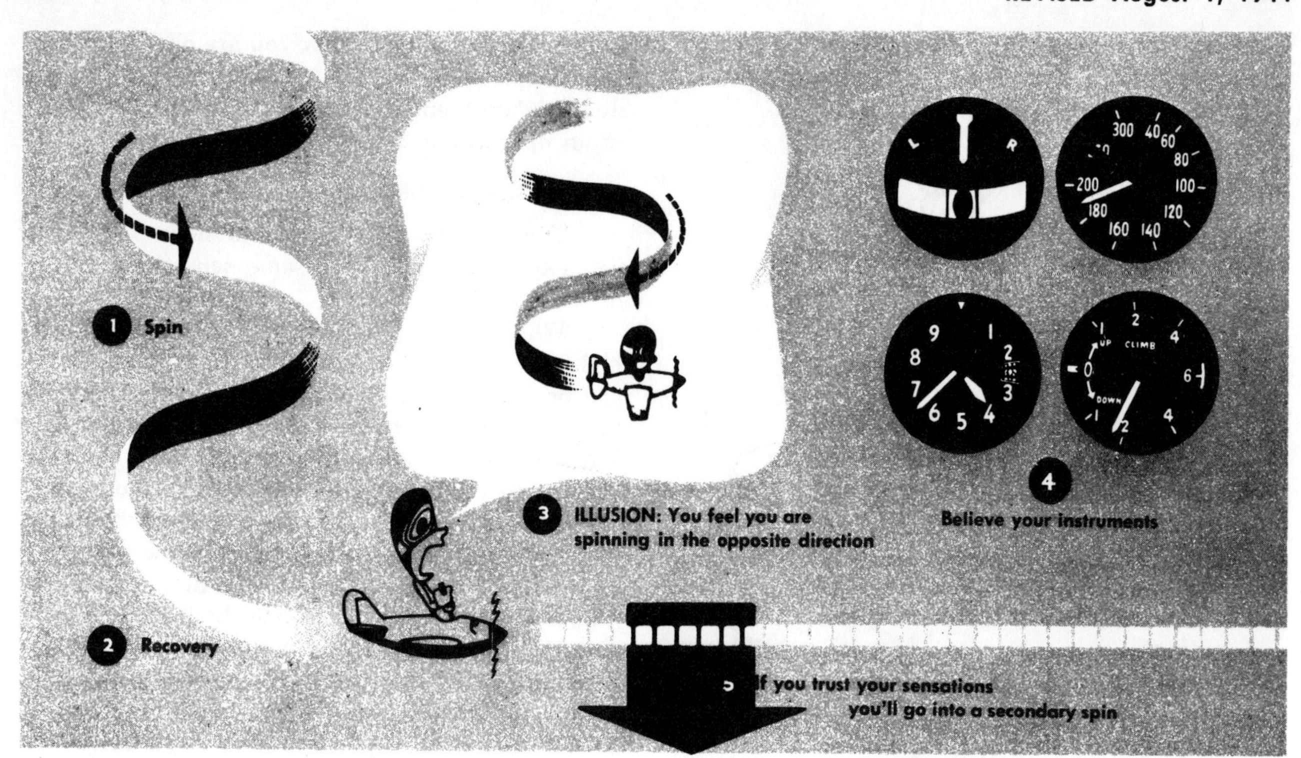


in the opposite direction. In attempting to correct for this sensation, you may fall into a secondary spin. **Check spin recovery by your instruments, and trust them.**

## NIGHT FLYING

Many of the accidents in night flying result from the fact that pilots rely too much on vision, rather than on their instruments.

At night the inexperienced pilot may look around continually to find some light on the ground by which he can orient himself. Unless he is flying near a large city where there are enough lights to make a good pattern, this habit is extremely hazardous.

Any experienced pilot can tell how he has mistaken a star for a light beneath him and how he thought lights were moving past him, when actually he was turning about the lights. A pilot can easily get so confused that he doesn't know which way is up, or whether the airplane is turning, diving, rolling, or climbing.

Most confusing and dangerous of all is the situation in which the pilot attempts to look backward to see lights behind him, such as after a takeoff into unlighted territory.

You may become confused easily in night flying because your eyes deceive you. You have no definite horizon to use as a plane of orientation; you have only isolated points of light. Your senses may mislead you if you depend solely on these lights for orientation, and you may become greatly confused in handling your airplane.

The one solution for this is to watch the instrument panel, with only occasional glances at the lights. If you get the habit of using your instruments as a major reference and of using lights only as a secondary reference, you will not get into trouble.

## AIRSICKNESS

Airsickness is one form of motion sickness, like sea sickness and train sickness.

Contributing factors are: rough air, fatigue, poor ventilation, prolonged cold or heat, vibration, emotional tension, and poor physical condition.

The following aids in preventing or minimizing airsickness have been suggested by the Air Surgeon. Pass them on to your crew members:

1. Keep the horizon in full view except when tactical consideration, such as searching the sky for enemy fighters, dictates otherwise.

2. Sit as near as possible to the center of gravity of the airplane, where least motion will be felt.

3. Concentrate thoroughly on your mission to its termination.

Reference: Technical Order 30-100A-1, Section 2.

Every pilot is familiar with vertigo — "dizziness" in flight. As it is a definite hazard to safety, you should understand what causes it, and what precautions to take when you are affected by it.

### Cause of Dizziness

The seat of the sense of balance is the inner ear where there are three semi-circular canals at right angles to each other. These canals correspond to the three axes of rotation.

They are filled with fluid. Sensitive hairs extend from the walls of the dilated ends of the canals into the fluid. When the head tips or moves from side to side, the liquid in the canals, because of inertia and the force of gravity, attempts to stay in its original position.

Move or tip a glass of water and you will see that the water tries to stay "put" in its relation to the earth and to the points of the compass.

The same thing happens to the liquid in the inner ear and any movement of the head sets this fluid in motion and in turn tips the hairs in the liquid so that the movement and the resulting position are telegraphed to the brain. If the movement is slow, there may be no sensation of turning at all. This fact is of extreme importance in blind flying.

If a rapid movement is suddenly slowed or stopped, the inertia of the liquid causes pressure against the opposite set of hairs and thus a false sensation of movement in the opposite direction is produced.

Take a glass of water, rotate it rapidly, then stop the glass and watch the liquid continue to rotate.

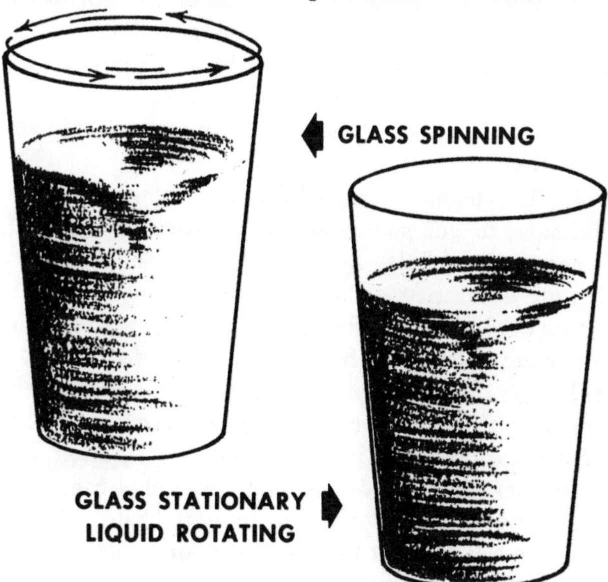


The same thing goes on in your inner ear. That is one reason why you feel dizzy after whirling, spinning, or rolling. It is the basis of vertigo, but not the only cause. Your other senses also contribute to it.

The sense of balance is a combination of what you see and what you feel (pressure on your feet, joints, or muscles, etc.) and what your inner ear tells you. If what you see and what you feel contradict what your inner ear tells you, your brain becomes confused — it doesn't know which to believe. That confusion is the cause of vertigo.

### Results of Vertigo on Flying Technique

The confusion resulting from vertigo naturally affects your flying technique, particularly when you are on instruments. Since you cannot trust your sense of "feel" (balance) you are inclined to do many queer and hazardous things. Here are a few of them:

1. During a steep turn you will feel that you are climbing. The steeper the bank the more you will feel it. Your instinct will tell you to get out of that climb before you stall.

**Don't trust that feeling, check with your instruments and trust them.**

2. When you return to level flight after a steep turn, you are likely to feel that you are diving or losing altitude. That is because centrifugal force, acting on the inner ear liquid and "the seat of your pants" has suddenly "let go" and you feel lighter. Your "feeling" is overcompensating.

**Again check with your instruments and trust them.**

3. In a perfect turn it is impossible for your sense of balance to tell you your angle of bank. Centrifugal force keeps the pressure in the inner ear normal so it feels like a direct pull instead of a bank. (You know what happens when you whirl a pail of water.)

If you can see the horizon, your eyes will set you right; if not, again **check with your instruments and trust them.**

4. "The leans" is a peculiar sensation which you are sure to get sometimes in instrument flying. It consists of the false feeling — often amounting to a definite assurance — that you are turning (or climbing or diving) when your instruments show that you are in level, straight flight. The reason for the false feeling is that in the many small movements of the airplane about its three axes, particularly in rough air, the liquid of the inner ear becomes agitated, and moves against the hairs in such a way that they telegraph the wrong information to your brain.

**Learn not to trust the "seat of your pants."**

5. Another case similar to "the leans" often arises when your plane has turned slowly off course, and you have brought it back quickly. The slow turn makes no impression on your inner ear, but the sharp turn does, and you get the strong feeling that your compass has "gone haywire."

**Trust your compass rather than your sense of direction.**

6. If you move your head sharply in any direction during a sharp turn or dive, you may find you have a feeling amounting almost to conviction that something has gone wrong with the maneuver. Your instinct tells you to correct your attitude.

**But don't do it until you have checked with your instruments.**

7. Flying in reduced visibility is especially hazardous if you attempt to fly partly on instruments and partly on "feel" and attempted contact. It is particularly dangerous if you let your eyes jump from the instruments to what you can see of the ground and the horizon, and back again to instruments, etc. Your eyes and ears and sense of balance are likely to become hopelessly confused.

**Fly reduced visibility wholly on instruments, or make visual contact only occasionally for check. Don't fly contact and instruments at the same time.**

8. In formation flying in reduced visibility, you may find yourself affected by some phase of vertigo. It may occur when the formation has been flying in the vicinity of city lights and then turns toward a dark quadrant of the sky where you cannot see the horizon or orient yourself by sight. Wingmen are more often affected because of a certain amount of skidding and maneuvering which is necessary to maintain formation. You may feel that there is something radically wrong with the leader and the rest of the flight. Your normal reaction is to correct what you think is wrong. You feel an uncontrollable urge to get out of formation and straighten out. **Force yourself to trust your formation leader. Don't pull away from the formation. Take a quick glance at your instruments. If you are in level flight, shake your head. But if you are in a turn or maneuver of any kind, wait until you are in level flight before you shake your head, or it will make matters worse.**

9. If you are nervous, tired, or tense, you will be more likely to suffer from vertigo than if you keep yourself relaxed and rested. Obviously that is not always possible, particularly on long operational flights. But if you know what you may expect, and understand its cause, you can guard against most of the hazard of vertigo.

10. **Finally: In any difference between your sense of balance (your "feel") and your instruments, keep your eyes in the cockpit, trust your instruments, and you can minimize the effects of vertigo.**

### Carbon Monoxide

Be aware of the possible danger of carbon monoxide in the airplane and on the ground. Carbon monoxide is extremely dangerous and the hazard is greatest at high altitudes. Although carbon monoxide has no odor, its presence should be suspected whenever you smell exhaust gases.

There is especial danger in single-engine aircraft where a short exhaust stack is directly in front of the cockpit.

Enemy gunfire not necessarily disabling the airplane may open up exhaust collector rings and holes in the fuselage and fire wall, causing dangerous amounts of carbon monoxide to filter into the cockpit. Never use an exhaust heater during combat.

In poorly ventilated buildings, hangars, and tents in cold regions, carbon monoxide becomes a hazard when gasoline or oil stoves, lamps or motors are being used. **Therefore guard against carbon monoxide by insuring proper ventilation under such circumstances.**

The symptoms of carbon monoxide intoxication are similar to those of oxygen lack, as would be expected since carbon monoxide prevents oxygen from being absorbed by the blood. Headaches, nausea, dizziness, dimming of vision, unconsciousness and even death follow exposure to concentrations of carbon monoxide of varying degrees. But don't rely on any symptoms to warn you.

If the odor of exhaust gases is present and carbon monoxide is suspected:

1. Open all cockpit hoods, windows, tent flaps and doors and obtain as much cross ventilation as possible.
2. In the airplane, drop to a lower altitude and **put on your oxygen mask with the Automix lever in the "OFF" position so that you breathe pure oxygen.**
3. Turn off exhaust heaters, lamps, other heaters or motors if they are in use.
4. **See your Flight Surgeon as soon as possible.**

### Poison Gas

If you are exposed to poison gas and do not have your gas mask with you the oxygen mask and flying goggles can be used as a substitute. To make the oxygen mask function in this capacity turn the Automix lever to the "OFF" position. This excludes any poison gas in the atmosphere since 100% oxygen is supplied even at ground level.

# SECTION 5 POWER PLANT

# GROUND OPERATION OF ENGINES

## Starting of Aircraft Engines

### First

Make sure ignition is off, then turn engine over four or five revolutions if engine has been idle for over two hours, or if excessive priming has been used during starting attempts.

### Second

Open cowl flaps.

### Third

Move carburetor air heat control to "COLD" position.

### Fourth

Make sure fuel valve is turned on to a tank that contains fuel. On multi-engined airplanes, set fuel cock to supply fuel to all engines.

### Fifth

Be certain that the cross-feed fuel shut-off cock is in "OFF" position on twin-engined airplanes. This cross-feed fuel cock permits the fuel pump on one engine to supply fuel to both engines in case of a fuel pump failure on the other engine.

### Sixth

Move throttle almost to closed position—600 to 800 rpm. With low-pressure carburetor, put Fuel Mixture control in "RICH." With high-injection carburetor keep in "IDLE CUT-OFF" until engine fires.

### Seventh

Make sure that the propeller controls of two position and Hamilton Standard propellers are set in the "LOW" rpm position, where they should be upon stopping the engine. About one minute after starting shift into "HIGH" rpm as this reduces the load on the engine and improves cooling.

Other constant speed propellers should be in the "HIGH" rpm position when stopped, so they will be properly set for starting and warm-up.

### Eighth

Prime engine while energizing starter by using either hand type priming pump or electric type priming pump. Three or four strokes are usually adequate but up to 10 strokes may be used on large radial engines when cold. (One second on electric primer is equal to one stroke of the hand primer.) Avoid excessive priming as it washes the oil off of the cylinder walls causing scoring of the barrels and seizing of the pistons. **Throttle priming is not authorized for any engine** except where special instructions issued in Technical Orders pertaining to a particular engine advise this method.

### Ninth

Turn ignition to "BOTH" position. Energize the starter, then hold meshing switch "ON" until engine fires. Open the throttle slowly to the desired warm-up speed.

On installations using the combination inertia direct cranking starters **engage both the meshing switch and energizing switch at the same time** after the starter has been energized.

### Supercharger Control

Engines equipped with turbo-driven superchargers should be started with the waste gate open and supercharger regulator in "OFF" position.

Engines equipped with two-speed superchargers should in all cases be started and take-off made with supercharger in low blower.

Engines equipped with two-stage superchargers should be started in accordance with the flight operating instructions for the specific engine involved.

### Warm-Up Procedure and Oil Pressure Check

Aircraft engines will always be warmed up on the ground until proper lubrication and engine operation for the take-off and flight are assured. Excepting in an emergency, engines will not be run on the ground for warm-up or test unless chocks are placed in front of the wheels of the airplane regardless of whether or not the airplane is equipped with brakes.

Be sure to watch the oil pressure gage. If it does not indicate pressure within 30 seconds, shut off the engine and have an investigation made.

After the oil gage indicates pressure, shift propeller to take-off setting and run engine at 600-800 rpm until oil pressure is normal (30-65 lb sq in.) for this speed.

Use both magnetos except on engine check.

Do not permit engine rpm to exceed ½ of the maximum permissible ground rpm during the warm-up period. When the engine maintains at least ⅔ of the minimum full power oil pressure specified for the particular engine, and the oil temperature gage shows a definite increase in oil temperature, the engine rpm may be increased to check for proper functioning of the engine and engine instruments, except that the maximum permissible ground rpm will not be maintained for periods in excess of 20 to 30 seconds on the ground.

### Ignition System

Switch from one magneto to the other and note the loss of revolutions or manifold pressure. The normal loss in rpm on one magneto should not exceed 100 rpm.

NOTE: Whenever an engine is operating on only one magneto, the manifold pressure must not exceed maximum cruising manifold pressure to avoid detonation when firing on only one set of plugs.

If the airplane is not equipped with a manifold pressure gage, the magneto check will be made with not more than 85 per cent of the rpm developed when the throttle is opened to the throttle stop.

If the engine is equipped with an "Idle Cut-Off" carburetor, it is necessary to check the "OFF" position of the ignition switch to assure the proper connection of the ground wires. This check, made at idling speeds, need only be made at the start of the day's flying.

### Controllable Propellers

Check the operation of the controllable propeller. During ground operation, the manifold pressure must not exceed the specified maximum cruising manifold pressure, to prevent overheating of the engine while operating in the low rpm (high pitch) position.

### Fuel Supply

Check the functioning of all fuel tanks during the warm-up by switching the fuel valve to each tank long enough to insure that the fuel from that tank has a chance to flow to the engine.

### Superchargers

Check the operation of the two-speed supercharger by shifting the clutches immediately following each engine warm-up.

1. Set propeller governor in low pitch (high rpm).
2. Set engine speed at 1,500 rpm.
3. Move supercharger control lever to the "HIGH" position and lock.
4. Open throttle to obtain not more than 30 in. Hg.
5. Check manifold pressure.
6. Immediately shift to "LOW" position. A sudden decrease in manifold pressure shows the two-speed supercharger drive is operating properly.

Check flight operating instructions for specific engines for check on two-speed and two-stage superchargers.

Supercharger controls.

### Taxiing

Taxi from the line for take-off or to the line after landing with propeller controls in "HIGH" rpm (low pitch) position.

Set engine and cowl flaps "FULL OPEN" to prevent overheating the engine.

### Changing Power Condition

To avoid excessive pressures within cylinders with resultant detonation and possibility of failure, use the following procedure when changing condition of power:

#### Increasing Engine Power

1. Adjust mixture control to obtain fuel-air ratio specified for power condition desired.
2. Adjust propeller control to obtain desired rpm.
3. Adjust throttle control to obtain desired manifold pressure.
4. Readjust mixture control, if necessary.

#### Decreasing Engine Power

1. Adjust throttle control to obtain desired manifold pressure.
2. Adjust propeller control to obtain desired rpm.
3. Readjust throttle controls, if necessary.
4. Adjust mixture controls to obtain desired fuel-air ratio.

### Stopping Aircraft Engines

This preliminary procedure applies to all engines regardless of the type of carburetor or fuel system installation:

1. Mixture control in "FULL RICH."
2. Throttle in normal idling position.
3. Propeller controls of two position and Hamilton Standard will be in "LOW" rpm; other constant speed propellers in "HIGH" rpm.
4. Nose cowl or radiator shutters (if installed) fully opened at normal idling speed until engine temperature has cooled below cruising temperature.

**The following procedure applies to engines equipped with float type carburetor WITHOUT idle cut-off.**

1. Turn the fuel valve to "OFF" position.
2. Idle engine at 800-1,000 rpm until fuel pressure drops to zero.
3. As engine dies, move throttle slowly forward. When engine stops, cut ignition.
4. Turn fuel valve "ON" after engine has stopped, and operate wobble pump until fuel pressure gage indicates pressure showing that carburetor and fuel lines are filled with fuel.

**The following procedure applies to engines equipped with float type carburetor WITH idle cut-off.**

1. Leave fuel valve "ON."
2. Idle engine at 800-1,000 rpm.
3. Set mixture control at "IDLE CUT-OFF."
4. Set throttle at "FULL OPEN."
5. Cut ignition after engine stops.

**The following procedure applies to engines equipped WITH fuel injectors (no carburetor).**

1. Leave fuel valve "ON."

**NOTE: Fuel cock will never be shut off except in emergency.**

2. Idle engine at 600-800 rpm.
3. Set mixture control to maximum lean without moving the throttle.
4. Cut ignition after engine stops.

**CAUTION**

On high output engines it may be necessary to idle at a higher speed than normal to prevent overheating and fouling of plugs.

### Oil Dilution

To operate oil dilution system before stopping the engine:

1. Turn oil dilution control "ON," when a cold weather start is anticipated, for four minutes at 1,000 rpm, then stop the engine.
2. Repeat the oil dilution after approximately 15 minutes.

NOTE: Dilution is ineffective while oil temperatures are above 70°C (158°F).

If oil temperatures are too high, stop engine and wait for it to cool to 40°C to 50°C (104°F to 122°F), then restart it and proceed with the oil dilution.

A normal start should be made without regard to the oil dilution system. If a heavy viscous oil is indicated by oil pressure too high, or by fluctuating oil pressure when engine rpm is increased, push dilution control "ON." This method should be used only if time and extreme temperature conditions do not permit engine warm-up in the usual manner. Over-dilution is likely to occur under these conditions because of low oil flow and a cold engine which

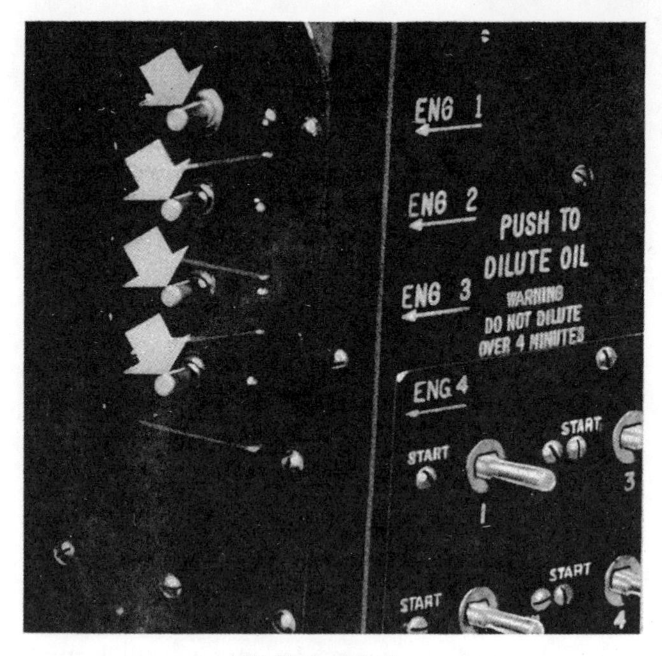


Oil dilution switches.

results in little evaporation until engine warms up. If dilution is used, close observation of oil pressure is necessary to determine whether or not the oil has been over-diluted, resulting in low oil pressure.

NOTE: If necessary, immediate take-off may be made after oil dilution without the normal warm-up, provided there has been a rise in oil temperature and the oil pressure is steady.

Over-dilution will not occur usually if the diluting operation is done immediately after flight, while the engine and oil are warm. Over-dilution has no serious effect on engine bearings if the oil pressure remains normal.

### To Eliminate Engine Failure

Excessive ground running time of airplanes is one of the main reasons for early failure of the engines. Reduce the ground running time by the following procedures:

1. Hold ground tests and warm-up periods to a minimum.
2. Do not taxi away from the line for take-off until the control tower operator gives the necessary instructions.
3. Shut off the engine when it appears that take-off will be delayed due to other planes landing or other causes.
4. Use proper judgment in taxiing to position to prevent long periods of ground running of engines.

**REFERENCE:** Technical Order 02-1-29, dated Nov. 23, 1942

# FLIGHT OPERATION OF

### Operating Limits Depend Upon Grade of Fuel Used

Military aircraft engines designed to use 100/130 grade of fuel will be operated in combat and on certain missions with the grade of fuel for which they were designed; but much of your flying within the domestic area will be on a lower grade fuel than that for which the engines were designed.

You must be absolutely certain of the grade of fuel you are using and accommodate the operating range of your engines to it.

To conserve 100 octane aircraft fuel, 91 octane is now specified for most military aircraft operating in continental United States on training or routine flights.

### Missions On Which 100 Octane Will Be Used

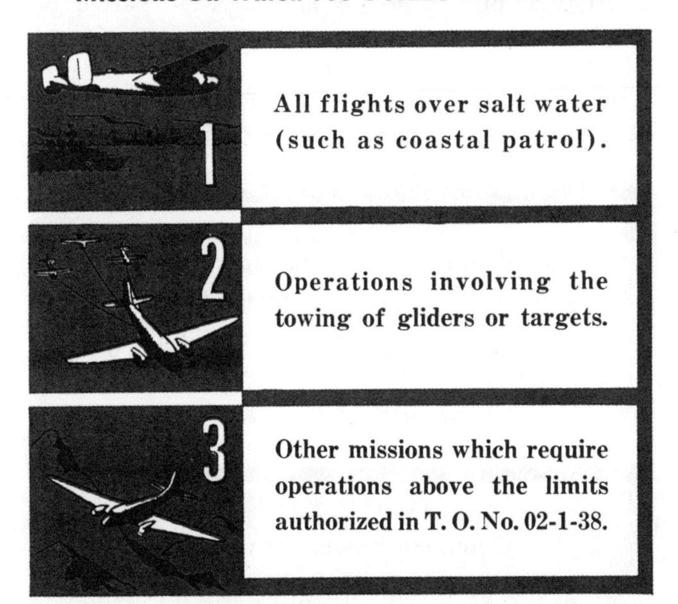


All aircraft which require and are serviced with 100 octane fuel will be operated at all times using the most economical power and cruising conditions.

### Mixed Grades of Fuel

On any mission named above, never mix 100/130 octane with a lower grade fuel. If your tanks contain any 91 octane, have them fully drained before servicing with the higher grade fuel.

It is permissible when occasions demand to mix fuels of grade 91 and grade 100/130. But when these grades are mixed, you must observe operating limits for grade 91.

### Reduction of Full Rated Power

When fuel grade of Spec. No. AN-VV-781, amendment 5, AN-F-27, or AN-F-28 is used, full rated military power is allowed. When fuel grade of Spec. No. AN-VV-781, amendment 4, is used, you will reduce maximum boost by 10% below full rated power.

### Watch Your Operating Limits

**In using the lower grade fuel, pilots must use special care in flight operations of engines to avoid any condition that may lead to detonation,** which is always a danger in engines operating on a grade of fuel lower than that for which they were designed.

All aircraft serviced with the lower grade fuel will have attached to the control column or ignition switch a red warning tag bearing the following notation:

> **Warning To Pilot**
>
> This airplane serviced with fuel, grade 91, Specification No. AN-F-26. Operate engine within the maximum limits prescribed in T. O. No. 02-1-38.

### Detonation in Aircraft Engines

Detonation may be indicated by unusual engine roughness, cylinder temperature increase, erratic fuel-air ratio, and by the exhaust flame.

Engine roughness does not necessarily indicate that there is detonation, but when unusual roughness is present it may be due to detonation.

Cylinder temperature increase cannot be relied on as a definite indication of detonation but since detonation liberates an unusual amount of heat to the cylinder walls an increase in cylinder temperatures may be due to detonation.

An erratic fuel-air ratio reading may indicate detonation. If, as the mixture is leaned out, the needle does not show a leaner mixture or backs up the scale towards the rich side, detonation has probably occurred and the mixture should be richened.

Exhaust flame is the most reliable indication of detonation when exhaust stacks are visible. Intermittent puffs of dense black smoke, often accompanied by sparks or glowing carbon, indicate detonation. Black puffs occurring with great regularity indicate severe detonation. A rich mixture is indicated by full red flames with steady black smoke. Black smoke and red flame may also be caused by high oil consumption, ice in the carburetor, or other causes of poor fuel distribution. Light blue flames, almost invisible, indicate a mixture setting which produces maximum power.

Detonation may be caused by use of fuels of low octane number, low fuel-air ratio, and engine operating conditions.

Use of fuels of low octane number will cause detonation since **fuels are rated in terms of their resistance to detonation.** The greater their resistance to detonation the higher the octane number of the fuel. A higher octane number fuel than that specified may safely be used but a lower octane number must be used only when specified in technical orders or in an emergency. In this case the engine instructions covering the use of a lower grade fuel must be strictly followed.

The greatest tendency for fuel to detonate occurs near the fuel-air ratio for best power. Leaning the mixture increases the detonation, while rich mixture increases anti-knock value and cools the cylinders.

A change in engine operating conditions which increases the pressure or temperature in the cylinder increases the chance of detonation. Therefore, detonation may be caused by the following:

1. Increasing the manifold pressure.
2. Advancing the ignition timing or by operating on one spark plug.
3. Increasing intake air temperature by use of the carburetor air heater.
4. Increasing fuel-air mixture temperature, as by using high blower gear ratio instead of low blower to obtain a given manifold pressure.
5. An increase of cylinder temperature.
6. Building up of engine deposits which tend to decrease the rate at which heat is conducted away from the combustion chambers.
7. Operating engine for an excessive period on the ground.
8. Leaning fuel-air mixture.

Results of detonation are loss of power, overheating, preignition, and damage to the engine.

**To obtain detonation-free operation immediately reduce the manifold pressure, richen the mixture, and reduce carburetor air heat to the minimum at which icing may be prevented.**

### Changing Power Condition

To avoid excessive pressures within cylinders with resultant detonation and possibility of failure, use the following procedure when changing condition of power:

#### Increasing Engine Power

1. Adjust mixture control to obtain the fuel-air ratio specified for power condition desired.
2. Adjust propeller control to obtain desired rpm.
3. Adjust throttle control to obtain desired manifold pressure.
4. Readjust mixture control, if necessary.

Propeller, throttle, and mixture control quadrant.

## Decreasing Engine Power

1. Adjust throttle control to obtain desired manifold pressure.
2. Adjust propeller control to obtain desired rpm.
3. Readjust throttle controls, if necessary.
4. Adjust mixture controls to obtain desired fuel-air ratio.

## Fuel-Air Ratios

The fuel-air ratio values are the absolute minimum with which dependable engine operation can be expected. Use fuel-air ratios somewhat richer than the minimums specified except when extreme ranges are essential. This increase will help prevent sticking and burning of piston rings.

## Air Speeds During Climbs

To obtain proper cooling, maintain a calibrated IAS of at least the best climbing speed specified for the particular airplane, with at least a 10 per cent greater IAS desirable unless the tactical situation prevents, or unless maximum altitudes are desired.

NOTE: **Under no circumstances** will airplanes be climbed for protracted periods at calibrated IAS less than the best climbing speed when using rated engine power.

The true climbing speeds for all later model airplanes appear in Flight Characteristics Section of the Handbook of Flight Operating Instructions for each type of airplane.

## Overspeeding of Aircraft Engines

When manifold pressures and engine speeds have exceeded those specified, follow this procedure, without exception.

Report the following information to the Squadron, Station, or Sub-Depot Engineering Officer, as the case may be, immediately upon landing.

1. The maximum rpm and manifold pressure which was obtained during flight.
2. Duration in minutes of the overspeed and overpower operation.
3. Reason for overspeed and overpower, if known.
4. Total engine time and time since last overhauled when overspeed or overpower operations occur.

REFERENCE: Technical Orders 02-1-29, dated November 23, 1942, and 02-1-69, dated May 22, 1943.

# ENGINE INSTRUMENTS

### Manifold Pressure Gage

Manifold pressure gages should be checked before every flight and after every 50 hours of service:

**Preflight**—Check the instrument by gunning the engine with the supercharger off, if it is of the turbine type. The needle should move freely toward the high side of the scale. Check against barometer, if practical. Be sure manual, automatic, or balance line drain has been operated properly during first warm-up. Drain for 30 seconds at 800 to 1,000 rpm to obtain proper setting.

In the case of any sign of malfunction, all connections should be checked for leaks.

**50-Hour**—The reading of the instrument should be checked against the station barometer every 50 hours. If the reading differs more than 0.4 inches Hg, the instrument should be removed for a bench test and resetting of the pointer.

### Fuel Mixture Indicator

The fuel mixture indicator shows the fuel-to-air ratio and is used as a visual guide for setting the mixture control. When making mixture adjustments, allow at least one minute to elapse before reading the instrument. This will provide time for the change in the carburetor setting to evidence itself.

The fuel-air indicator has a uniform degree of accuracy over a wide range, provided there is no detonation. If there is detonation, the indicated fuel-air ratio will be higher than the actual value.

If, as the mixture is leaned out, the indicating needle does not show a leaner mixture, or backs up toward the rich side of the scale, **detonation is taking place** and the mixture should be made richer.

If, as the mixture control is shifted from position to position, the indicating pointer remains near the "A" point (air point), the instrument is inoperable.

### Oil-Pressure Gages

During the operation of aircraft in temperatures below 32°F, engine oil congealing in the oil-pressure gage line may cause the gage to react sluggishly. When this condition exists, the oil-pressure gage line should be drained and refilled with petroleum base hydraulic fluid.

Normally, the fluid will mix with the engine oil very slowly. From 60 to 90 days may elapse before the gage will begin to operate sluggishly again.

# MARKING OF INSTRUMENTS

The operating limits and ranges of engine instruments and the airspeed indicator are marked on the face of the instruments to facilitate reading these instruments.

**The engine instruments are marked with short radial lines in red to indicate limits which should not be exceeded.** Operating ranges are shown by green arcs. A white radial line at the bottom of the cover glass is used as an index to show rotation of the glass.

**The maximum permissible indicated airspeed is marked with a red line** extending from the center of the dial and passing directly over the point corresponding to the maximum permissible indicated airspeed.

REFERENCE: Technical Order 05-1-17, dated November 3, 1943

### Standard Fuels

The standard fuels of the Army Air Forces are:

| | |
|---|---|
| Grade 62 | Grade 87 |
| Grade 73 | Grade 91 |
| Grade 80 | Grade 100 |
| Grade 130 | |

You'll find the correct grade to be used marked on your airplane.

### For Combat Use Only

100 octane fuel is needed in combat. Therefore, fuel, Grade 91, conforming to Spec. AN-F-26 is authorized for use, within the Continental United States, in practically all aircraft engines originally designed for 100 octane fuel. This substitution must be made only under operating conditions specified in T. O. 02-1-38. **A Red Warning Tag, noting the substitution and listing specific operating conditions, will be placed on the control column or ignition switch of the airplane.**

### Commercial Fuels

When buying commercial fuels for normal operation, grades must be equal to or higher than required AAF grade, to obtain safe engine operation. Fuels having higher grade ratings than that specified for a given engine may be used as a substitute, if necessary. **If commercial fuel is procured, obtain only enough to enable you to reach a source of AAF specification fuel.**

### Aromatic Fuels

All fuels of grade 100 or higher are aromatic. In using them observe carefully the markings on all fuel tanks. **Do not use aromatic fuels in tanks not treated for their use.**

### Detonation

In using lower grade fuel you must take every precaution to avoid detonation.

Be sure that all spark plugs are functioning, particularly when using a fuel of lower grade than that for which the engine was designed. One dead spark plug may cause detonation.

Remember that operation at higher manifold pressure and lower rpm is more likely to cause detonation than lower manifold pressure with higher rpm —in the same range of operation.

Detonation can be reduced by placing the mixture control in "MANUAL RICH" as no compensation is made for altitude and the mixture will become richer as altitude is gained. If the carburetor is equipped with automatic mixture control, richer mixtures can be obtained by having control in "MANUAL RICH" than in "AUTOMATIC RICH" position. **Never fly the airplane in "AUTOMATIC LEAN" position when using low octane fuel except when specific Technical Orders so direct.**

### Fuel System Operation

Fuel system failures and forced landings will result if you do not know the operating characteristics of the fuel system, and the proper operations to be performed to insure its normal functioning.

**IMPORTANT**

Before flying any type of airplane, be sure you know:

1. Fuel consumption of engine, at cruising and full throttle.
2. Capacity of each fuel tank in airplane.
3. Location of each fuel valve in pilot's compartment and its function with reference to operation of the engine.
4. Location of the sight gages, if so equipped.

### Reserve Fuel Supply

Reserve fuel supply is provided by two methods: a standpipe in the main fuel tank, or by auxiliary tanks. **When a standpipe is used as a reserve supply, and it is necessary to take off or land on a reduced fuel supply, use the "RESERVE" position.** All current model airplanes are designed to provide a reserve fuel supply of approximately 20 minutes of full throttle operation. As this requirement was not in effect at the time some of the older airplanes were built, **determine before take-off the amount of reserve supply carried.**

### Fuel Tank Change-Over

**Do not permit fuel tanks to run dry at any time,** particularly in the case of airplanes equipped with gravity systems or high pressure carburetors. The change from one tank to another will be made at the pilot's discretion but must be made before any loss of pressure is apparent.

Change-over from one source of supply to another will be made as follows, while the engine is still running:

1. Hold the airplane in level flight position, throttle unchanged.
2. Mixture control "FULL RICH."
3. Switch fuel valve to required position.
4. Operate wobble pump slowly, if any decrease in pressure is noted.

**If the engine has stopped, nose the airplane down to maintain flying speed and follow the same procedure.**

Check fuel cock settings by "click" or "feel."

Some airplanes may be equipped with cross-feed valves to permit one fuel pump to supply pressure to both engines, in case fuel pressure fails or becomes low on one engine. If this happens, turn cross-feed valve "ON."

Airplanes having independent fuel systems for each engine have a cross suction line provided with a shut-off valve. Normally this valve is "OFF" with engines operating from their respective tanks. To operate both engines from the right wing tank, for instance, open the shut-off valve and close the left wing tank selector valve.

**Don't trust fuel cock settings.** Determine either by the "click" or "feel" method and not merely by the position of the control handle of the valve.

Operate wobble pump slowly.

### Fuel Gages

Fuel level gages usually are designed to show the quantity of fuel in the tanks only when the airplane is in flight position. **Gages should only be read in this position.** When gages having scales showing amount for both ground and level flight position are used, **do not confuse ground readings and flight readings.**

There are several types of fuel level gages: sight gage, float and lever gage, and the electric liquidometer gage. The electric liquidometer gage makes possible accurate measurement of the fuel in any number of tanks on a single indicator dial in the cockpit. By turning the switch connected to the gage, the amount of fuel in any tank may be read directly from the gage.

Read fuel gages in flight position only.

### Fuel Booster Pumps

Use fuel booster pumps:

1. At all times during take-off and landing. If pumps have two speeds, use lower speed.
2. At all altitudes below 1,000 feet. If pumps have two speeds, use lower speed.
3. At all altitudes above 10,000 feet.
4. Between 1,000 feet altitude and 10,000 feet altitude when conditions warrant their use.

On airplanes having separate booster pump connections to each tank, the pump will be used only for the tank from which fuel is being drawn. When using fuel from a tank which has no booster pump connection, turn all pumps "OFF."

### Carburetor Mixture Controls

Mixture controls will be used in conjunction with specific instructions for various engines. They are manually adjusted when a fixed pitch propeller is used. A constant speed propeller requires the use of a fuel-air ratio gage for manual adjustment of the mixture. Automatic mixture controls do not require manual adjustment unless specified in engine operating instructions.

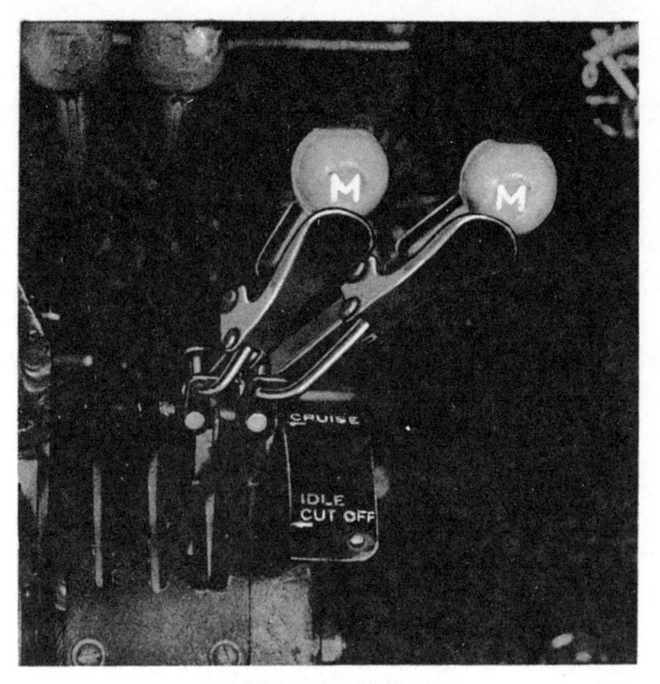


Mixture control.

The one-position automatic mixture control automatically controls mixture with changes in altitude and temperatures without regard to throttle setting, and in no way controls the power output of the engine. The need for manual control is done away with when it is set in "AUTOMATIC" position. If it becomes damaged, mixture may be operated manually by shifting lever from "AUTOMATIC" to the desired manual position.

The two-position automatic mixture control may be set in "MANUAL," "AUTOMATIC RICH," or "AUTOMATIC LEAN" position.

In "MANUAL" position, mixture is adjusted manually.

In "AUTOMATIC RICH" position, mixture is automatically adjusted for safe engine operation up to normal rated power, manifold pressure, and speed.

Use "AUTOMATIC LEAN" position only if manifold pressures and speeds are below those specified, and only with specified fuel.

**Caution**

For operation at maximum cruising power and above, always set mixture in "AUTOMATIC RICH" position or adjust manually for proper mixture.

**Caution**

When operating an engine with mixture control in "MANUAL LEAN" position at high altitudes, it is important that the mixture be richened during descent. When airplane is leveled off, readjust mixture control for proper operating conditions.

Engines equipped with fixed pitch propellers should be run up for take-off as follows:

1. As near take-off manifold pressure and rpm as possible.
2. Mixture leaned enough to increase speed 50 to 70 rpm.

Engines equipped with constant speed propellers cannot be adjusted on the ground, since fuel-air ratio gages are not sensitive enough to permit adjustment of mixture during run-up and take-off. They should be flown at the approximate altitude of the field from which take-off will be made as follows:

1. At take-off manifold pressure and rpm.
2. Mixture leaned enough to eliminate rough engine operation or adjusted for proper fuel-air indicator reading.
3. Mark setting on the mixture quadrant for use in take-off from that field.

Engines equipped with automatic controls will be set for take-off above 3,500 feet as follows:

1. One-position automatic type set in "AUTOMATIC."
2. Two-position automatic type set in "AUTOMATIC RICH."
3. Power control type set in "EMERGENCY" and adjusted manually.

REFERENCE: Technical Orders No. 06-5-1, dated August 18, 1942; No. 01-1-118, dated June 29, 1942; No. 03-10-15, dated May 19, 1942; No. 03-10-28, dated September 17, 1942, and No. 03-10G-1, dated February 6, 1939.

# OIL SYSTEMS

## Oil Grades

No one grade of oil can be specified for use in all airplanes under all climatic conditions. Oil becomes cold and viscous in cold weather and it is difficult to start and obtain proper oil circulation in the engine prior to take-off. This can be eliminated by using oil dilution.

The grade of oil most satisfactory at low temperatures begins to break down at high temperatures. When this happens it will be difficult to maintain oil pressures within prescribed limits, and oil consumption will become excessive. These factors in high temperature operation necessitate a maximum limit for oil temperatures. The exact point at which oil film will break down has not been accurately determined, but maximum allowable "oil-in" temperatures specified are selected within safe limits. **Avoid operation for long periods above maximum temperatures.**

The oil grades under present and future specifications, now being procured, and the equivalent commercial grades are shown in table I.

**Table I**

| Grade Designation Spec. AN-VV-O-446 Present | Grade Designation Spec. AN-VV-O-4462 Future | Commercial SAE No. |
|---|---|---|
| 1080 | 1080 | 40 |
| 1100 | 1100 | 50 |
| 1120 | 1120 | 60 |

No one grade of oil will meet all conditions, but most operations fall within the ground temperature range of 4°C (40°F) and above, for which grade 1120 is specified. This grade will be considered standard except where starting or warm-up difficulties are found during operations at low temperatures. For engines using oil dilution, grade 1100 will be used.

## Commercial Oil

Whenever it becomes necessary to procure engine lubricating oil during cross-country flights, the best available quality of commercial oil of one of the well-known brands, equal in viscosity to the specification oil required, and containing no "additives," will be obtained. The SAE viscosity ratings corresponding to the specification grade numbers are indicated in table I.

## Temperature Ranges

Table II lists three grades of lubricating oil with recommended operating temperature ranges. Overlapping of ground temperatures is to avoid setting up a requirement for changing from one grade to another when frequent temperature changes occur locally.

Temperatures of 4°C (40°F) and above, cover generally the range prevailing in the summer throughout the country.

Temperatures of −7°C to +27°C (20°F to 80°F), cover generally the range encountered during winter months in the southern part of the United States.

Temperatures of 10°C (50°F) and below, cover the winter range encountered in the northern portion of the United States.

Airplanes using oil dilution for starting in cold weather may operate in most temperature conditions with grade 1120 oil, but use grade 1100 when ground temperatures are below 4°C (40°F). **Oil lighter than grade 1100 never should be needed for airplanes using the dilution system.**

## Oil Levels

Filler openings and level cocks (when the latter are used) are located on aircraft oil tanks and indicate proper oil levels for corresponding fuel loads.

Oil level for the maximum fuel load is indicated when oil overflows from the filler neck when the tank is being serviced.

When airplanes use level cocks, the lowest one indicates the oil level for the smallest fuel load and the highest one indicates the oil level for the greatest fuel load.

**Table II**

| | 1120 GRADE OIL | 1100 GRADE OIL | 1080 GRADE OIL |
|---|---|---|---|
| Air Temperature at Ground | 4°C (40°F) and above | −7° to +27°C (20° to 80°F) | 10°C (50°F) and below |

Low ground temperatures listed for each grade are sufficiently high that starting and warm-up difficulties should not ordinarily result. If airplanes are stored in heated hangars, the various grades may be used at still lower temperatures. Starting difficulties may occur below —15°C (5°F), unless the engine is started immediately upon removal from a heated hangar or after being serviced with warm oil.

| | 1120 GRADE OIL | 1100 GRADE OIL | 1080 GRADE OIL |
|---|---|---|---|
| Safe Max. "Oil-in" Temp. | 95°C (203°F) | 85°C (185°F) | 76°C (167°F) |

Temperatures above those listed should not be the cause of forced landings **unless they are accompanied by low oil pressure.**

| | 1120 GRADE OIL | 1100 GRADE OIL | 1080 GRADE OIL |
|---|---|---|---|
| Safe Min. "Oil-in" Temp. | 20°C (68°F) | 10°C (50°F) | 0°C (32°F) |

Temperatures listed are conservative, **but avoid continued operation below limits specified.** A steady oil pressure reading is a more reliable indication of proper lubrication at low temperatures than oil temperature readings.

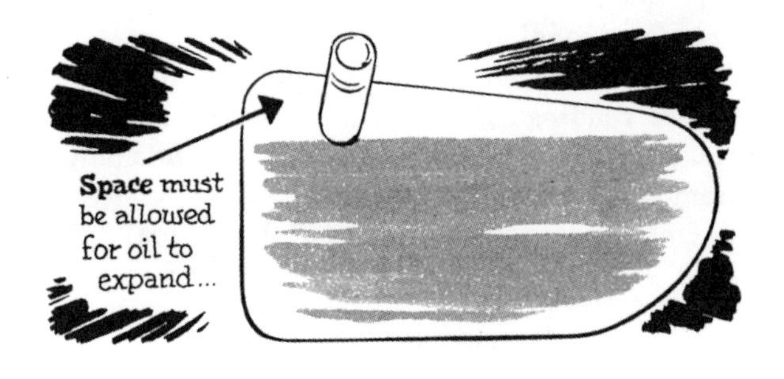


Agitation of the oil under normal operating conditions results in foaming of the oil in the tank. Therefore, when checking the oil supply when there is foam present in the oil tank, use the utmost caution to ascertain the true level of the oil. Using the measuring stick to determine the oil supply will be inaccurate under these conditions as it is difficult to tell where the true oil level begins. Use the measuring stick only when no other means of judging the oil level is possible.

**CAUTION**

In no case will aircraft be serviced with fuel without servicing with oil, if necessary.

On most AAF aircraft the best way to fill the oil tank is to add oil until the oil level is apparent at the neck of the filler opening into the tank.

When the oil is heated to engine operating temperature, it expands. Space in the tank must be allowed for this expansion.

### Oil Temperature Regulators

Increased engine power and bearing loads, higher cooling temperatures, and other factors cause increased oil temperatures. Therefore, it is necessary to provide oil coolers with sufficient area to cool the oil under conditions of high temperature operations. For low temperature conditions, regulation is obtained by a valve that by-passes the oil cooler. This by-passed oil flows around the cooler jacket to keep the core from freezing up.

Adjust regulator shutters, as necessary, during cold weather operation to restrict the air flow through the regulator. If oil congeals in the regulator core, the cooling effect of the core is lost and high oil temperatures will result.

Variations in cooling capacity, regulator sizes, and oil flow in different installations vary the conditions of shutter closings. Shutter adjustments are dictated by experience with individual airplanes, and will reduce to a minimum difficulties resulting from excessive oil temperatures.

REFERENCE: Technical Orders No. 06-10-1, dated November 27, 1942, and No. 01-1-25, dated April 29, 1941.

# HYDRAULICS

## Operation

The principle of the hydraulic system is that fluid is used to transmit a force to operate a mechanism. This mechanism may be the bomb bay doors, landing gear, cowl flaps, wing flaps, brakes, or some accessory.

The fluid used is either a specially prepared mineral oil containing red dye, or a fluid consisting of a mixture of alcohol and castor oil containing blue dye. This dye is placed in the fluid for identification purposes and also to facilitate the detection of leaks.

The system designed for mineral oil utilizes a different packing material than the system designed for alcohol and castor oil. **Interchanging of fluids will cause malfunctioning of the systems because of detrimental effect on packings and gumming of the fluids.**

Emergency hydraulic hand pump.

In operation, force is developed either by a power or hand pump and is transmitted through a column of fluid, confined in tubing, to a piston against which the force is directed. In general, the piston is attached to a mechanism so that movement of the piston will operate the mechanism.

Hydraulic pressure gage.

Position indicator.

The Handbook of Flight Operating Instructions for each airplane shows where the selector valves are located that control the different mechanisms. Move the selector valve to the desired position for operation of the mechanism.

The force of the fluid is measured in pounds per square inch by a pressure gage, mounted on the instrument panel in most airplanes.

**At the first sign of malfunctioning, look at the pressure reading.** If the gage reads zero, it may be because of malfunctioning of the gage and not the system. To check:

1. Operate wing flaps by use of the normal system.
2. Observe the wing flap position indicator.

## Emergency Operation

To lower landing gear, if there is no hydraulic pressure:

1. Operate the selector valve and let the landing gear drop down by its weight.
2. Observe the landing gear indicator to see if the gear is down and locked.

To operate a mechanism if the hydraulic system has failed:

1. Move selector valve to operating position.
2. Use the hand pump to operate the system.

If the power pump will not operate a mechanism because of the low level of the fluid in the reservoir, use the hand pumps as there will be enough fluid in the standpipe.

Remember that the Handbook of Flight Operating Instructions contains detailed instructions in case of hydraulic system failure. Be familiar with these instructions for the different types of airplanes flown.

### General

The blade angle of a fixed pitch propeller cannot be changed. The blade angle of an adjustable pitch propeller can be changed only on the ground. The useful blade angle of both of these propellers is a compromise between the high rpm needed for takeoff and the low rpm that is best for cruising. These propellers are used only on training airplanes.

**A two-position propeller** provides both a high rpm setting for takeoff and a low rpm setting for cruising efficiency.

**The constant speed propeller** permits an engine to develop full horsepower for takeoff, and allows automatic readjustment of the blade angle during flight for particular power and altitude conditions.

### Feathering

Some constant speed propellers incorporate a feathering feature. Feathering a propeller permits the stopping of a disabled and vibrating engine, decreases the drag of the propeller and increases the performance of the airplane with the remaining good engine.

A feathered propeller on a twin-engined airplane increases the single engine ceiling appreciably over that which can be maintained with the useless propeller windmilling.

If necessary to maintain level flight, increase the propeller and throttle settings of operating engine, but don't exceed maximum permissible operating limits of the engine except in an emergency.

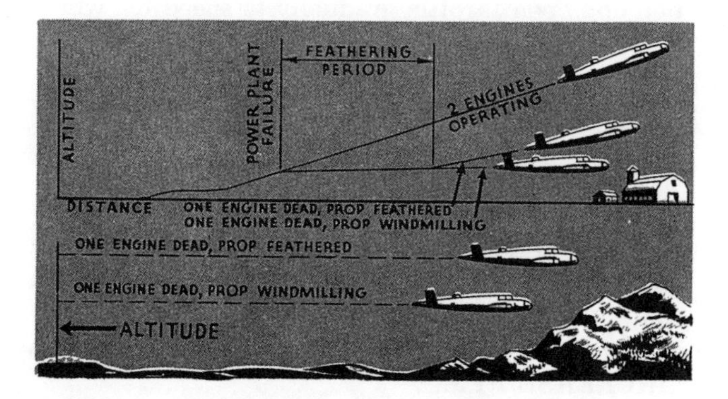


**Maximum performance at all altitudes with a feathered propeller is obtained by trimming the flight controls for level flight.**

### Operating Tips

1. Observe engine operating instructions.
2. Increase rpm first, then manifold pressure.
3. Decrease manifold pressure first, then rpm.
4. Keep throttle closed when unfeathering.

**For maximum endurance** select airspeed and reduce rpm to lowest permissible value for the manifold pressure required to maintain selected airspeed. Adjust mixture as specified by engine operating instructions.

**For economical cruising** reduce manifold pressure to approximate cruising value then reduce rpm to value specified by engine operating instructions. Readjust manifold pressure to specified limits.

### Featheritis

Don't jump for the feathering switch the minute you think you have engine trouble.

**Check these things first:**

1. It may be that your tachometer is registering incorrectly.
2. Look at your airspeed.
3. Listen to sound of suspected engine.
4. Look at the exhaust.
5. Make sure the trouble is not caused by inverter failure (See PIF 5-8-3).

**Remember, unless you are practicing feathering you don't want to lose an engine if you don't have to.**

# CURTISS ELECTRIC PROPELLER

### Starting Engine

1. Before starting engine, set circuit breaker to "ON" position. Leave it "ON" at all times during flight.

2. Set selector switch to "AUTOMATIC". If propeller is not already in low pitch, the electric motor will decrease the angle when the engine is started until the minimum is reached.

3. Start engine.

### Preflight Check

**Manual Operation:**

1. Safety switch "ON" and selector switch "OFF" (center).

2. With engine running at approximately 2000 rpm check the manual propeller operation with the selector switch first in "DECREASE" then in "INCREASE RPM," noting that the engine speed responds accordingly, **but do not decrease rpm excessively (200 rpm maximum).**

**Automatic Operation:**

1. Safety switch "ON" and selector switch to "AUTOMATIC."

2. Propeller control to takeoff rpm.

3. Open throttle to obtain 2000 rpm.

4. Pull propeller control back until you notice approximately 200 rpm reduction in engine speed.

5. Return propeller control to takeoff position noting that you again attain 2000 rpm.

6. Close throttle.

If your airplane is equipped with a feathering Curtiss Electric Propeller, immediately following the manual and automatic operation checks, and before you close throttle, perform the following **feathering check:**

1. Feathering switch to "FEATHER" until you notice approximately 200 rpm reduction in engine speed. (This takes only one to two seconds. Don't reduce rpm excessively.)

2. Return feathering switch to "NORMAL," noting that you again attain 2000 rpm (because of automatic operation).

3. Close throttle.

These propeller checks can be made in sequence and concurrently with the magneto check without shutting the engine down between each check.

### Takeoff, Climb, and Level Flight

1. Set selector switch to "AUTOMATIC."

2. Adjust propeller for takeoff rpm.

3. Adjust throttle to obtain desired manifold pressure. During climb and level flight, with selector switch in "AUTOMATIC", the governor holds engine speed constant by varying the pitch of propeller to suit different engine powers or flight conditions.

If a different engine speed is desired during flight, adjust the propeller slowly until tachometer registers the proper speed.

Any combination of engine rpm and manifold pressure is obtained within the operating limitations of the engine by independent adjustment of throttle and propeller.

When cruising altitude is reached, set throttle for cruising, adjust propeller to cruising rpm.

If your propeller governor fails, you may still maintain satisfactory performance by placing the selector switch in "FIXED PITCH."

### Feathering

1. Close throttle.

2. Set feathering switch to "FEATHER."

3. Move mixture control to "IDLE CUT-OFF."

4. Shut "OFF" fuel supply to engine.

5. Leave ignition switch "ON" until propeller stops, then turn switch "OFF."

### Unfeathering

1. Turn ignition switch "ON" with throttle closed.

2. Set propeller control to "DECREASE RPM."

3. Turn "ON" fuel supply.

4. Move mixture control to "FULL RICH."

5. Set feathering switch to "NORMAL."

6. Hold selector switch to "INCREASE RPM" until tachometer reading reaches 800 rpm; then release switch.

7. Allow engine to operate at this rpm until required operating temperature is obtained. Otherwise you may damage the engine.

8. Set selector switch to "AUTOMATIC" position and open throttle gradually to speed for which governor is set.

9. Adjust mixture, throttle, and governor to desired power and engine rpm.

### Approach and Landing

**Place propeller in maximum cruising rpm position to prevent overspeeding of engine if throttle is opened in an emergency.** If necessary to interrupt the glide and make another landing approach, adjust throttle first, then place propeller control in "INCREASE RPM."

**After landing, before stopping engine,** place control in "INCREASE RPM."

### Run-Away Propeller

In case of a run-away propeller, that is, one which allows engine to overspeed, check:

1. Circuit breakers.
2. Switch settings.
3. Try to reduce rpm by propeller control.
4. Hold selector switch in the "DECREASE RPM" position.

If this fails to reduce the rpm, turn the feathering switch on, then quickly off again to reduce rpm to the correct setting. Be careful not to reduce the rpm too much when using this method.

### Power Failure

The battery and generator are your sources of electric power. If the generator fails in flight place the selector switch in "FIXED PITCH" to conserve electric energy. Switch it back to "AUTOMATIC" before landing, so that you will have automatic operation if you have to go around again.

### Selective Fixed Pitch Operation

If you desire selective fixed pitch operation, or if the constant speed control fails:

1. Move selector switch to "FIXED PITCH".
2. Hold switch momentarily on "INCREASE RPM" or "DECREASE RPM" (as required) to obtain the rpm setting you want.

### Reset Circuit Breaker

If the circuit breaker opens because of an overload on the propeller circuit let it cool for 10 or 15 seconds, then reset it to "ON". Otherwise, the propeller blades will remain at a fixed angle setting.

In emergencies hold the circuit breaker button full in if a continuous overload exists. It will carry extremely high loads.

REFERENCE: Technical Order .03-20B-3 and Technical Manual 1-412

## HAMILTON STANDARD COUNTERWEIGHT TYPE PROPELLER

### Starting Engine

Before starting engine, place propeller governor control in "DECREASE RPM" position. Then start engine and run for at least 1 minute. This is done to avoid starving the engine thrust bearing of oil immediately upon starting. **This is opposite to the setting used for all other propellers.**

### Warm-up

Shift propeller control to "INCREASE RPM" position for warm-up in accordance with operating instructions for the particular type of engine.

### Takeoff and Climb

Place governor control in "INCREASE RPM" position for takeoff and climb. As airplane accelerates, engine speed increases until it reaches governor setting. Setting has been adjusted previously and should be within 25 rpm of the red line on the tachometer. From this point on, rpm is held constant by the governor. As soon as safety permits, reduce manifold pressure and rpm. **Always reduce manifold pressure and then rpm. Move throttle and governor controls slowly.**

### Level Flight

Set rpm for normal level flight. It will be held constant by the governor within close limits.

### Engine Failure

In event of engine failure, place governor control in "DECREASE RPM" to reduce drag.

### Approach and Landing

**Place governor control in maximum cruising rpm position.** If necessary to interrupt glide and make another landing approach, first adjust throttle, then place governor control in "INCREASE RPM."

**Before stopping engine make sure that propeller has shifted to low rpm to prevent exposure and corrosion of the propeller piston.**

REFERENCE: Technical Order 03-20CB-1 and Technical Manual 1-412.

# HAMILTON STANDARD HYDROMATIC PROPELLER

### Preflight Check of Feathering System

1. After engine is warmed up set governor control to approximately 1,400 rpm.
2. Open throttle until governor is functioning, i.e., engine operating at constant rpm with slight increase in manifold pressure.
3. Depress feathering button until the engine speed drops to 1,200 rpm.
4. Pull feathering button out. Engine speed should return to original governing value in a few seconds.
5. If propeller does not respond as above, feathering system should be checked immediately.

### Takeoff and Climb

Place propeller governor control in "INCREASE RPM" for takeoff and climb.

As the airplane accelerates, engine speed increases until it reaches the governor setting. This setting has been previously adjusted and should be within 25 rpm of the red line on the tachometer. Thereafter, rpm is held constant by governor. As soon as safety permits, reduce manifold pressure and rpm. **Always reduce manifold pressure, then rpm. Move throttle and governor controls slowly.**

### Level Flight

Set rpm desired for normal level flight. The governor will hold it constant within close limits.

### Approach and Landing

**Place governor control in maximum cruising rpm position for approach and landing.** If necessary to interrupt the glide and make another landing approach, adjust throttle first, then place governor control in high rpm position.

### Feathering

1. Close throttle.
2. Depress feathering button.
3. Move mixture control to "IDLE CUT-OFF."
4. Shut "OFF" fuel supply to engine.
5. Leave ignition switch "ON" until propeller stops, then turn switch "OFF."

In case of practice feathering while flying, don't leave propeller feathered for more than 15 minutes. Longer periods of feathering may allow enough oil to seep past the piston rings of the lower cylinders to damage the engine when you unfeather the propeller and restart the engine.

**Practice feathering only between 5000 and 10,000 feet above terrain over which you are flying.**

### Unfeathering

1. Turn ignition switch "ON" with throttle closed.
2. Set propeller control to the minimum rpm position.
3. Turn "ON" fuel supply.
4. Depress propeller feathering button. Keep closed until tachometer reading reaches approximately 1000 rpm; then release.
5. Adjust mixture control.
6. Allow engine to operate at this rpm until required temperature is obtained. Then open throttle gradually, causing engine to speed up to the minimum rpm or the speed for which governor is set.
7. Adjust throttle, mixture, and governor setting to desired power and engine rpm, and synchronize.

In flight, the propeller begins to windmill and crank engine as soon as it starts to unfeather. Engine speed increases rapidly as power is applied and automatic unfeathering proceeds.

When unfeathering a propeller after engine has cooled, idle at slow speed until engine is thoroughly warmed up. Otherwise serious damage may result. Windmilling action of unfeathering acts as a powerful cranking force and may overspeed engine beyond safe idling speed unless unfeathering is stopped while propeller still is at a high angle.

If the engine idles at too high a speed, you can reduce rpm by depressing the feathering button briefly. Pull out button when desired rpm is reached.

If you don't pull the button out, the propeller will

continue to the full feathered position at which time the button pops out.

If you hit the feathering switch accidentally, pull it out at once. **Do nothing more** as the propeller will return shortly to governor control.

REFERENCE: Technical Order 03-20-14 and Technical Manual 1-412.

# AEROPROP

### Starting Engine

Start engine with propeller control in "INCREASE RPM." This reduces the load or drag of the propeller and results in easier starting and warm-up of the engine. Also, this is normally the position of propeller prior to stopping engine.

### Takeoff and Climb

Place propeller in high rpm position for takeoff and climb. As the throttle is advanced, engine rpm increases until it reaches the amount for which the high rpm stop has been set. From this point on, rpm is held constant by the governor.

As soon after takeoff as safety permits, reduce both manifold pressure and rpm. Always reduce manifold pressure first and then rpm. Move throttle and propeller controls slowly.

### Level Flight

Set rpm desired for normal level flight. The governor will hold it constant within close limits.

### Approach and Landing

**Place propeller in maximum cruising rpm position for approach and landing.** If necessary to interrupt the glide and make another landing approach, adjust throttle first and then place propeller control in "INCREASE RPM."

**Stop engine with control in "INCREASE RPM."**

### Aeroprop Purging

If you suspect poor governing, purge the propeller by changing the manifold pressure from 20" Hg to 37" Hg (for the P-39Q-10) several times in succession. Always consult the Technical Order for the airplane operating instructions.

### Dual Rotation Propellers

The operating instructions for the dual rotation Aeroprop are the same as for conventional rotating propellers; move propeller control forward for increased rpm, backward for decreased rpm.

REFERENCE: Technical Order 03-20E-2.

### Description

The Electrical Power System consists of a Generator, Voltage Regulator, Reverse Current Relay, Battery, Battery Switch, Generator Switch, Ammeter, and Voltmeter. In multi-engined airplanes, this complete set of equipment is duplicated for each engine.

1. The Generator is the primary source of power for all electrical equipment.
2. The Voltage Regulator keeps the generator voltage constant at all times.
3. The Reverse Current Relay automatically disconnects the generator from the main circuit when the engine stops or when the generator ceases to operate.
4. The Battery is a **small** power reserve to supply extra power for momentary peak loads. **Don't use it as a primary source of power** except in emergencies.
5. The Battery Switch connects and disconnects the battery from the main circuit.
6. The Generator Switch connects and disconnects the generator from the main circuit.

### Starting Engines

When an external power source is available, always use it to start engines.

If no external power is available and the airplane is equipped with an auxiliary power unit, start the auxiliary unit and use it to start the engines. Stop the auxiliary unit as soon as the engines are running normally.

> **Caution**
>
> Don't use the battery to start engines except when auxiliary source is not available.

If necessary to start an engine on the battery, crank it for short, intermittent periods only. This allows the battery to recover slightly between starter loads.

On multi-engined airplanes, start only one engine on the battery. Allow that engine to warm up sufficiently, then run it at 1,600 rpm to 1,800 rpm while cranking the other engines with the starter. In this way, the generator of the one engine supplies current for starting the others.

### Checking Electrical System

Run the engine at 1,600 rpm to 1,800 rpm and check the generator output voltage on the ship's voltmeter. On multi-engined airplanes, repeat this process with each engine.

NOTE: Voltage of each generator must be between 13.5 and 14.5 volts on 12-volt systems, and between 27.0 and 29.0 volts on 24-volt systems, with not more than 0.3 volts variation between any two generators on multi-engine airplanes. If greater variation is shown, don't take off until qualified maintenance personnel make necessary adjustments.

On multi-engined airplanes, check equalizing of generator loads. With all engines running at 1,600 rpm to 1,800 rpm, turn on an electrical load equal to 10% to 20% of the total capacity of all generators. Then compare the reading of all ammeters. Readings must not vary more than 10%. If the variation is greater than 10%, don't take off until maintenance personnel make necessary adjustments.

On multi-engined airplanes, also check the batteries by turning the battery switches to "ON," one at a time. If the total of all ammeter readings increases more than 60 amperes when any one battery is turned "ON," have that battery checked, and replaced if discharged.

### Ground Instructions

While on the ground, keep all "ON-OFF" switch-controlled electrical units turned to "OFF," unless the unit is being used or tested.

**Always use external power source when available** for testing electrical equipment to conserve battery energy.

**After each flight, first turn off all electrical equipment. After this is done turn the battery switches off.** This procedure is necessary to protect the battery switches, as they are not designed to interrupt the entire load.

### Flight Instructions

During flight, whenever the voltmeter reading is greater than 15 volts for 12-volt systems, or more than 30 volts for 24-volt systems, turn electrical accessories "OFF," and land at the first opportunity.

In multi-engined airplanes, turn the generator switch "OFF" on any generator whose ammeter indicates excessively high or low current in comparison with others in the airplane.

When flying a multi-engined airplane with one or more engines inoperative, always turn enough electrical equipment "OFF" to make sure that the total load does not exceed the capacity of the operating generators.

In case of generator failure (usually indicated by zero ammeter reading), the battery will supply energy for the electrical system for only a few minutes of normal flying.

Under these conditions:

1. **Turn all electrical units "OFF"** and operate them manually or mechanically as far as possible.

2. **Avoid use of the radio except when absolutely necessary,** and then only for brief, intermittent periods.

### Conservation of Electrical Power

Conserve electric power in every way possible by observing such economies as these:

1. Always start engines and test electrical equipment with an external power source when one is available. Exceptions to this rule may be made only in emergencies.

2. Avoid unnecessary use of electrical current at all times by keeping every unit turned "OFF" unless it is needed at the moment or is being tested. This will conserve the generator and other electrical equipment for their maximum life. For example, do not turn landing lights "ON" until actually coming in for a landing.

3. Take particular care to conserve electrical power at high altitudes. Low air pressure at these altitudes reduces cooling efficiency so greatly that the generator should not be operated at full load.

4. Do everything possible in cold weather to hold the **battery load** to the very minimum. The efficiency of any battery falls sharply as temperatures go down. Since the battery of an airplane is not designed to carry the normal electrical load even when conditions are most favorable, you cannot rely on it without the generators for more than a few minutes' operation, particularly at sub-zero temperatures.

When airplanes are to stand idle, in sub-zero temperatures, for more than a few hours, the batteries should be removed and stored in as warm a location as possible (normal room temperature preferred). This procedure is necessary because the efficiency of cold batteries is low and a partially discharged battery cannot be effectively charged while flying in an airplane at sub-zero temperatures.

#### Caution

1. Keep all generator switches "ON" at all times during flight except when failure of a generator makes it necessary to turn its switch "OFF." Turning one generator "OFF" tends to overload the remainder of the system.
2. Pay particular attention to generator readings when flying at high altitudes, since generator brush life is extremely short under these conditions.
3. Reduce the electrical load whenever ammeter indicates that the capacity of operating generators is being exceeded.
4. **Always** turn "OFF" all battery switches before attempting a crash landing to reduce danger of fire.

REFERENCES: T. O. No. 03-5B-1, dated February 5, 1942.
T. O. No. 01-20EF-2, dated December 1, 1942.
T. O. No. 01-1-61, dated October 24, 1942.

# ELECTRICAL INSTRUMENT FAILURE

Inverters are used in multi-engine airplanes to change direct current (delivered by the battery system) to alternating current suitable for use in the Autosyn instrument system. Autosyn instruments, which are electrically powered, include certain engine instruments as well as landing gear and flap indicators.

Always check batteries and make sure generators are working before take-off. Failure of either will cause inverter failure, which will cause Autosyn instruments to fail, since they are dependent on inverter power.

Two inverters are supplied in most airplanes. (Exceptions are: The B-25, B-26 and C-47A.) Power can be drawn from either one.

**CAUTION**

**Don't flick the switch from one inverter to another in flight. Run on one inverter only. Save the spare until you need it.**

**If instruments do not function normally, or cease to function, flick the inverter switch to the spare inverter.**

**In the event your Autosyn instruments still will not function—even after trying the spare inverter—do not become alarmed. You will be able to make a normal landing after reaching a suitable landing field if you will:**

1. Reduce power and set engines by eye and ear, using flight instruments, which are not dependent on the inverters, to maintain approximate cruising conditions.
2. Extend landing gear and flaps and check them visually.

In case of inverter failure, remember your engine instruments will remain in the same position as at time of failure, or will indicate "INCREASE" or "DECREASE" if affected by vibration. For instance, on take-off with 2700 rpm, reduction of rpm would not be indicated, although it is actually occurring each time the "DECREASE RPM" control is moved. This is also true of the manifold pressure gauges and other instruments. The sound of your engine will tell you whether you are having merely inverter failure or actual engine trouble.

**DON'T BAIL OR FEATHER BECAUSE YOUR INSTRUMENTS FAIL - YOU CAN STILL MAKE A NORMAL LANDING!**

# ELECTRICAL INSTRUMENT FAILURE

Inverters are used in multi-engine airplanes to change direct current (delivered by the battery system) to alternating current suitable for use in the Autosyn instrument system. Autosyn instruments, which are electrically powered, include certain engine instruments as well as landing gear and flap indicators.

Always check batteries and make sure generators are working before take-off. Failure of either will cause inverter failure.

Failure of an inverter will cause Autosyn instrument failure. Since these instruments are dependent on inverter power, two inverters are supplied in most airplanes. (Exceptions are: The B-25, B-26 and C-47A.) Power can be drawn from either one.

**CAUTION**

**Do not use your inverters alternately in flight. Instead of switching from one to the other, save your spare inverter in case the normal inverter fails.**

**If instruments do not function normally, or cease to function, flick the inverter switch to the spare inverter.**

**In the event your Autosyn instruments still will not function—even after trying the spare inverter—do not become alarmed. You will be able to make a normal landing after reaching a suitable landing field if you will:**

1. Reduce power and set engines by eye and ear, using flight instruments, which are not dependent on the inverters, to maintain approximate cruising conditions.

2. Extend landing gear and flaps and check them visually.

In case of inverter failure, remember your engine instruments will remain in the same position as at time of failure, or will indicate "increase" or "decrease" if affected by vibration. For instance, on take-off with 2700 rpm, reduction of rpm would not be indicated, although it is actually occurring each time the "decrease rpm" lever is moved. This is also true of the manifold pressure gauges and other instruments. The sound of your engine will tell you whether you are having merely inverter failure or actual engine trouble.

THIS IS A TEMPORARY PAGE TO BE INSERTED FACING PAGE 5-8-3. IT WILL BE REPLACED IN AN EARLY REVISION.

# SECTION 6 THE AIRPLANE

Tricycle landing gear has several advantages over the conventional type:

1. It will not let the airplane ground loop.
2. It gives the airplane an approximately normal flight attitude on the ground.
3. It protects the airplane from nosing over when the pilot applies the brakes too heavily.
4. It provides better visibility while taxiing.

Because the tricycle gear has to be handled differently from conventional gear in ground operations, you should become familiar with its characteristics.

The nose wheel is not "steerable," that is, it is not connected with the rudder control. It tends to roll straight like a bicycle ridden "hands off."

The suggestions which follow are of a general nature and are not to be construed as conflicting with the Handbook of Flight Operating Instructions for the particular type airplane being operated.

### Taxiing

When an airplane equipped with tricycle gear moves very slowly, it has little stability. Difficulty, therefore, may be encountered in maneuvering at low speed.

**Caution**

**Get the airplane rolling before attempting a turn. Don't use brakes and engines to turn the airplane from a position at rest.**

### Take-Off

Because the center of gravity is forward of the main wheels and the airplane is in a flat attitude on the ground, it tends to hug the ground regardless of speed, particularly if no flaps are used. Therefore, it is necessary to "pull" the airplane off when sufficient airspeed for take-off is attained.

At minimum take-off speed the angle of attack of the wings is just at the stalling angle. This angle feels unsafe and the natural tendency is to wait until sufficient speed is attained to produce full lift at a lesser angle of attack. Consequently, take-offs often are made at speeds much higher than necessary, imposing abnormal stress on tires and gear.

### Best Take-Off Procedure

The most desirable setting of the center of gravity is such that the nose wheel can be raised from the ground at approximately 80 per cent of the take-off speed. At this speed you can raise the nose sufficiently to produce a convenient angle of attack less than the stalling angle, and the airplane will break away from the ground freely when the proper velocity is reached without giving you an "uncertain feeling."

### Landing

**Don't land unduly fast.** Airplanes with tricycle landing gear may be landed at various speeds from normal landing speed to 150 mph or more. When the airspeed indicator registers more than the normal landing speed, unnecessary strains are placed on landing gear.

The best landing is one where the nose of the airplane is well up, and the main wheels touch the ground before the nose wheel. This landing attitude is equivalent to that of a landing with conventional gear in which the main landing gear wheels touch while the tail wheel is approximately 6 inches in the air. **Don't land so "tail low" that the emergency skid or tail bumper strikes, as this produces excessive shock loads on the empennage.**

Immediately after landing, the airplane has a large amount of momentum. Due to aerodynamic drag this momentum is quickly dissipated during the initial part of the ground run. This is especially true if the nose wheel is held off after landing, as the angle of attack is greater, thus producing more drag. **Let the airplane roll some distance before applying brakes.** This is much easier on the brakes and will save them for the time you need them most.

#### Caution

**Don't apply brakes before the nose wheel is on the ground.**

### Taxiing, Landing, or Taking Off on Soft Ground

1. When taxiing on soft ground or sand, be sure you know which way the nose wheel is headed. For example: if it is headed to the left when stopped, don't try to steer it either straight ahead or to the right until it is moving. Doing so would cause the nose wheel to dig in deeper.
2. Keep as much weight off the nose wheel as possible, as the nose wheel may not run true.

### Slippery Terrain

Sliding the wheels on ice or very slippery sod may cause the main wheels to skid and the airplane to turn sideways exactly as an automobile would with locked wheels on icy pavement. This is not a ground loop. Control can be regained almost instantaneously by releasing the brakes.

#### Caution

Don't slide your wheels. You can stop quicker under any conditions, slippery or not, by braking just short of the point where wheels will slide.

REFERENCE: Technical Order No. 01-1-33, dated December 16, 1942.

# LANDING WHEEL *Brakes*

Repeated excessive application of brakes without allowing enough time for cooling between applications causes temperatures to increase dangerously. This may result in brakes freezing, complete breakdown of the brake structure, failure of the brake drums and wheels, blowing out of tires and tubes, and in extreme cases destruction by fire of wheel and brake installations.

Excessively short stops from high speeds and dragging brakes for an appreciable distance while taxiing at low speeds will produce the same results.

### To Maneuver on Ground

Control multi-engined airplanes by use of the engines and single engine airplanes by steerable tail wheel, when provided. If necessary to use brakes to steer airplane, use them as little and as lightly as possible.

### Precautions

1. Don't drag brakes while taxiing. Avoid applying toe pressure inadvertently while operating rudder pedals.

2. Don't use the brakes for speeding up turns unless a sharp turn is necessary to avoid obstructions.

3. Don't stop the airplane as quickly as possible when it is unnecessary. Take advantage of the full length of the runway during the landing roll, using brakes as little and as lightly as possible.

4. Apply brakes firmly but without jamming them.

5. **Don't set parking brakes while brakes are hot.** They may "freeze."

6. **Don't set and lock brakes during flight;** you may land with them on. However, vibration caused by rotation of landing wheels after take-off may be stopped by applying brakes lightly for a short period.

7. If making successive landings, leave the landing wheels extended in the slipstream for at least 15 minutes between landings to allow adequate cooling. If landing wheels are retracted, allow 30 minutes for cooling.

> **Caution**
>
> Don't overheat your brakes. Brakes which get too hot during taxiing may "freeze" after take-off.

### Emergency Brakes

Know the location of your emergency brake control before take-off. You may need it in a hurry.

Refer to the Handbook of Flight Operating Instructions for instructions on their use.

REFERENCE: Technical Order No. 01-1-27, dated April 24, 1942.

# FLAPS

## General

Using flaps increases the lift of a wing by changing its effective camber, or increasing its area, or both; and also reduces length of roll upon landing by providing a resisting surface to the passing air.

The increase in lift is accompanied by a large increase in drag and by an increase in virtual angle of attack. If the airplane remains in the same attitude when flaps are used, it will approach dangerously near to the stall.

## Use

**Correct use of flaps in flight is to steepen the gliding angle and *not to decrease the gliding speed.***

A depressed flap increases the wing diving moment, generally neutralized by increased downwash on the tail. Simultaneous use of elevator trim tabs may be necessary to maintain longitudinal balance.

### Take-Off

Using flaps for take-off increases drag, tending to increase the run, while the increase in lift tends to reduce the run. These two effects combine to give a minimum take-off distance for a definite flap setting.

A split or bent type flap may be lowered from one-fourth to one-half (usually 10 to 20 degrees) to improve take-off. Within these limits the exact setting will vary with the individual type of airplane and can be obtained by experiment.

In general, an airplane with flaps in take-off position will clear an obstacle by a greater margin than it will if flaps are not used. The take-off run itself may be increased. There is no general rule which can be applied. Information determined by test will be in the operation instructions for each type of airplane.

In event of engine failure following take-off in a maximum performance climb using flaps, the danger of stalling is greater than when flaps are not used due to more abrupt stalling characteristics.

**CAUTION**

If engine fails when flaps are down, **keep flaps down.** Sudden raising may result in an excessive loss of lift.

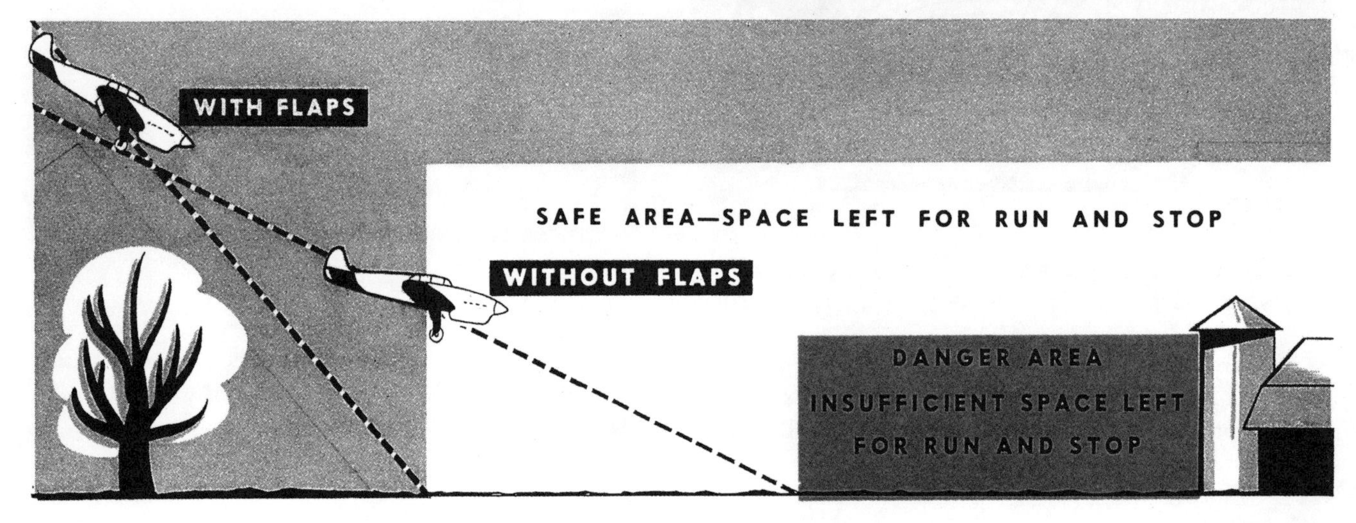

## Forced Landings

In the event of a forced landing made from an altitude which permits maneuvering and where there is insufficient or unsuitable terrain for rolling to a stop, **have landing gear retracted and flaps extended.**

Flaps will provide normal drag and lift with which you are familiar in normal landing routine. Don't worry about the damage caused by landing with flaps extended, since it will be relatively slight compared to what might ensue if they were not used.

## Landing

Split or bent type flaps limit maneuverability of the airplane and eventually stall it more abruptly. The following recommendations are made for landing:

1. With flaps down, **glide at a speed not less than 10 to 15% above the stalling or landing speed** listed in the operation instructions of the airplane to provide sufficient margin of safety against inadvertent stalls and to provide sufficient kinetic energy to force the tail into position for landing.

2. **Be sure to maintain gliding speed to the point of leveling off for landing. Using flaps causes an airplane to stall rapidly when leveled off.** Consequently you must level off at a lower altitude than when flaps are not used.

3. The effect of a cross wind on a landing airplane is to apply a yawing force which tends to swing the nose into the wind. This effect is greater when using flaps. The resultant effect upon the airplane is to increase lift on the up-wind wing and to decrease it on the down-wind wing. This tends to tip the airplane over in a down-wind direction. Since both wings stall rapidly, the tipping force, though strong, is of short duration. This effect is especially critical in a short coupled airplane.

Unless extreme caution is used, you may overcontrol and lose control. Use caution and sufficient excess speed to insure adequate control. In crosswinds of appreciable velocities, use no more flap than is necessary.

## Precautions

1. **Don't raise flaps immediately after take-off** or at other times when near the stalling speed. An appreciable interval of time may be required for speed and angle of attack to become adjusted to the new condition.

2. **Raise flaps gradually** to avoid a sudden loss of lift.

3. When controllable pitch propellers are installed, **use the maximum cruising rpm position whenever the flaps are depressed.** This makes available the maximum thrust under the resulting low speed and high drag condition.

4. **Don't lower flaps at an indicated air speed higher than the limiting air speed for flaps** posted in the cockpit.

5. After flaps have been lowered, keep air speed below the limiting speed to avoid overstressing the structure.

6. **Avoid rapid or otherwise violent maneuvers,** including steep banks, when flaps are extended.

7. Use flaps only for normal maneuvers necessary for landing or take-off.

8. Avoid long flight or glides with flaps lowered.

9. Be cautious when using flaps on low winged airplanes on rough ground or on fields that are covered with mud, water puddles, or slush. Don't taxi such airplanes with flaps down because of the limited ground clearance and the possibility of damage to the flaps by objects being thrown against them by the wheels and propeller.

10. Avoid water puddles and slush wherever possible when making landings or take-offs with flaps extended. If water or slush may be encountered on the ground when landing, extend flaps only partially if landing conditions permit.

REFERENCE: Technical Order 01-1-60.

# SURFACE CONTROL LOCKS

**GET THESE CONTROL LOCKS OFF BEFORE YOU REV HER UP**

Several cases of serious damage to aircraft control surfaces, etc., have been caused by failure to use control locks when the airplane was exposed to high winds or propeller blasts from other airplanes.

To minimize the danger of an airplane taking off with such damage undetected, and possibly losing or impairing the surface control mechanism in flight, observe the following precautions:

**Check all flight controls for free movement before take-off to insure that all control locks are released or removed.**

1. **Don't run the engines at high speeds during ground check or taxiing with the surface controls locked.**

2. If possible, **keep the movable tail surfaces in a neutral position during engine test.**

3. On airplanes having internal surface control locks, **engage the locking devices whenever the controls are not manned.** On airplanes having external surface control locks, install the locks whenever the airplane is unattended. Follow this procedure regardless of whether or not there are high winds.

4. Some older types of airplanes may have neither internal nor external locks. Some method of locking the surface controls will be improvised locally and applied in accordance with instructions given in (3) above.

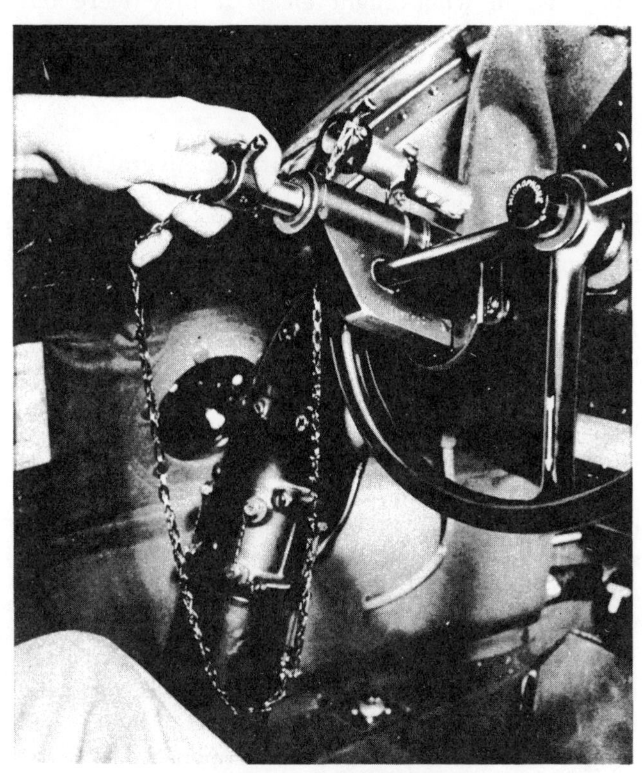

Internal surface control lock.

### Inspection of Surface Controls

If the control surfaces are caught by the wind or propeller blasts and moved violently against their stops or to the limit of their travel when taxiing, awaiting take-off, maneuvering to a parking position, etc., a special pre-flight inspection will be made before the airplane is flown again.

Inspect all control surfaces, hinges, hinge brackets, control horns, attachment of surfaces to torque tubes, etc.

A similar pre-flight inspection will be made, if during the preceding night or since the last flight, unusually violent or gusty wind conditions have occurred in the vicinity of the parked airplane.

REFERENCE: Technical Order 01-1-29.

# HOW TO TRIM YOUR PLANE

Because the loading of an airplane is never uniform—the load changing with the number and placing of the crew, the placing of the cargo, the power used, and the continually changing weight of the gasoline in the tanks—some means must be employed to keep the ship flying naturally in a straight line with its wings level and its nose neither up nor down. This aerodynamic counterbalancing for load change is accomplished by offsetting the natural flight positions of the control surfaces through the use of "trim tabs." The number of trim tabs varies with the size of the airplane, ranging from a single tab on the elevator of a small airplane to tabs on the ailerons and rudder as well as the elevator of larger airplanes.

### For Take-Off

**MAKE SURE THAT ALL TRIM-TAB ADJUSTMENTS ARE IN THE TAKE-OFF POSITION.**

### In Flight

When trimming a small airplane fitted only with a "nose up" and "nose down" trim-tab adjustment:

**FIRST**—climb to the desired altitude. If there is too much forward pressure on the control column during a normal climb, adjust the elevator trim tab to bring the nose up.

**SECOND**—adjust the engine to its rated cruising power.

**THIRD**—move the wheel or control stick back and forth and determine by feel if the ship is "nose heavy" or "tail heavy." If continual back pressure is needed to keep the rate-of-climb indicator at zero, the airplane is "nose heavy." If forward pressure is required, it is "tail heavy."

**FOURTH**—adjust the trim-tab adjustment control to counteract this tendency, being careful to follow the direction of turn indicated to bring the nose "down" or "up."

To trim a large airplane equipped with tabs on ailerons and rudder as well as elevator:

**FIRST**—climb to desired altitude and adjust engine to its rated cruising power.

**SECOND**—adjust for "nose up" or "nose down" as indicated for small ships as above.

**THIRD**—adjust for low left or low right wing by adjusting aileron trim-tab control.

**FOURTH**—adjust for right or left yaw (turn) by adjusting the rudder trim-tab control.

**FIFTH**—recheck the "nose up—nose down" and low-wing trim-tab adjustments.

### Landing

Most airplanes must be trimmed for the best gliding speed when coming down for a landing.

> **Warning**
>
> The placing of the trim-tab adjustment controls and the direction in which the tab controls are turned for a desired trim-tab reaction varies with the airplane. Check the indicated directions before taking off.

### Trim-Tab Don'ts

Don't adjust the trim tabs too rapidly. Cases have been reported where a too rapid adjustment of the tabs has thrown the airplane into a snap roll.

Don't forget that as your gasoline supply decreases the balance of the airplane may change and further tab adjustments may be necessary.

**REMEMBER**—the main function of the trim tabs is to ease the pressure on the controls and save the pilot work. If that is not accomplished the adjustments are not correct.

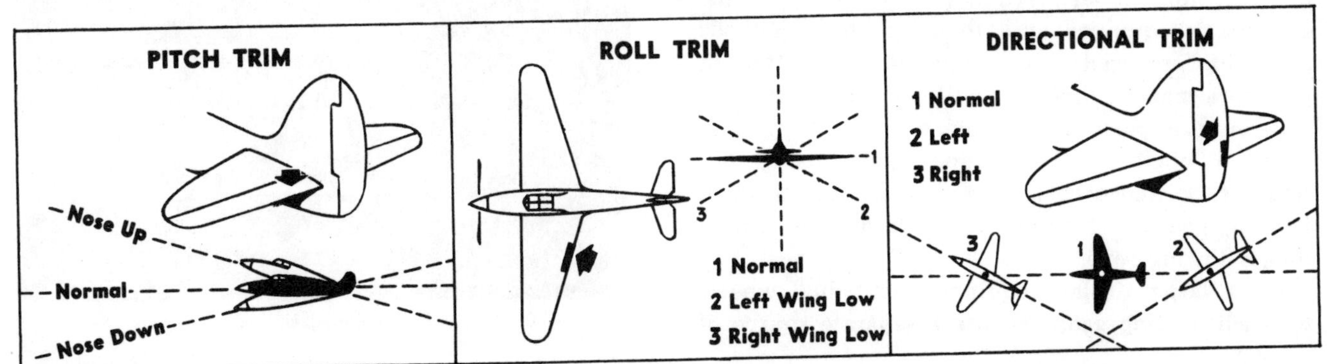

# CARE OF THE AIRPLANE

Make it a point to see that the airplane is properly handled by the ground crew and be familiar with the routine of handling the plane on the ground. This is necessary when using emergency fields where no qualified ground crew is available.

Be certain that the ground crew gives the exterior finish of the airplane proper attention. If operating near the sea, **make sure that all spray and salt deposits are washed away promptly.** See that all oil and soot accumulations are removed frequently. Have all Plexiglas surfaces properly cleaned. Keep the cockpit clean and in good order at all times.

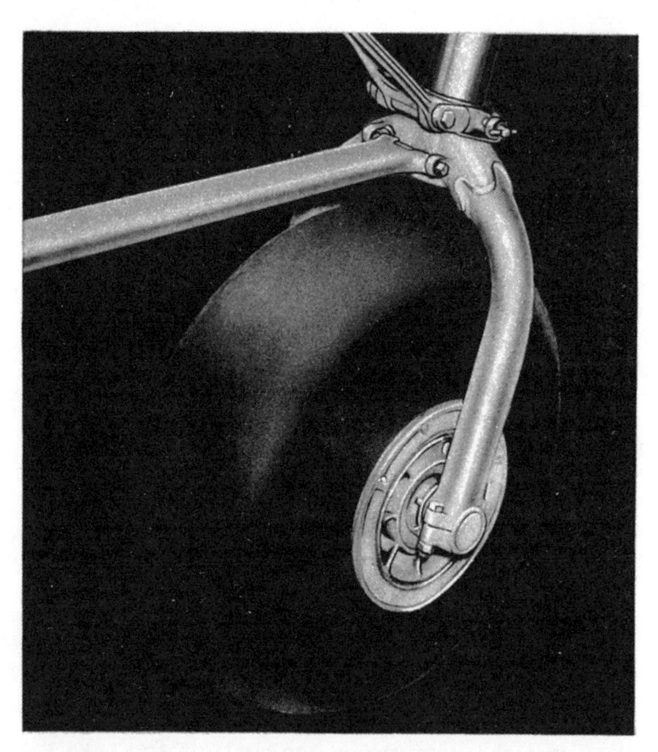

## Towing

See that the ground crew uses towing rings, towing bars, or similar devices for towing when the airplane is equipped with them. If other methods of towing are used, towing lines must be attached in such a manner that they will not damage streamlines or other parts of the airplane.

Use a length of towing rope approximately 3½ times the distance between the wheels of the main landing gear. If towing over soft ground, use a longer towing rope.

On a large airplane equipped with dual wheels on each landing gear leg, use a separate source of power to operate the brakes, and station one or more men at the tail wheel steering handle. When maneuvering near hangars, airplanes on the ground, or other obstacles, assign a man to each wing tip. **A man will go in front of and facing the airplane when it is being towed.**

When a tow bar is attached to the nose wheel assembly, always check to be sure that the tow bar pin is in place and locked, and that the nose wheel snubber is unlocked. Pay particular attention to this precaution if the tow is to be made over soggy or rough ground.

If in doubt as to the method of towing to be used, consult the Handbook of Flight Operating Instructions for the particular airplane. **Avoid high speeds and sudden stops when towing.**

## Mooring

When mooring an airplane, use the following sizes of ropes or cables according to the weight of the airplane. If the airplane is under 5,000 lbs., use ½-inch rope or 3/16-inch cable; from 5,000 to 16,000 lbs., use ¾-inch rope or ¼-inch cable; and if over 16,000 lbs., use 1-inch rope or ⅜-inch cable. Leave enough slack in ropes to prevent them from putting a heavy strain on the airplane's structure in case they tighten because of absorption of moisture.

**Lock the surface controls with locks when they are available.** If there are none, lock the controls in the cockpit by the use of the built-in locks. Place elevators parallel to the ground, rudder and ailerons in neutral position, and automatic pilot "OFF," if the airplane is so equipped. Lock the tail or nose

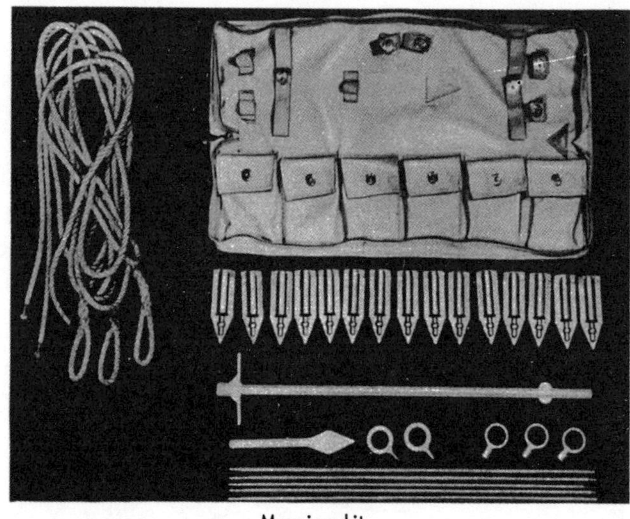

Mooring kit.

wheel in fore and aft direction, and head the airplane into the wind with chocks in front and back of each wheel. Apply parking brakes on all but B-18 series airplanes.

### Emergency Moorings

For emergencies, or when warnings of storms or high winds are received, take the following additional precautions in mooring:

Clamp a spoiler on the wings to avoid damage to fabric covering of the wing or to the controls. The spoiler may consist of a wooden two-by-four with a length equal to approximately 75 percent of the wingspan. Locate it 10 to 15 percent of the average chord aft and parallel to the leading edge of the wing. Put the four-inch dimension perpendicular to the wing surface, with little or no space between the spoiler and the wing surface.

Install felt-padded wooden clamps to lock all movable control surfaces securely. These clamps can be made on the spot and will be used in addition to inbuilt surface control locks.

### Special Mooring Equipment

The type D-1 mooring kit provides mooring points for temporary purposes. Instructions for assembling and inserting these anchor rods in the ground are included in each kit.

A "dead-man" mooring anchor may be used for semi-permanent and temporary mooring in the field.

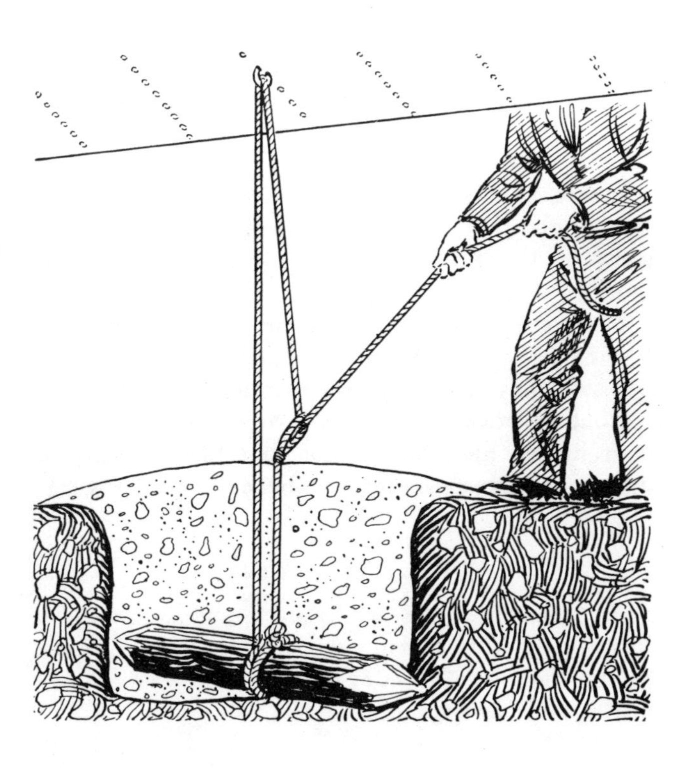

1. Secure a cable, by splicing or with a cable clamp, around a wooden block at least 12 inches long and 3 inches by 3 inches square (if wood is not available, other material of at least equal strength may be used).
2. Form free end of the cable into a loop.
3. Bury the "dead-man" 3 to 6 feet deep (depending on the firmness of the soil).
4. Attach mooring ropes to the cable loop.

Stakes similar to tent pegs may be used to moor the airplane. They should be 2 inches to 3 inches in diameter, and 3 feet to 5 feet long. Drive them into the ground at an angle not greater than 30 degrees from vertical. Position each stake so that the mooring rope will tend to pull it into a vertical position.

### Precaution When Lifting Tail Section

The underside of metal-covered stabilizers may be damaged if mechanics put their backs or shoulders under the stabilizer to lift the tail. Never permit this method. of indiscriminate lifting. **Insist that the ground crew lift only at points provided in the structure of the airplane which are clearly marked by the builder.**

### Plexiglas

Plexiglas is used for gun turrets, windshields, nose pieces, cabin windows, blisters, etc., on all combat planes of the Army Air Forces. **Cover Plexiglas with canvas covers when exposed to the sun,** as extreme heat will soften and bend the Plexiglas.

Clean all Plexiglas surfaces frequently with plenty of pure water. The bare hand is preferable as a wiper, since a cloth might contain gritty particles that will scratch the plastic. Kerosene will remove grease and oil. Do not apply strong soaps or any type of solvent as they will "cloud" this material.

In sandy and windy regions, cover Plexiglas surfaces when parking the airplane to prevent scratches which will reduce the transparency.

An automobile type wax, free from any abrasive material, will remove the majority of light scratches. Several applications with a small grit-free cloth may be necessary. Do not rub too hard or too long on one spot as the heat generated may be enough to soften the plastic.

REFERENCE: Technical Order 01-1-50, dated November 30, 1942.

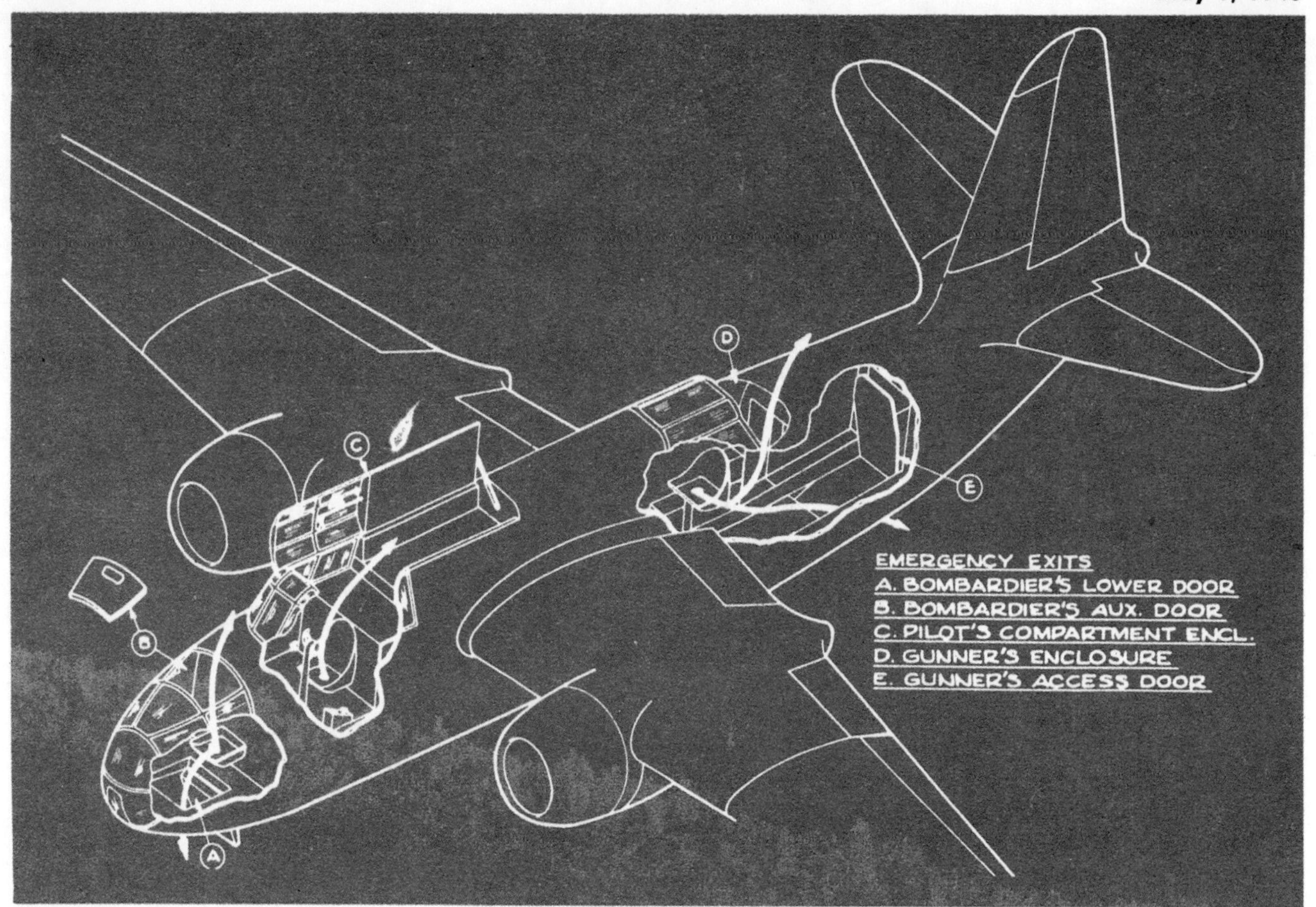


# Emergency Exits

All Army Air Forces airplanes contain means for quick exit, in the air, on the ground, or water.

Before you fly, be sure everyone aboard knows:

1. What exit to use.
2. How to use it.
3. When to use it.

**Hold frequent practice drills on the ground.** Teamwork and speed mean a lot in an emergency.

### What Exit?

In flight, upper exits are dangerous due to the possibility of being caught by a propeller or of striking the tail. Use lower or side exits whenever possible. Study Handbook of Flight Operating Instructions to learn how to bail out of your particular airplane.

On the ground or water, fasten all lower hatches before landing. Dump all upper ones. They may jam upon impact and delay is dangerous. In an emergency, you can knock a hole in the skin of the airplane. If a handaxe is provided on the airplane, know where it is and how to remove it.

### How To Escape

Emergency exits are provided with quick release red handles. Usually the door or hatch will be blown away by the windstream, after release and a light push.

**For a crash landing on land or water, don't dump lower hatches.** Dump the upper hatches, but remember that they may damage the tail assembly.

**Hold practice drills before flight.** Be sure each man knows his exact duties and the meaning of emergency signals by interphone, call light, or warning bell.

### What To Escape

Any emergency is unexpected and unusual, so keep your wits about you. Be deliberate, even though hurried. Consider the situation; make a decision; then give definite orders.

# Safety Belts

Fasten your seat belt on entering the plane and be sure that belts of all passengers are properly secured. The safety belt has two purposes:

1. To keep you in the airplane.
2. To protect you in case of a crash.

Keep belts fastened at all times except when duties require free movement about the plane. Replace the belt at once upon resuming seat.

Before acrobatics, check your belt and **see that the belt is tight.** The catch may have loosened. Always check with your passenger just before performing acrobatics. He may have a loose belt or a loose catch.

**Always put belt over, not through, parachute harness.** You may want to bail out.

Don't wear a twisted safety belt. It cuts.

## Safety Shoulder Harness

Always wear the safety shoulder harness when it is available.

It can be adjusted in two ways: to give freedom of movement (unlocked), or to prevent the wearer from being thrown forward on crash impact (locked). **Have it locked during all take-offs and landings (crash or otherwise), and in acrobatics.** Upon crash impact the shoulder straps prevent your head and chest from being thrown forward against the cowling, instrument panel, or gunsight. Experienced pilots and medical officers are convinced that a properly used shoulder harness has prevented many serious injuries and saved many lives. The shoulder harness will not break your neck.

**To adjust, be sure straps are over crossbar at the top of seat where a crossbar is installed.** Insert metal ends into tongue of seat belt lock. Pulling the latch handle to the right in the usual way releases both belt and harness.

Harness is locked by adjusting lever at left of seat base. When lever is upright (or to rear), harness is unlocked and allows you to lean forward. Pressing down on the top of the lever and shifting it forward sets a catch which allows the harness to lock when the wearer leans back against the seat. **To adjust the harness for fit, wear your parachute and adjust the straps to fit snugly in the locked position.** Otherwise the straps may be too short to allow the lock to catch when you lean back.

REFERENCE: Technical Order 03-1-2, AAF Regulation 62-18.

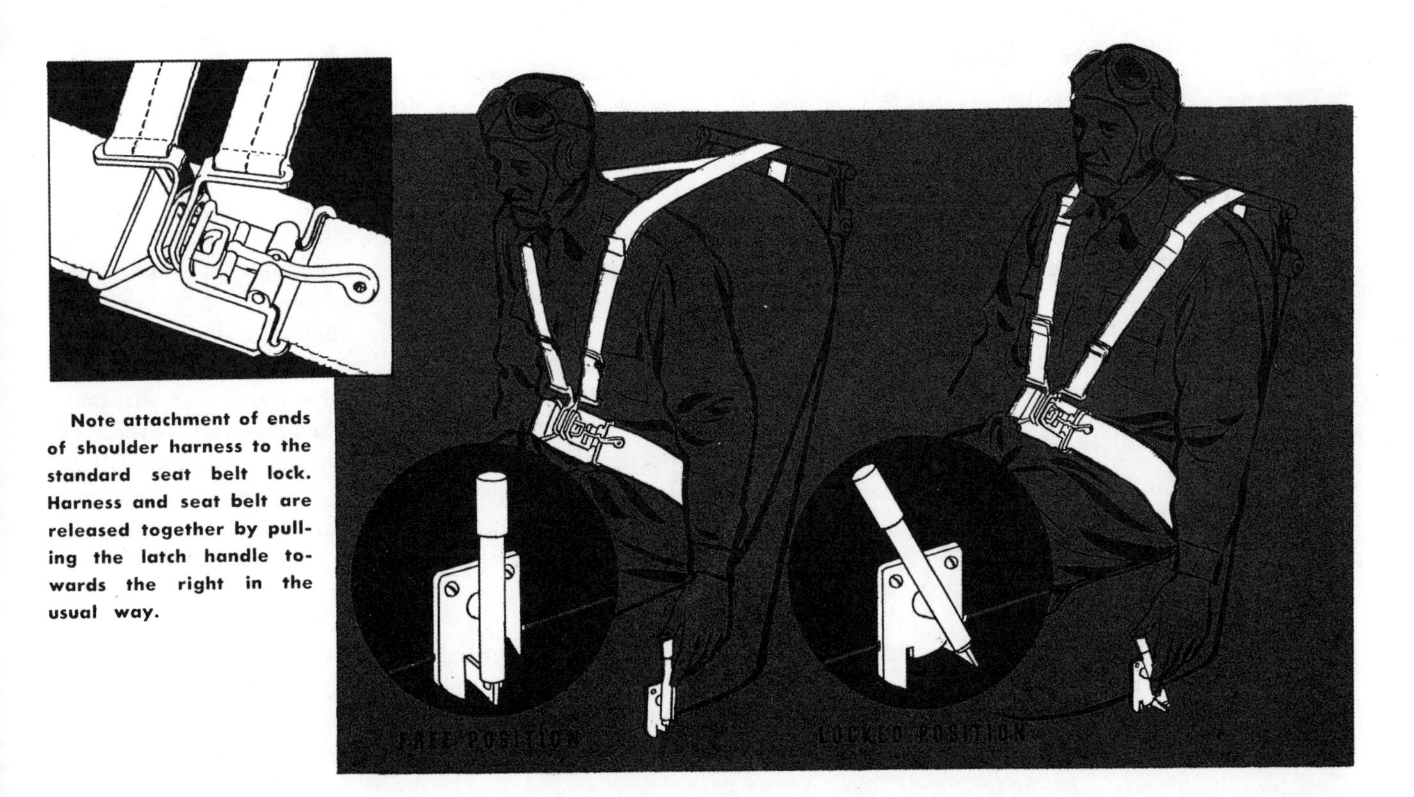


Note attachment of ends of shoulder harness to the standard seat belt lock. Harness and seat belt are released together by pulling the latch handle towards the right in the usual way.

# PRECAUTIONS WITH BOMB BAY DOORS

### On the Ground

The mechanism (hydraulic, mechanical, or pneumatic) operating bomb bay doors is powerful. Before opening doors on the ground, have a man on watch signal "Clear." **Don't open bomb bay doors blindly.**

**Close the doors with equal care. Be sure they are clear.**

**Always close bomb bay doors before starting engines on the ground, unless it is desired to blow gasoline fumes from the bomb bay.**

### In Flight

**Bombardier must get pilot's permission before opening bomb bay doors.** Certain airplanes use a trailing antenna which may catch on open bomb bay doors. Pilot or bombardier will notify radio operator to reel in antenna.

After take-off from a muddy or slushy field in cold weather, operate bomb bay doors through several cycles to prevent freezing closed.

Doors open automatically when salvoing bombs. **Don't drop bombs in the normal manner unless you know the bomb bay doors are open.** There usually is an automatic safety device to prevent this, but don't trust it.

You should know exactly how long it takes to open doors, salvo bombs, and close doors. In an emergency there may be just enough time to do this. **Don't crash land with the doors open.**

For emergency exit in flight go through open bomb bay doors. Study your Handbook of Flight Operating Instructions for the particular airplane you are flying. Learn each emergency method of opening bomb bay doors. **Be sure you know how long emergency means will keep doors open and watch that time.** You may have only a minute or less. If the doors remain open only a short time, have parachutes on and be ready to go before opening the doors.

Study emergency methods of bomb release on your airplane. **If there are several emergency methods, learn to use each, how long it takes, and whether the bombs will be armed or safe.**

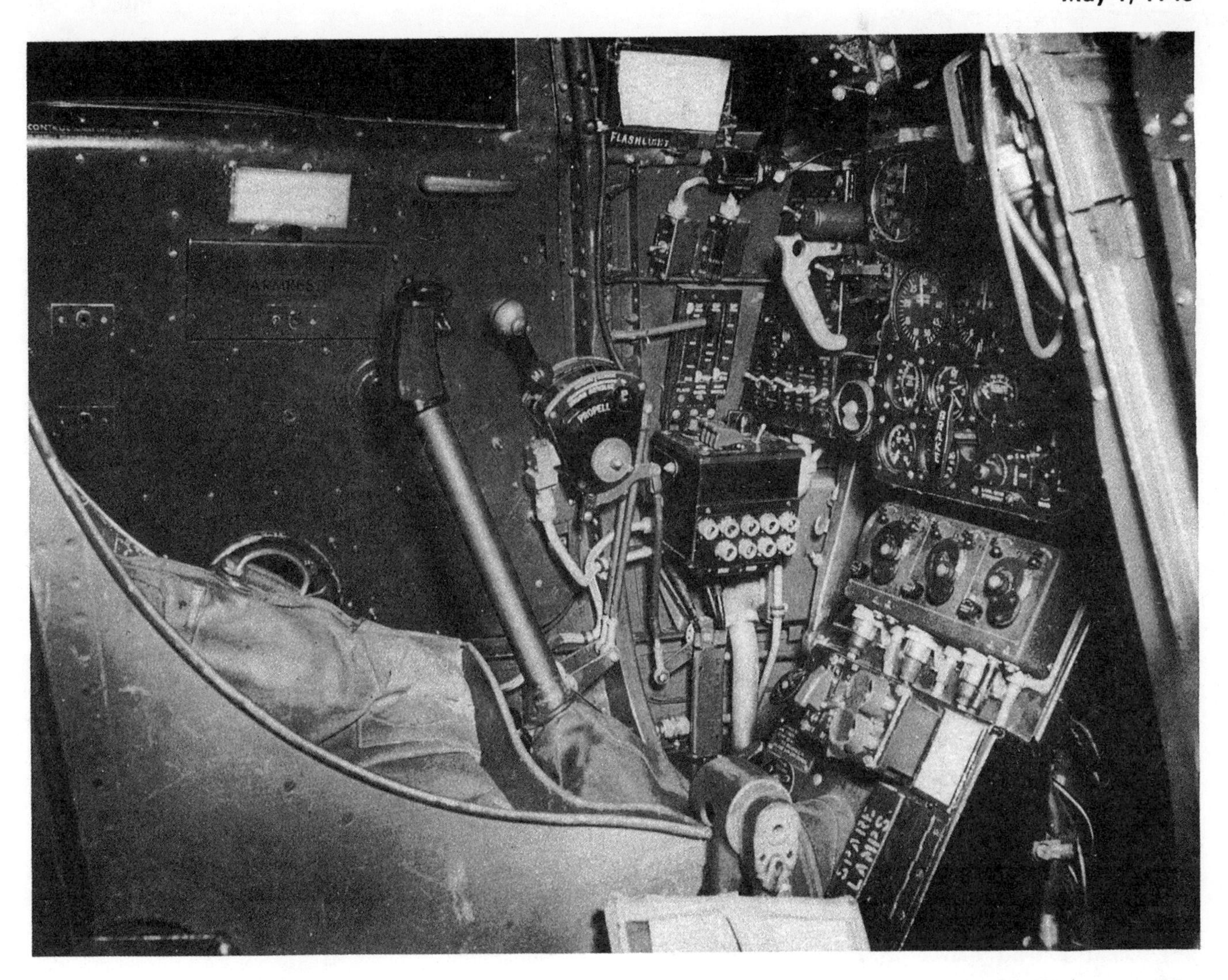


# Precautions against fouling controls

A serious hazard may be created by the flight controls becoming jammed whenever such articles as microphones, flashlights, oxygen mouthpieces, etc., are dropped to the floor and come in contact with the controls.

1. **Always replace accessories** and be sure that they are securely seated in the carrying hooks or other receptacles provided.
2. After using microphones or similar articles, tape the extra length of wire to a convenient part of the airplane to prevent their contacting any part of the controls in case they are dropped.
3. Oxygen breathing tubes may be held in place by retaining clips and loop-type clips which are installed in many airplanes. One retaining clip, used at or near the mouthpiece, should suffice for each tube, in most cases.
4. **Keep front cockpit free of loose equipment.**

REFERENCE: Technical Order 01-1-109, dated December 25, 1942.

## SAFETY HINTS

### High-Speed Aircraft

Pilots who have become accustomed to performance characteristics of training types of aircraft must be especially careful when they begin to fly high-speed aircraft. Many practices that are relatively harmless in trainers are hazardous in higher speed airplanes.

1. Don't extend hands or arms—or anything—out of windows or above the canopy.
2. Never attach streamers or other objects to any flight or control surfaces for identification purposes.
3. Make certain that all control surfaces, trim tabs, and hinges to them are clean. Mud, waste, or even ravelings of cleaning cloths or mooring lines will seriously affect the flight characteristics of the airplane.
4. Cockpit windows left partially opened by accident will often throw the airplane out of trim.
5. Use particular care when operating on sandy or gravelly fields. Scratches or dents made by sand, gravel, or high weeds and grass may injure control surfaces, air foils or propellers, and seriously affect the airplane's performance.
6. Remember that for any maneuver in high-speed aircraft you need lots of room. Altitude is absolutely essential for recovery from dives, spins, and stalls.
7. You need all the runway the field affords for landing and take-off.
8. Keep your landing speed 20% above stalling speed. If you miss your first attempt, you must be above stalling speed to recover and go round again.
9. Don't attempt to lower your flaps or landing gear at high speeds. Follow instructions in cockpit or Handbook of Flight Operating Instructions for the airplane.

### General

1. Be sure to check before you get into the cockpit that all surface control locks have been removed.
2. Check, as soon as you get into the cockpit, that the controls are unlocked.
3. Never start your engines until a crewman is standing by with a fire extinguisher.
4. Pay particular attention to your engine instruments during the warm-up period.
5. Make sure wheels are chocked when warming up on the apron.
6. Never taxi within 100 feet of aircraft on the line, or of buildings, without a crewman to guide you.
7. Don't rely too heavily on your brakes for making turns when taxiing.
8. Don't leave tools loose in cockpit. Don't allow extra lengths of wire to dangle. Fouled controls result from carelessness.

# THE LOAD ADJUSTER

The balance of an airplane is much more important than the total load. **Any pilot may refuse to fly any airplane if he is not satisfied with the loading or balance of the airplane.** A load adjuster is a device for computing balance without the aid of charts or graphs. It amounts to "test flying" the airplane on the ground without risk to crew or equipment.

Load adjusters vary for different model airplanes. Load adjusters for specific airplanes of the same model are alike except for the **INDEX. This index will vary with each individual airplane.**

You will find the load adjuster mounted in the pilot's compartment of the airplane. Check and see that the serial number on the case and on the airplane are the same. **If so, the index marked on the case may be used.**

The following illustrations show how balance of an A-20B was computed. All load not part of the basic airplane was computed with the load adjuster. **Don't leave out any part of the load, as every item of weight affects the balance.**

This is a sample problem:

| ITEM | SUB-TOTAL | TOTAL |
|---|---|---|
| Basic Airplane | | 15,935 |
| Gasoline: 394 U.S. gal (328 Imp. gal) | | 2,364 |
| Oil: 44 U. S. gal (36.6 Imp. gal) | | 330 |
| Bombs: | | |
| Wing Racks (2 ea 500 lb) | | 1,600 |
| Forward Bay | | 900 |
| Rear Bomb Bay | | 900 |
| Nose Compartment: | | |
| Bombardier | 200 | |
| Ammunition (1,000 rds .30 cal) | 65 | |
| Special Equipment | 50 | 315 |
| Pilot's Compartment: | | |
| Pilot | 200 | |
| Handbook Data | 25 | |
| Special Equipment | 50 | 275 |
| Nacelle Ammunition (2,000 rds .30 cal) | | 130 |
| Upper Gun Compartment: | | |
| Gunner | 200 | |
| Special Equipment | 200 | |
| Ammunition (1,000 rds .30 cal) | 65 | 465 |
| Lower Gun Section: | | |
| Special Equipment | 200 | |
| Ammunition (500 rds .30 cal) | 33 | 233 |
| GROSS WEIGHT | | 23,447 |

*To Find:* If the load distribution brings the airplane balance within permissible cg limits as indicated on the load adjuster "Loading Range" scale.

VIEW A—LOAD ADJUSTER AND CASE. CHECK TO SEE THAT THE SERIAL NUMBER ON THE CASE AND ON THE AIRPLANE ARE THE SAME.

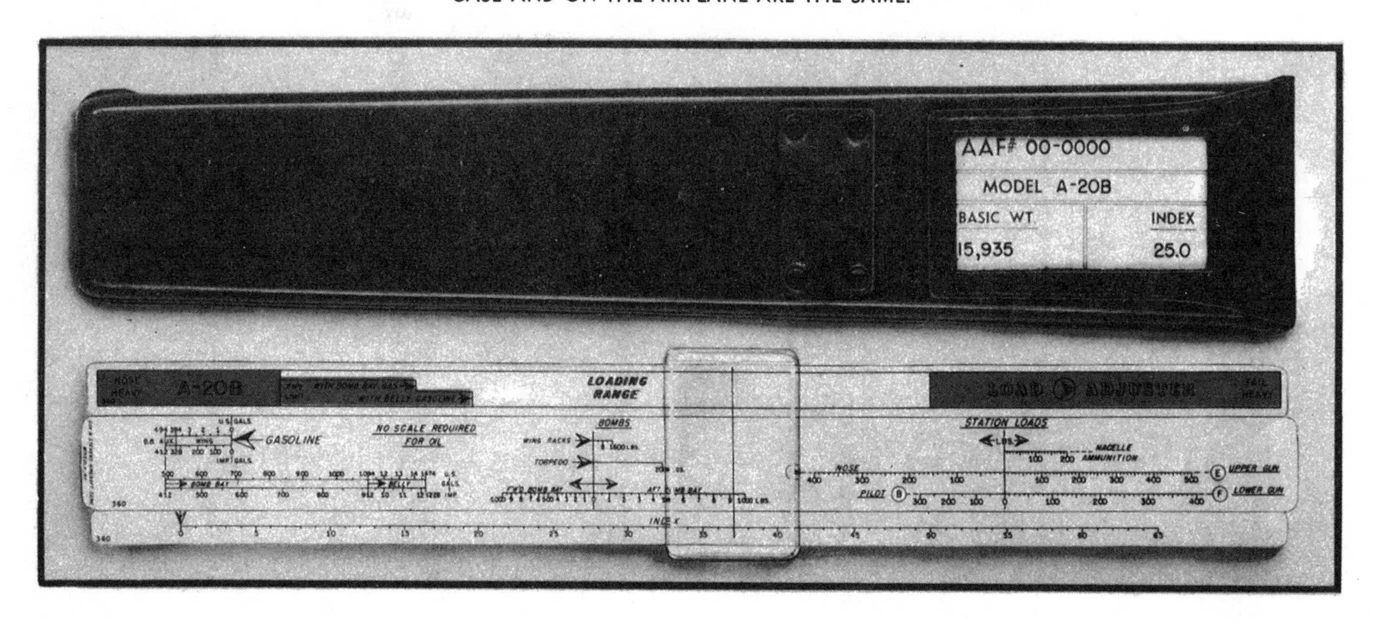

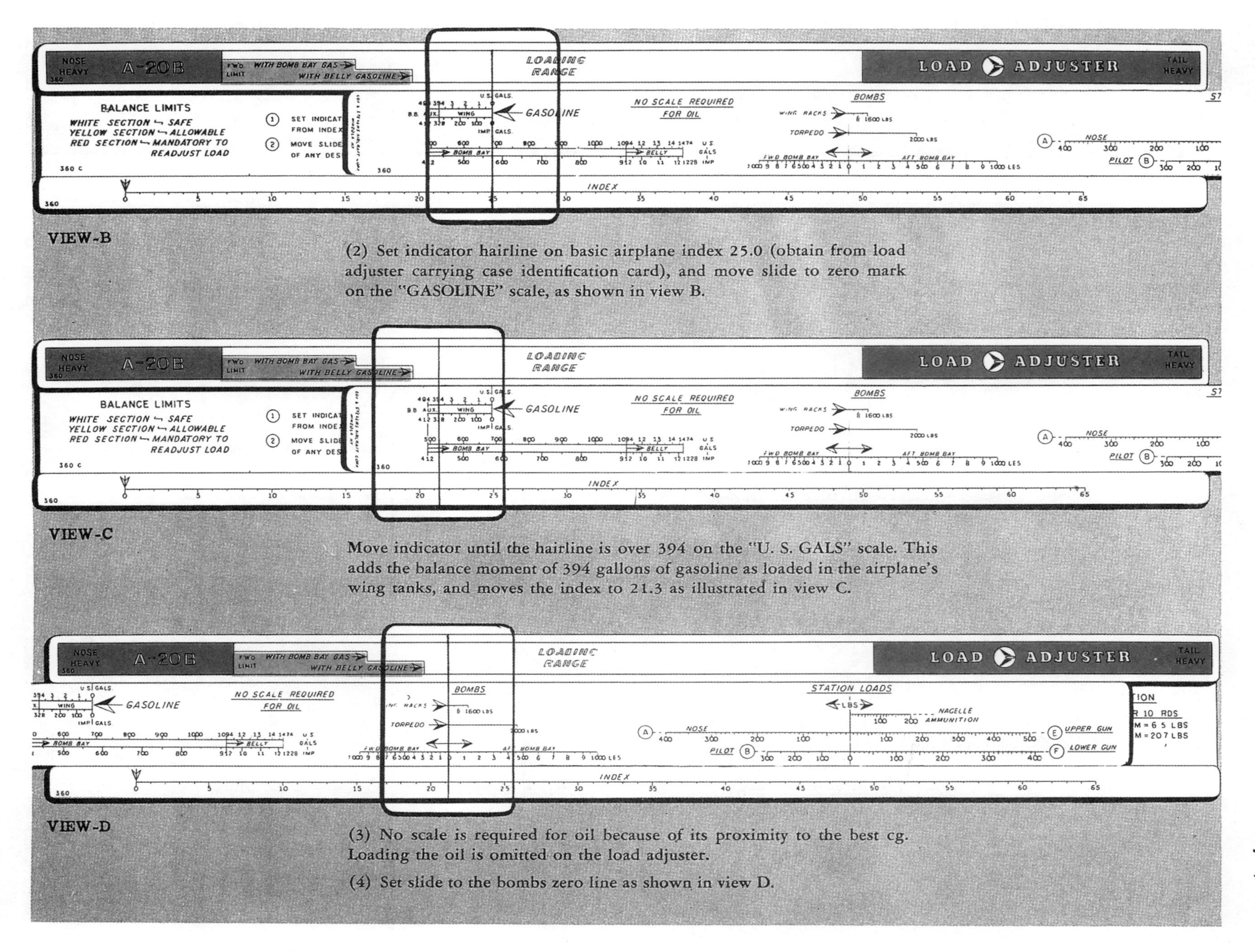


(2) Set indicator hairline on basic airplane index 25.0 (obtain from load adjuster carrying case identification card), and move slide to zero mark on the "GASOLINE" scale, as shown in view B.

Move indicator until the hairline is over 394 on the "U. S. GALS" scale. This adds the balance moment of 394 gallons of gasoline as loaded in the airplane's wing tanks, and moves the index to 21.3 as illustrated in view C.

(3) No scale is required for oil because of its proximity to the best cg. Loading the oil is omitted on the load adjuster.

(4) Set slide to the bombs zero line as shown in view D.

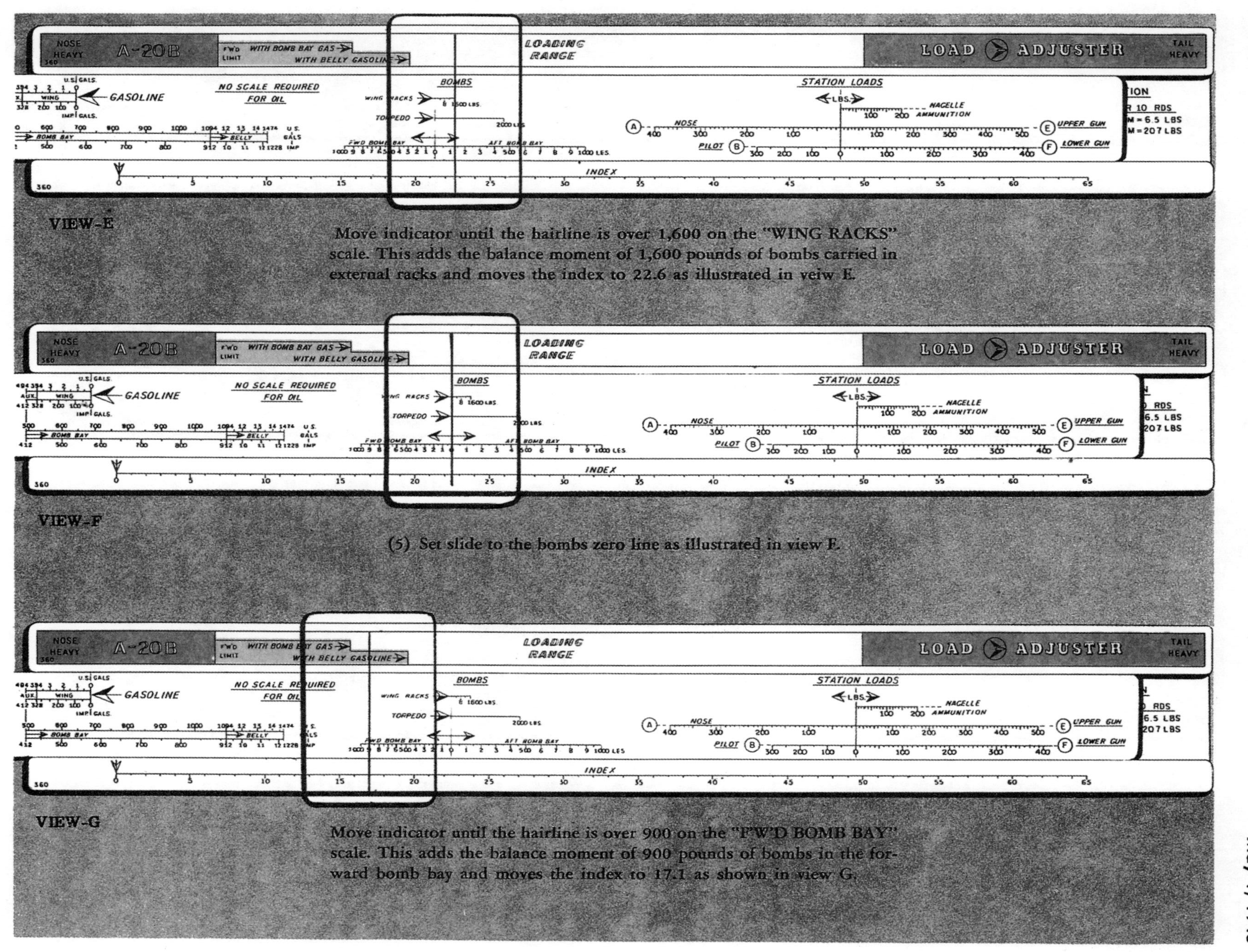


VIEW-E

Move indicator until the hairline is over 1,600 on the "WING RACKS" scale. This adds the balance moment of 1,600 pounds of bombs carried in external racks and moves the index to 22.6 as illustrated in veiw E.

VIEW-F

(5) Set slide to the bombs zero line as illustrated in view F.

VIEW-G

Move indicator until the hairline is over 900 on the "FW'D BOMB BAY" scale. This adds the balance moment of 900 pounds of bombs in the forward bomb bay and moves the index to 17.1 as shown in view G.

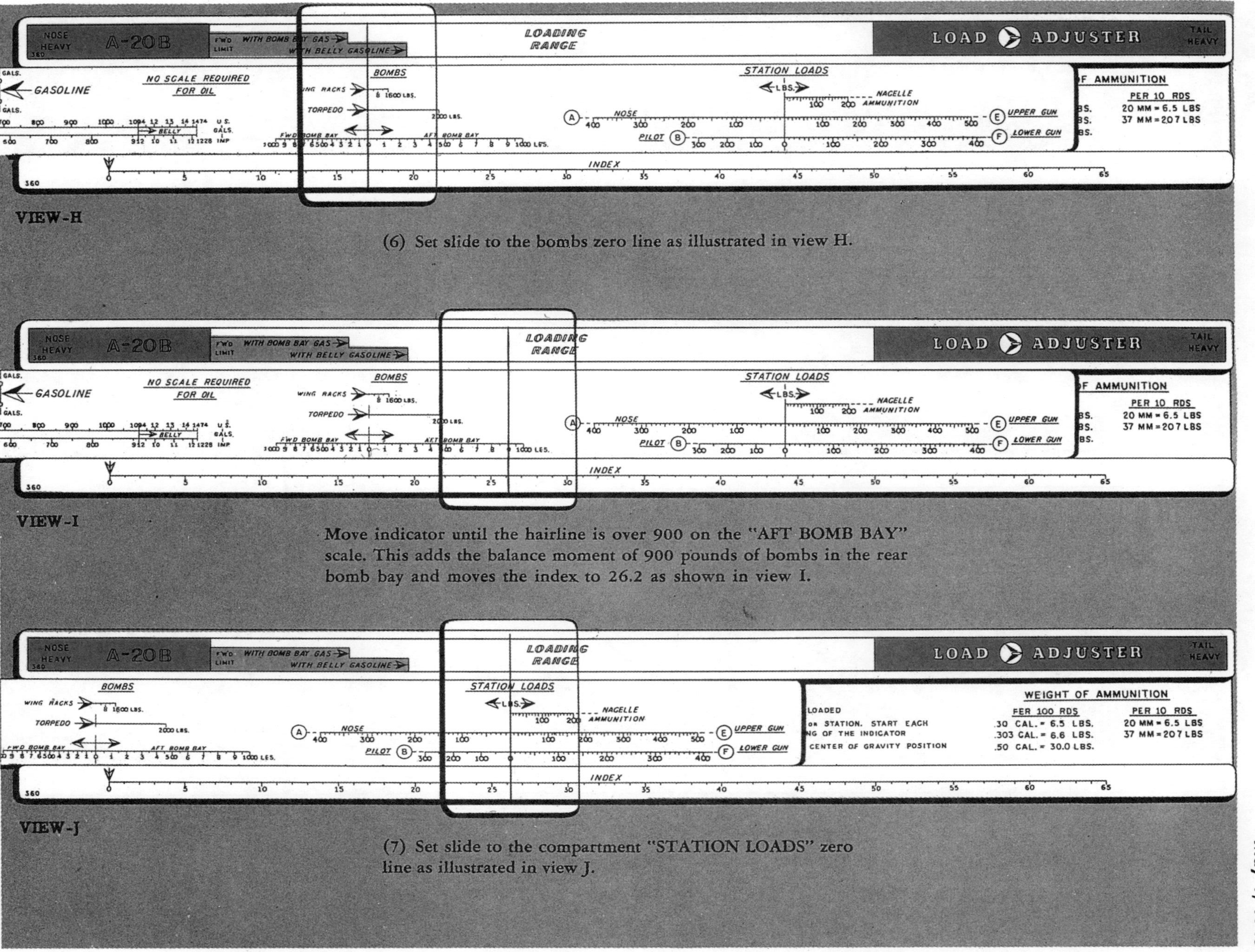


(6) Set slide to the bombs zero line as illustrated in view H.

Move indicator until the hairline is over 900 on the "AFT BOMB BAY" scale. This adds the balance moment of 900 pounds of bombs in the rear bomb bay and moves the index to 26.2 as shown in view I.

(7) Set slide to the compartment "STATION LOADS" zero line as illustrated in view J.

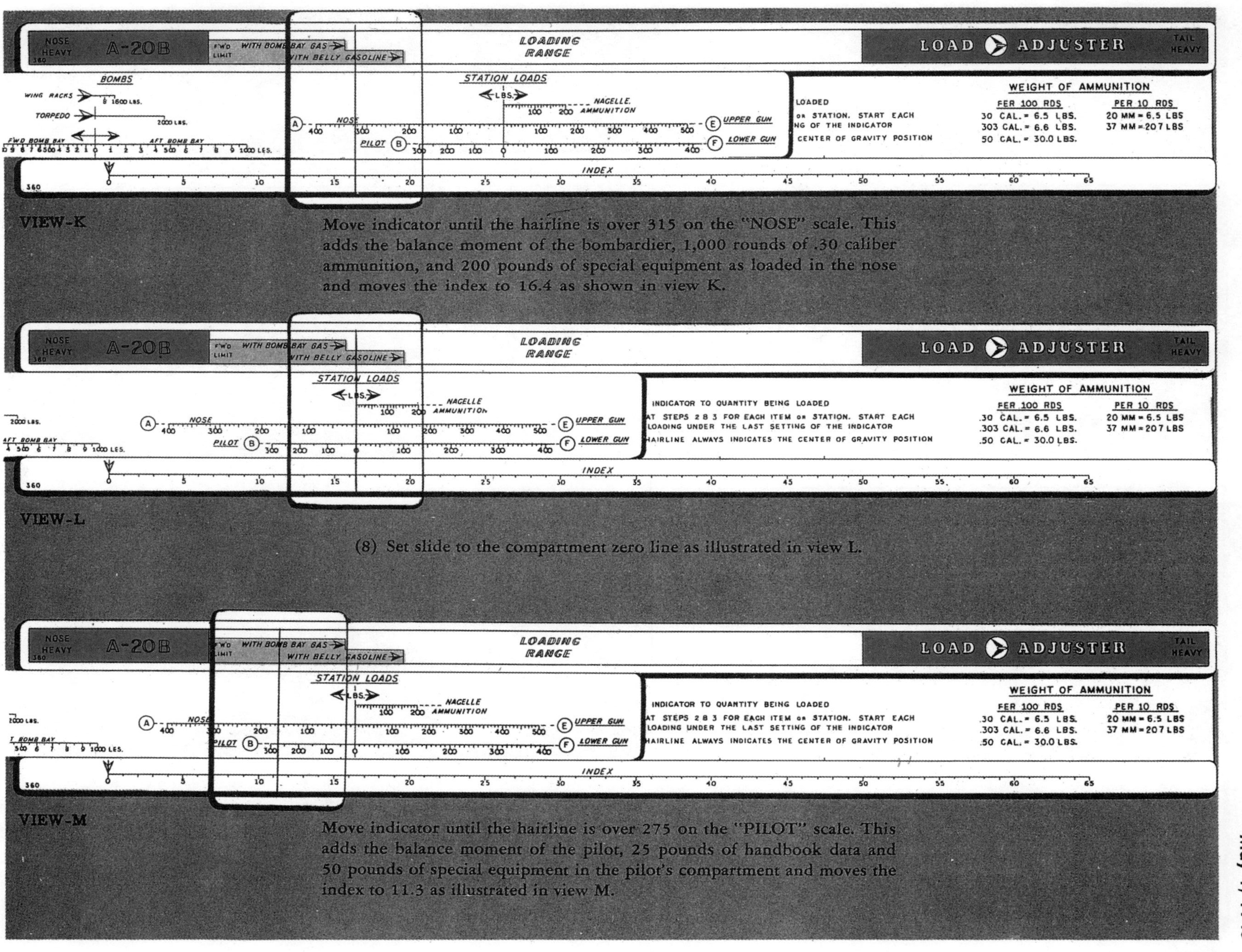


VIEW-K

Move indicator until the hairline is over 315 on the "NOSE" scale. This adds the balance moment of the bombardier, 1,000 rounds of .30 caliber ammunition, and 200 pounds of special equipment as loaded in the nose and moves the index to 16.4 as shown in view K.

VIEW-L

(8) Set slide to the compartment zero line as illustrated in view L.

VIEW-M

Move indicator until the hairline is over 275 on the "PILOT" scale. This adds the balance moment of the pilot, 25 pounds of handbook data and 50 pounds of special equipment in the pilot's compartment and moves the index to 11.3 as illustrated in view M.

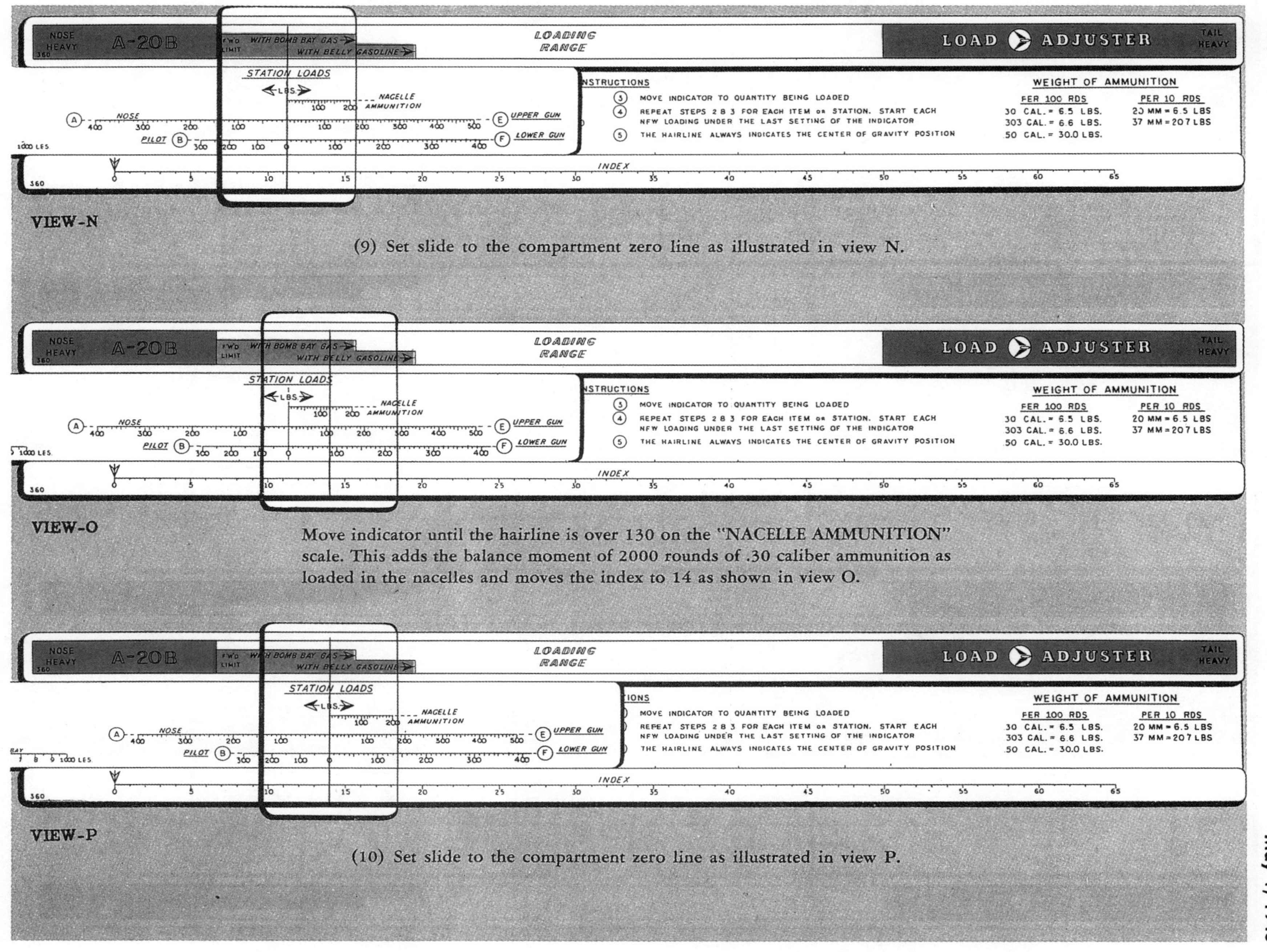


VIEW-N

(9) Set slide to the compartment zero line as illustrated in view N.

VIEW-O

Move indicator until the hairline is over 130 on the "NACELLE AMMUNITION" scale. This adds the balance moment of 2000 rounds of .30 caliber ammunition as loaded in the nacelles and moves the index to 14 as shown in view O.

VIEW-P

(10) Set slide to the compartment zero line as illustrated in view P.

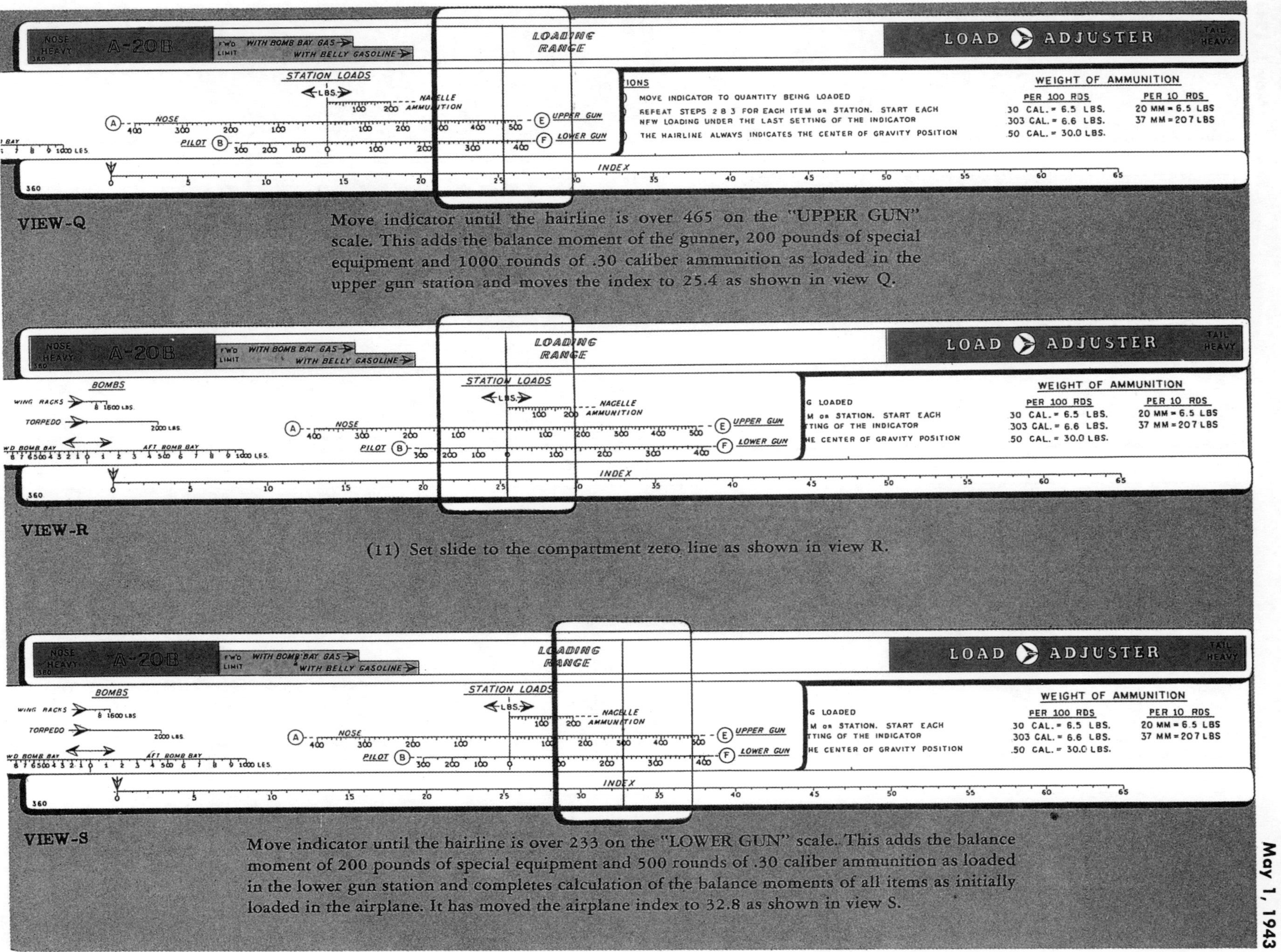


Move indicator until the hairline is over 465 on the "UPPER GUN" scale. This adds the balance moment of the gunner, 200 pounds of special equipment and 1000 rounds of .30 caliber ammunition as loaded in the upper gun station and moves the index to 25.4 as shown in view Q.

(11) Set slide to the compartment zero line as shown in view R.

Move indicator until the hairline is over 233 on the "LOWER GUN" scale. This adds the balance moment of 200 pounds of special equipment and 500 rounds of .30 caliber ammunition as loaded in the lower gun station and completes calculation of the balance moments of all items as initially loaded in the airplane. It has moved the airplane index to 32.8 as shown in view S.

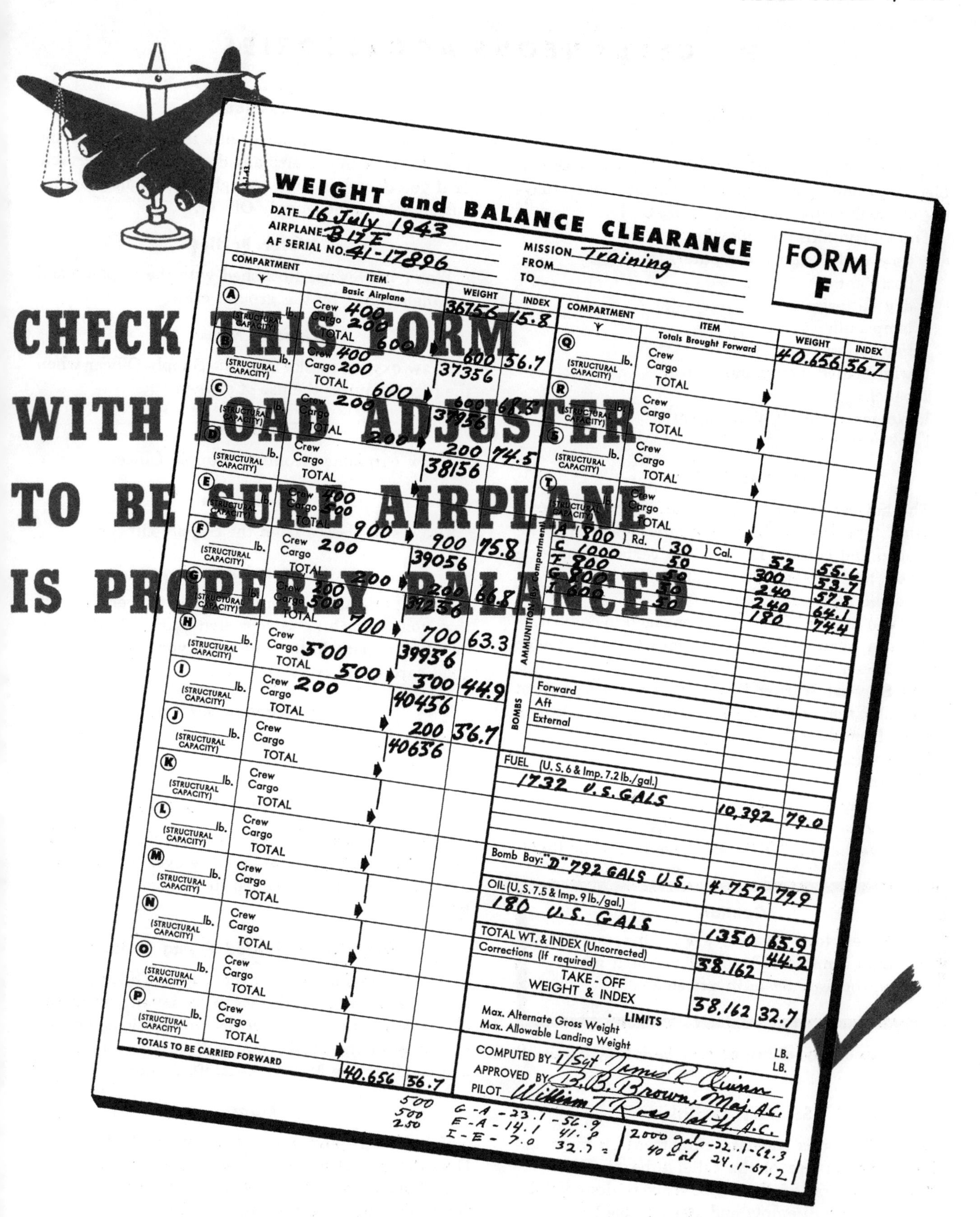


**WEIGHT and BALANCE CLEARANCE** — FORM F

DATE 16 July 1943
AIRPLANE B17F
AF SERIAL NO. 41-17896
MISSION Training
FROM
TO

| COMPARTMENT | ITEM | WEIGHT | INDEX |
|---|---|---|---|
| | Basic Airplane | 36756 | 15.8 |
| A (STRUCTURAL CAPACITY) lb. | Crew 400 Cargo 200 TOTAL 600 | 600 | 56.7 |
| B (STRUCTURAL CAPACITY) lb. | Crew 400 Cargo 200 TOTAL 600 | 37356 / 600 | 68.3 |
| C (STRUCTURAL CAPACITY) lb. | Crew 200 Cargo TOTAL 200 | 37956 / 200 | 74.5 |
| D (STRUCTURAL CAPACITY) lb. | Crew Cargo TOTAL | 38156 | |
| E (STRUCTURAL CAPACITY) lb. | Crew 400 Cargo 500 TOTAL 900 | 900 | 75.8 |
| F (STRUCTURAL CAPACITY) lb. | Crew Cargo 200 TOTAL 200 | 39056 / 200 | 66.8 |
| G (STRUCTURAL CAPACITY) lb. | Crew 200 Cargo 500 TOTAL 700 | 39256 / 700 | 63.3 |
| H (STRUCTURAL CAPACITY) lb. | Crew Cargo 500 TOTAL 500 | 39956 / 500 | 44.9 |
| I (STRUCTURAL CAPACITY) lb. | Crew 200 Cargo TOTAL | 40456 / 200 | 36.7 |
| J (STRUCTURAL CAPACITY) lb. | Crew Cargo TOTAL | 40656 | |
| K (STRUCTURAL CAPACITY) lb. | Crew Cargo TOTAL | | |
| L (STRUCTURAL CAPACITY) lb. | Crew Cargo TOTAL | | |
| M (STRUCTURAL CAPACITY) lb. | Crew Cargo TOTAL | | |
| N (STRUCTURAL CAPACITY) lb. | Crew Cargo TOTAL | | |
| O (STRUCTURAL CAPACITY) lb. | Crew Cargo TOTAL | | |
| P (STRUCTURAL CAPACITY) lb. | Crew Cargo TOTAL | | |
| TOTALS TO BE CARRIED FORWARD | | 40.656 | 56.7 |

| COMPARTMENT | ITEM | WEIGHT | INDEX |
|---|---|---|---|
| | Totals Brought Forward | 40.656 | 56.7 |
| Q (STRUCTURAL CAPACITY) lb. | Crew Cargo TOTAL | | |
| R (STRUCTURAL CAPACITY) lb. | Crew Cargo TOTAL | | |
| S (STRUCTURAL CAPACITY) lb. | Crew Cargo TOTAL | | |
| T (STRUCTURAL CAPACITY) lb. | Crew Cargo TOTAL | | |
| AMMUNITION (By Compartment) | A (800) Rd. (30) Cal. | | |
| | C 1000 50 | 52 | 55.6 |
| | F 800 50 | 300 | 53.7 |
| | G 800 50 | 240 | 57.8 |
| | I 600 50 | 240 | 64.1 |
| | | 180 | 74.4 |
| BOMBS | Forward | | |
| | Aft | | |
| | External | | |
| FUEL (U. S. 6 & Imp. 7.2 lb./gal.) | 1732 U.S. GALS | 10,392 | 79.0 |
| | Bomb Bay "D" 792 GALS U.S. | 4.752 | 79.9 |
| OIL (U.S. 7.5 & Imp. 9 lb./gal.) | 180 U.S. GALS | 1350 | 65.9 |
| TOTAL WT. & INDEX (Uncorrected) | | | 44.2 |
| Corrections (If required) | | 38.162 | |
| TAKE-OFF WEIGHT & INDEX | | 58.162 | 32.7 |

LIMITS
Max. Alternate Gross Weight LB.
Max. Allowable Landing Weight LB.

COMPUTED BY T/Sgt James R. Quinn
APPROVED BY B. B. Brown, Maj. A.C.
PILOT William T. Ross 1st Lt. A.C.

500 G-A-23.1-56.9
500 E-A-14.1 41.9
250 I-E-7.0 32.7
2000 gals-32.1-62.3
40 oil 24.1-67.2

# MISCELLANEOUS ACCESSORIES

## Navigation Equipment

A pilot's navigation kit may be procured, if available, through the Supply Officer. This kit is assembled to provide you with instruments for use in laying out and correcting flight courses.

The kit contains the following:

1 time and distance computer, type D-4
1 flashlight
4 dry batteries
1 lamp bulb
1 navigation case
1 white ruled memorandum pad
2 pencils
1 Weems, or equal, aircraft plotter
1 box of thumbtacks

## Heating Systems

**The hot air heater** is usually an exhaust or combustion type heater. In combat, the heater may be damaged and cause a fire or allow carbon monoxide to enter the fuselage with fatal results.

To avoid this:

1. Turn heater "OFF" when engaged in combat.
2. Turn heater "ON" after combat, if the heater has not been damaged.

**The Stewart Warner Heater** obtains its heat from the combustion of a fuel-air mixture and the electric current from the 24-volt system.

To operate:

1. Turn heater "ON" during warm-up.
2. Turn heater "OFF" before take-off.
3. After take-off turn heater "ON" again.
4. Leave heater "ON" until completion of flight.

**Caution**

Never turn heater "OFF" for more than 5 minutes during flight, or it may not operate again in extreme cold.

**A steam or glycol heater** may be left "ON" at all times.

## Sanitary Facilities

A relief tube is standard equipment on most airplanes. Report use of the relief tube on Form 1A after completion of the flight. This is done so the ground crew will empty and sterilize the tube.

## Cushions

Cushions are furnished with the airplane. If the ones that are in the airplane do not feel comfortable or if you do not sit high enough, draw some different cushions from the Supply Officer.

## Water Bottles

Water bottles are furnished with the airplane and are maintained by the ground crew.

## Oxygen Masks

Draw oxygen masks from the Supply Officer when the flight requires the use of oxygen.

## Earphones

Draw earphones from the Supply Officer.

## Sun Glasses

Draw sun glasses from the Supply Officer.

## Parachutes

Draw parachutes from the Supply Officer. Be sure that your parachute is repacked every 60 days. **Remember a parachute may save your life, so take good care of it.**

## Emergency Kit

Draw from the Supply Officer. These kits are fully explained in the Emergency Section.

## Parachute Type Life Raft

Draw the parachute life raft from the Supply Officer when the flight will be made over water. These life rafts are fully explained in the Emergency Section.

## Electrically Heated Flying Suits

Draw heated suits from the Supply Officer. These suits are designed to be worn as a unit, consisting of the suit, shoes, and gloves. Instructions for the operation of the heated suit will be found in "Heat and Cold" in the **Man Goes Aloft** Section.

## Maps

Draw maps and radio facility charts from the Operations Officer when the mission requires the use of maps. **Be sure they are up to date and give the radio range that you will require.**

# MISCELLANEOUS ACCESSORIES

## Navigation Equipment

A pilot's navigation kit may be procured, if available, through the Supply Officer. This kit is assembled to provide you with instruments for use in laying out and correcting flight courses.

The kit contains the following:

1 time and distance computer, type D-4
1 flashlight
4 dry batteries
1 lamp bulb
1 navigation case
1 white ruled memorandum pad
2 pencils
1 Weems, or equal, aircraft plotter
1 box of thumbtacks

## Heating Systems

**The hot air heater** is usually an exhaust or combustion type heater. In combat, the heater may be damaged and cause a fire or allow carbon monoxide to enter the fuselage with fatal results.

To avoid this:

1. Turn heater "OFF" when engaged in combat.
2. Turn heater "ON" after combat, if the heater has not been damaged.

**The Stewart Warner Heater** obtains its heat from the combustion of a fuel-air mixture and the electric current from the 24-volt system.

To operate:

1. Turn heater "ON" during warm-up.
2. Turn heater "OFF" before take-off.
3. After take-off turn heater "ON" again.
4. Leave heater "ON" until completion of flight.

**Caution**

Never turn heater "OFF" for more than 5 minutes during flight, or it may not operate again in extreme cold.

**A steam or glycol heater** may be left "ON" at all times.

## Sanitary Facilities

A relief tube is standard equipment on most airplanes. Report use of the relief tube on Form 1A after completion of the flight. This is done so the ground crew will empty and sterilize the tube.

## Cushions

Cushions are furnished with the airplane. If the ones that are in the airplane do not feel comfortable or if you do not sit high enough, draw some different cushions from the Supply Officer.

## Water Bottles

Water bottles are furnished with the airplane and are maintained by the ground crew.

## Oxygen Masks

Draw oxygen masks from the Supply Officer when the flight requires the use of oxygen.

## Earphones

Draw earphones from the Supply Officer.

## Sun Glasses

Draw sun glasses from the Supply Officer.

## Parachutes

Draw parachutes from the Supply Officer. Be sure that your parachute is repacked every 60 days. **Remember a parachute may save your life, so take good care of it.**

## Emergency Kit

Draw from the Supply Officer. These kits are fully explained in the Emergency Section.

## Parachute Type Life Raft

Draw the parachute life raft from the Supply Officer when the flight will be made over water. These life rafts are fully explained in the Emergency Section.

## Electrically Heated Flying Suits

Draw heated suits from the Supply Officer. These suits are designed to be worn as a unit, consisting of the suit, shoes, and gloves. Instructions for the operation of the heated suit will be found in "Heat and Cold" in the **Man Goes Aloft** Section.

## Maps

Draw maps and radio facility charts from the Operations Officer when the mission requires the use of maps. **Be sure they are up to date and give the radio range that you will require.**

# SECTION 7 ARMAMENT

# SAFETY PRECAUTIONS WHEN RELEASING BOMBS

To safeguard the airplane and occupants from possible premature detonation of bombs caused by one bomb striking the fuse of another, observe the following precautions:

1. Don't release bombs containing fuses other than types M103, M110, and M106 in "SALVO" when in the "ARMED" condition.

2. **Keep bomb release handle (type L-21A or other types that may be employed) in "SAFE" position at all times while in flight excepting the actual time of approach and bombing of target.**

## Fuses

The recently developed fuses have a long arming delay feature which provides arming times and corresponding distances the bomb will fall, as follows:

The long arming delay feature of the M103, M110, and M106 fuses affords adequate safety only within the fall distances and functioning time shown. Detonation of the bomb may occur as a result of one bomb striking the fuse of another at distances greater than those listed below.

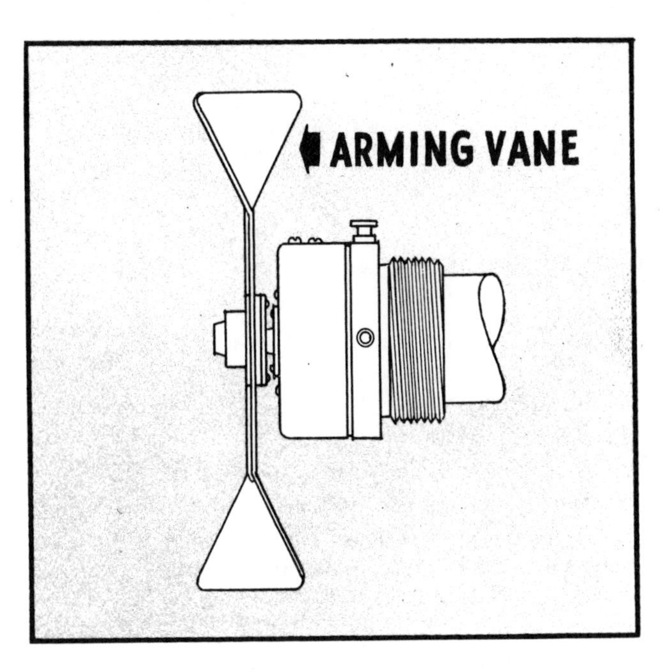


## "SALVO" Release

Since "SALVO" release is an inaccurate and undesirable method of releasing bombs and is **provided in airplanes only for emergency use,** the existent danger, though slight, does not justify indiscriminate dropping of bombs in "SALVO" and "ARMED." Therefore, salvo-armed release may be used at the airplane commander's discretion for bombs with types M103, M110, and M106 fuses only when no other system is possible.

REFERENCE: Technical Order 11-25-3.

| Bomb Fuse | Plane Speed (mph) | Arming Time (seconds) | Corresponding Fall Distance (feet) |
|---|---|---|---|
| M103 (nose) | 150 | 6.3 | 640 |
| | 300 | | * about 320 |
| M110 (nose) | 300 | 5.0 | 400 |
| M106 (tail) | Arms upon release, but will not function until 45 seconds after impact. | | |

* Actual test figures not available.

# MINIMUM ALTITUDE OF RELEASE AND PREARMING OF FUSES

The safe dropping altitudes shown in chart on the following page are known to safeguard both airplane and personnel from exploding bombs. Use them in all practice bombing.

On combat missions there is no definite line between safe and dangerous operation. **Bombing from lower altitudes within the danger zone is justified only when great tactical advantage is gained.**

### Arming of Fuses

Bomb fuses are armed by a rotating arming vane which is turned by air in the manner of a propeller. Therefore, a certain distance of air travel is required to turn the arming vane the preset number of revolutions necessary to arm the fuse.

Normal arming of bomb fuses, M103 (nose), and M100, M101, M102, M100A1, M101A1, M102A1, AN-M100A1, AN-M101A1, AN-M102A1 (tail), requires from 1,000 to 3,500 feet of air travel. Dive bombing and low altitude bombing over water have produced bomb failures because the bomb fuses were not completely armed at time of impact. When used for low altitude water impact bombing, **the above fuses will be partially armed only before dropping.**

Soft, normal, or frozen ground, water, and other objects which bombs can be expected to pierce will be considered as penetrable targets. Concrete roadways, concrete bridges, very rocky terrain, and similar objects will be considered as impenetrable targets. For mast head bombing at the altitudes shown, impact upon the target must be obtained during the first bounce of ricochet flight.

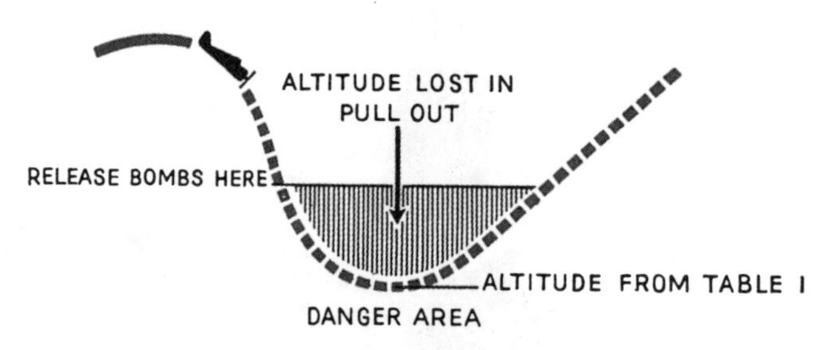


### Dive Bombing

For safety from bomb fragments in dive bombing, the altitude of release must be considerably greater than that required in horizontal bombing because of the altitude lost in pull-out. Therefore, **in dive bombing, release bombs at an altitude great enough that the airplane will not pass below the applicable altitude listed in table during pull-out.**

# SAFE ALTITUDES OF RELEASE FOR HORIZONTAL BOMBING

| Type or Size of Bomb | Nature of Target | Nose Fuse Setting | | Tail Fuse Setting | | | | Hydro-static Fuse |
|---|---|---|---|---|---|---|---|---|
| | | Instant | .1 sec. Delay | Non-delay or .01 sec. | .025 to .1 sec. | 3-5 sec. | 11-45 sec. | |
| Fragmentation Frag. Cluster | All | 500 (1) | | | | | | |
| Parachute Frag. and Cluster | All | 100 (2) | | | | | | |
| | Penetrable | 1500 (3) | 650 (3) | 1000 (6, 7) | 650 (6, 7, 13) | 400 (10, 11) | 100 (9, 11, 8) | 100 (14, 15) |
| 100 to 1000 lb. Demolition | Impenetrable | 1500 (3) | 1500 (3) | 1000 (6, 7) | 1000 (6, 7, 13) | 650 (10, 11) | 100 (9, 11, 8) | |
| | Mast Head Bombing | | | | | 100 (10, 11) | | |
| | Penetrable | 2000 (3) | 650 (3) | 1500 (7) | 650 (7, 13) | 400 (10, 11) | 100 (9, 11, 8) | 100 (15) |
| 2000 lbs. or less | Impenetrable | 2000 (3) | 2000 (3) | 1500 (7) | 1000 (7, 13) | 650 (10, 11) | 100 (9, 11, 8) | |
| | Mast Head Bombing | | | | | 100 (10, 11) | | |
| | Penetrable | 2500 (3. | 650 (3) | 2000 (7) | 650 (7) | 400 (10, 11) | 100 (9, 11, 8) | |
| Over 2000 lbs. | Impenetrable | 2500 (3) | | 2000 (7) | | | | |
| | Mast Head Bombing | | | | | | | |
| 325 lb. Depth Bomb | All | 750 (4) | | | | | | 100 (12) |
| Incendiary Cluster 4 lb. | All | 500* | | | | | | |
| Incendiary Cluster 2 lb. and 6 lb. | All | 1000* | | | | | | |
| Chem. Bomb M47 Series | All | 100 (5) | | | | | | |

*Distance of fall required to stabilize bombs. Lower altitude may be used if necessary but will cause an increasing number of duds.

(1) M110 Fuse
(2) M104
(3) M103, AN-M103
(4) AN-Mk19
(5) M108
(6) M100 M101, M100A1, M101A1, AN-M100A1, AN-M101A1, AN-M101A2, AN-M101A2.
(7) M102, M102A1, AN-M102A1, AN-M102A2.
(8) M-106
(9) M106A1
(10) M106A2
(11) M112, M113, M774.
(12) AN-Mk24
(13) AN-Mk28
(14) AN-Mk29
(15) AN-Mk30

# MACHINE GUNS AND CANNON

### Before Take-Off

For combat missions check with your armorer:

1. To make sure that your guns have been cleaned and inspected and are ready to fire.
2. That proper amount of ammunition is provided.
3. That guns are charged. This must be done on the ground because most installations do not provide for charging while in flight.

### After Take-Off

When safely off the ground on combat missions, turn safety switches "ON." **In the excitement of combat you might forget to turn them on.**

**Before landing, turn safety switches "OFF."**

### Firing Machine Guns

Do not fire a burst of more than 75 rounds maximum from a .50- or .30-caliber machine gun. Approximately 1 minute after firing a 75-round burst, firing may be resumed and 20 rounds fired and repeated each minute thereafter.

For synchronized guns where the heat from the engine preheats the gun, limit the initial burst to 50 rounds, and after approximately 1 minute, firing may be resumed and 15 to 20 rounds fired and repeated each minute thereafter.

The initial long burst will heat the barrel to the maximum permissible temperature, and repeated firing after a minute delay with a reduced number of rounds per minute will maintain the barrel at the high temperature. Thus the initial burst of 50 to 75 rounds or a 50- to 75-round burst followed by firing 20 rounds for each succeeding minute requires a cooling time or cessation of fire for approximately 15 minutes before the long burst can be repeated.

If long bursts are not fired, approximately 25 rounds may be fired each minute over long periods.

### Firing Cannon

Do not exceed a maximum of 20-round bursts when firing 20-mm cannon, and limit your average burst to 10 rounds.

Do not exceed a maximum of 15-round bursts when firing 37-mm cannon, and limit your average burst to 5 rounds.

### General

Machine guns and cannon must be properly maintained.

Before and after each combat flight, guns will be thoroughly cleaned and inspected.

If on flights other than combat missions, guns will be protected with oil preservative.

REFERENCE: Technical Manual 9-225, dated April 30, 1942

# GUN SIGHTS

## General

A machine gun in an airplane is fired normally at an indicated air speed of more than 150 miles per hour at a target which usually is moving at a high speed. Conditions and time do not permit firing long bursts with the hope of spraying the target sufficiently to insure a vital hit.

**Be ready to take immediate advantage of the few seconds available during which the target is exposed** to deliver accurate, effective bursts.

Aircraft machine gun sights are designed to enable firing with accuracy under all normal combat conditions. To use the sights effectively, **have a thorough knowledge of the proper use of the various types of sights.**

## Theory of Lead

To understand clearly the theory of lead it must be remembered that to obtain a hit, the bullet and the enemy airplane must arrive at the same point simultaneously. It takes a certain period of time for the bullet to travel the distance between the two airplanes. The center line of the bore of the gun, if extended, should not pass through the target, but should strike at some point along the apparent line of flight of the target at an appreciable amount in front of the enemy airplane.

The radius of the larger ring of a sight is such that the angle between the edge and center of the ring is equal to a definite lead angle. In the case of ring and bead sights the gunner's eye must be a definite dis-

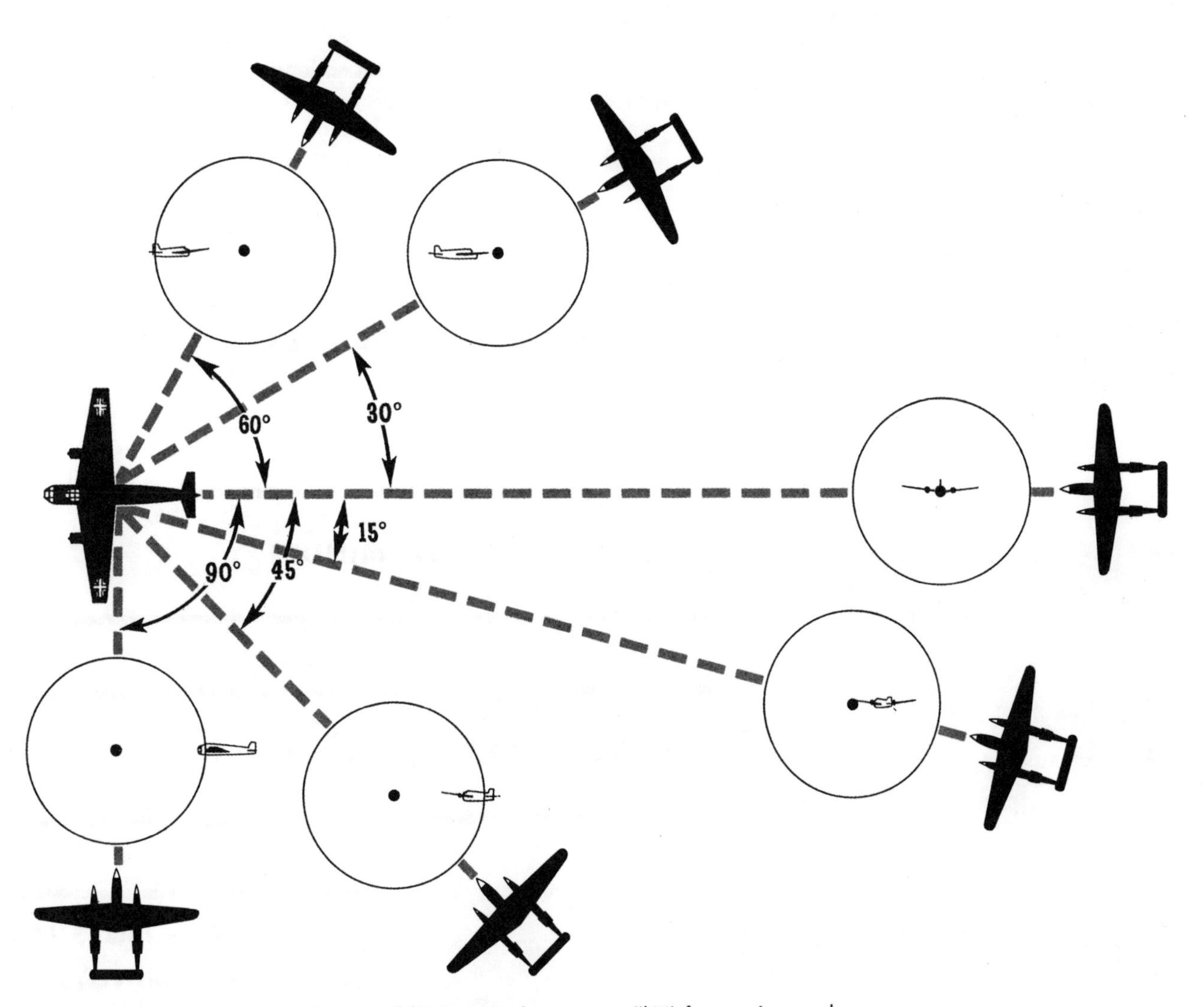


Amount of lead required to score a "hit" from various angles.

tance (usually 8 inches) from the ring. This angle may be expressed in terms of mils, or in terms of miles per hour of relative speed perpendicular to the line of sight. That is, a 150-mile-per-hour ring establishes the proper lead angle at the gunner's eye when the target is traveling at a relative speed of 150 miles per hour perpendicular to the line of sight. This is exactly true for one particular range only, usually 400 yards. However, if the relative speed perpendicular to the line of sight is constant, then the change in the lead angle is small as the range changes.

Lead required when target approaches from various angles.

### Reflex Sights

The basic principle of all optical sights is the projection of the image of a reticle of some form out to infinity so that the gunner can aim the gun by moving the sight until the reticle appears to be superimposed on the target. The reticle size is computed so as to be an aid in range estimation. With this sight it is not necessary for the gunner to hold his eye very steadily at a fixed distance from the sight while lining up a pair of reference points with the target. When the distance between the gunner's eye and the sight is too great, it is possible that part of the reticle will not be seen. This is purely a function of the size of the reticle and the physical diameter of the lens system used and does not affect the use of the sight so long as the reticle can be seen. When the complete reticle can be seen, it can be used to estimate lead or range.

**NOTE: Check your sight lamp before take-off.**

REFERENCE: Technical Order No. 11-35-1, dated September 18, 1942.

## DISTANCE TRAVELED BY TARGET AIRCRAFT BETWEEN SUCCEEDING PROJECTILES

(Armor Piercing Ammunition – 90° Deflection Shooting)
Computation based on 300 mph speed of target

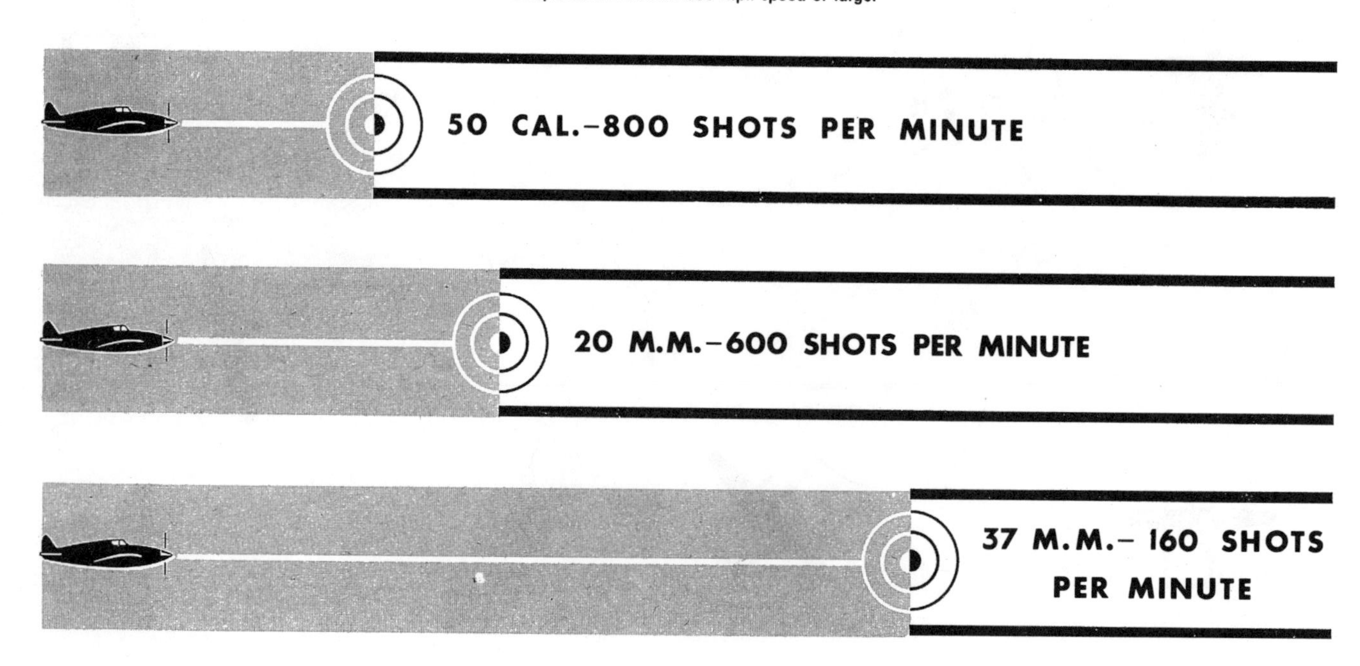

# SECTION 8 EMERGENCIES

# TO BAIL OR NOT TO BAIL

**If you lose an engine** in a heavily loaded multi-engine airplane, you may be able to continue under power to a safe landing place either by dumping your bombs and auxiliary tanks or by throwing cargo overboard.

If you are obviously too short of fuel to reach a safe landing, make a forced landing before your engines go dead. Choose a favorable spot and make a normal landing or a belly landing as circumstances indicate. **Remember,** a crash landing with the airplane under full control is much safer than an out-of-control crack-up, so **maintain adequate air speed for control.**

When caught in or above an overcast under conditions necessitating an immediate forced landing, bail out unless you are certain of at least a thousand-foot ceiling under the overcast and good terrain for landing.

If trouble develops over the jungle or desert or in the Arctic, crash land your plane if it is at all possible. The outline of your plane will help searching parties to locate you. It will also provide shelter and emergency equipment, and the remaining gasoline and oil can be drained and burned to provide warmth and smoke or flame for signalling. If it becomes necessary for you to trek out on foot, your plane will furnish materials for improvised signals, knives, tents, clothing, fish spears, and navigation instruments.

If you do bail out, DON'T FORGET YOUR MAPS, emergency rations, and emergency water cans. Tuck them in your pockets and inside your jacket.

### Remember

When leaving on any flight that will take you over the jungle, the desert, the Arctic, or any rugged country WEAR SHOES THAT YOU CAN WALK HOME IN, and wear clothes for the climate.

In case of a crash landing, any landing that you can walk away from is a good one. However, many forced landings are made in which little or no damage occurs to airplane or crew. They are the result of forethought, calm execution, and adherence to a few fundamental principles. The following suggestions may help you. Think them over. Plan in advance for that day when you are confronted with a forced landing.

1. Stay calm. This is the primary rule for any emergency.

2. Decide quickly what must be done and use all your skill in accomplishing the landing.

3. Dump bombs in "safe," fuel, other cargo or equipment that may endanger a successful landing, but be sure that open ground is beneath you. Be sure that no loose articles will be thrown about in the plane by the impact.

4. Warn crew members of the impending crash in order that they may assume positions which will result in the least personal injury.

5. Keep flying speed until the plane is on the ground. If you slow down while high in the air, you may stall or spin in.

6. Land as nearly up wind as possible, never over 90° from the wind.

7. Don't attempt turns near the ground; a stall usually results.

8. In case of forced landing immediately after take-off, land straight ahead. Turning back to the field has killed many good pilots.

9. Keep landing gear retracted unless the terrain is such that a lowered landing gear will not cause a nose-over. Don't fear a belly landing; it can be made with comparatively little damage.

10. Use flaps so that the plane will glide and land in the manner to which you are accustomed. Flaps also reduce your actual touching down speed.

11. If any power is available, use it to level the plane before striking.

12. Just before the crash, turn off the ignition switch and turn off the airplane master switch to avoid fire.

13. The main factor in escaping unhurt from a crash landing is the rate of stopping. For instance: if a solid obstruction is ahead, groundloop your plane to kill the speed and to get the wing in a position where it can absorb the blow.

14. Your position during a forced landing is important. Take the brunt of the crash through the thickness of your body rather than the length. If possible, just before crashing cushion your head in the crook of your arm and lean forward against a part of the plane towards the direction of travel. Make sure that your head is not free to lurch forward. **Brace yourself with a crash, not against it.** Never brace yourself with legs or arms rigidly extended. The bones are strong and you may be speared by your own skeleton.

## After a Forced Landing in Uninhabited Territory

1. Stick with your plane; the chance of being spotted is better.

2. Often you can attract attention with smoke. If nothing else is available, pour oil on rags and make a smudge of them.

3. Set out panels of metal or cloth so they can be noticed from the air. Make the designs big enough to be seen.

4. Make a shelter from airplane parts, or a tent from your parachute.

5. An excellent pack sack can be made from your parachute harness and shroud lines.

6. Inner tubes of tires may be removed and taken along to use as a raft if a collapsible boat is not provided.

7. When absolutely sure that the search for you has been given up, select the things that you will need and start out on a definite course.

8. Take the compass along with you.

9. If there are any containers, take along some motor oil for use as fuel for cooking or as a lamp or signal.

10. Purify water before drinking by boiling three minutes or with chlorine tablets.

11. Take along your Very pistol for signalling if you sight a plane or ship.

12. If lost, follow a stream downstream or follow the coast; you have a better chance of reaching habitation.

13. Travel slowly to avoid exhaustion.

**DON'T LOSE YOUR HEAD**

## Before take-off

You may be forced down at sea and now is the time to start preparing for such an emergency. Here is how to do it:

1. Know the proper procedure for ditching a land plane on water.
2. Know the use of emergency equipment provided for ditching purposes.

As pilot, it is your responsibility to make certain that your crew also has this knowledge. The best way to acquire it is through ground practice drills.

Ditching and dinghy drills will familiarize all personnel with the duties to be performed when the order "prepare for ditching" is given. If these drills are mastered well enough so that each crew member can carry them out in a darkened plane under unfavorable conditions, his education is at least partially completed.

However, before taking off on a long over-water mission, there are several additional points of importance to be considered. They are:

1. Be sure all emergency equipment functions properly and that you have all you may need in the airplane. Pay particular attention to $CO_2$ cartridges on the life raft and Mae West, water containers, Very pistol, emergency radio, and other signal devices.
2. Test escape hatches. Know—don't hope—that they will operate if the need arises.
3. Recheck life vest adjustments. If your vest does not fit you properly, waist and leg straps may bind when the vest is inflated in an emergency.
4. Drink as much water as you can hold comfortable. Contrary to popular belief, the body can store water. (During flight, fill up from water store which cannot be taken aboard the raft. This applies to food also.)

If these tips are followed, you and your crew will be prepared in advance if the plane must be ditched.

## Before ditching

At the first indication of trouble, it is your duty as pilot to notify the radio operator to transmit distress calls. They can be cancelled later if ditching becomes unnecessary.

Do not delay too long when it becomes evident that the plane must be landed on the water. Be certain there is enough gasoline left to make the water landing under power. Also, as soon as possible, give the order: "Prepare for ditching."

If the crew has been properly drilled, each man will respond to the order as practiced—"Co-pilot ditching" . . . "navigator ditching" . . . etc.—until all crew members have acknowledged the order.

Immediately, they will loosen shirt collars and remove ties to prevent strangulation. Oxygen masks will also be taken off, unless the plane is above 12,000 feet. In that event, the main oxygen supply or emergency oxygen bottle will be used until you give further instructions. Heavy flying boots, but not clothing, should likewise be removed.

This done, crew members will start jettisoning all loose equipment, guns, ammunition, etc. Upper hatches will be released to facilitate exit upon landing. Extra gasoline, not needed for a power landing, will also be jettisoned.

Bombs and depth charges should be salvoed if the plane has sufficient altitude. Bombs and depth charges, if not dumped, **must be placed on safe.**

All lower hatches and bomb doors must be securely fastened and landing gear retracted.

### Navigator and Radio Operator

Upon receipt of the ditching order, the navigator calculates the plane's position, course and speed. He passes this information to the radio operator. Then he destroys secret papers.

The radio operator tunes the liaison transmitter to MFDF and sends SOS, position, and call sign continuously.

He also turns IFF to distress and remains on intercommunication, clamping down the key on the pilot's order to "Take ditching posts."

### Pilot and Co-pilot

The pilot and co-pilot open the escape hatch through which they will exit. They make sure that safety belts and shoulder harness are secured.

The pilot also keeps up a running account of the ditching. Crewmen are either too busy or unable to observe what is taking place outside the plane.

### Crew

The duties of crew members obviously vary with the model of plane being flown. After performing such duties each man goes to, or remains at, his ditching post.

Equipment needed in the life raft is gathered and carried to the ditching posts by the personnel, and secured against the impact. Parachute pads, seat cushions, etc., are used to protect the face, head, and back against the shock of landing. When this is done the crew awaits the order to "brace for ditching." This order should be given by the pilot five seconds before the first impact.

### Landing Impacts

Regardless of the plane, two impacts will be felt—the first a mild jolt when the tail drags the water, the second a severe shock when the nose strikes. Crewmen will hold crash positions until the plane comes to rest.

Life vests should not be inflated inside the plane unless crew members are certain that escape hatches through which they will exit are large enough to accommodate both themselves and the inflated vests.

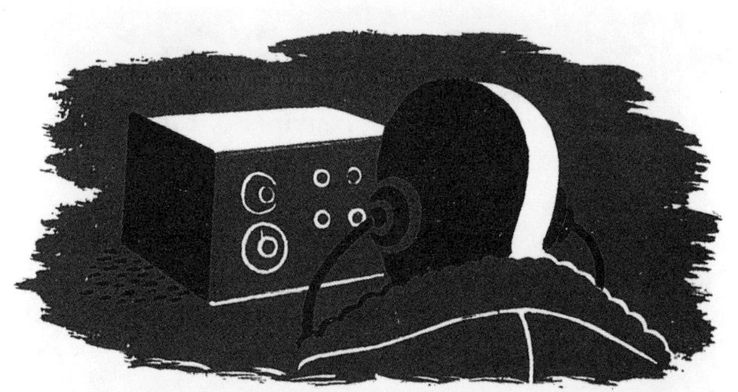

The radio operator tunes liaison transmitter to MFDF and sends SOS, position, and call sign continuously. He also turns IFF to distress and remains on intercommunication, clamping down key on pilot's order to "take ditching post." He relays information given him by navigator.

All loose equipment is jettisoned. Upper hatches are released to facilitate exit. Bombs and depth charges are salvoed if the plane has sufficient altitude; if not dumped, they are placed on SAFE.

Emergency equipment should be taken to the raft. Water is most important, then signal equipment. Remember to take the Very pistol, cartridges, signal mirror, emergency radio and sea marker. These will help in your rescue.

# Ditching the plane

The sea is a tricky landing strip. So it is important to determine the wind's **direction** and **velocity** by observing the surface of the water.

On a glassy sea it will be difficult to judge altitude on the approach. If the radio operator reels out the trailing antenna, the output meter will tell him when the antenna touches the water.

### Wind Direction

Wave formations determine wind direction.

Ripples form in a light breeze. They are bow shaped, varying in size from two inches to two feet. They move downwind.

As breeze freshens, small parallel waves build up with smaller ripples along tops of waves. If wind shifts, ripples form at an angle to parallel crests, indicating true wind direction.

As wind velocity increases, bubbles and foam will form. They will be blown to leeward and wind streaks will appear as straight lines parallel to the direction of the wind.

At still higher wind velocity, the waves will build up until small, white, foamy crests appear on them. The foam will stream out on the windward side of the wave.

Watch the surface carefully when you see these indications of sudden shifts in the wind:

1. A sudden calm or glassy surface.
2. The approach of a squall or a sudden summer shower. A rapid wind shift of as much as 180 degrees may be expected.
3. Curved wind streaks. Long, straight streaks indicate a steady wind.
4. A distant line appearing on the surface, as if caused by a rip tide.

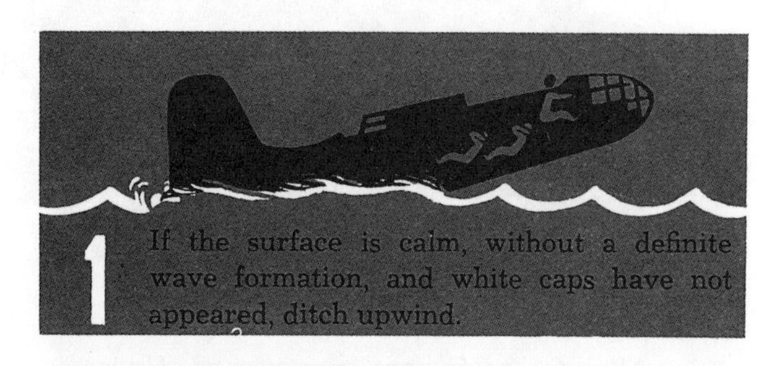

### Approach

If possible, maintain intercommunication with the crew until the last moment. Then warn them soon enough of the impending impact.

Lower flaps to reduce speed of approach and touch-down. Flap setting depends upon the type of aircraft flown. A steep nose-down approach is dangerous.

If your engines are functioning properly, use a normal approach glide to insure full control. Maintain a margin of excess speed after the round-out to allow you to choose the best spot for ditching. Hold off until you reach stalling speed, then set down tail first.

### Use of Engines

If one engine is available, a little power may be used to flatten the approach. But do not use so much power that the plane cannot be turned against the engine right down to the stall. Always have a margin of rudder control. **On no account should power be increased during the final stages of ditching.**

If two engines are available on one side only, the inner engine should be used.

If, for example, the inner left and outer right engines are available, it will be possible to use considerable power by adjusting the throttles so that little rudder is required.

The value of power in ditching is so great that the pilot should always ditch before fuel is quite exhausted.

Don't jump into the raft. Beware of jagged wing surfaces when launching it.

## DITCHING THE B-17

Speed in abandoning the B-17 is important, since the airplane will remain afloat for only a short time. However, equipment, which will be needed for survival in the open raft—especially water and signaling equipment—must not be forgotten.

After the plane has been successfully ditched and the crew abandons it, both rafts are automatically inflated by the bell turret gunner who pulls the dinghy release. The tail gunner, seated in the left raft, is first to exit from the radio compartment hatch. He secures equipment. The navigator, second to exit, is shown at the side of the raft after he has stowed the emergency radio. The radio operator, shown holding a line attached to the left dinghy, is fifth to exit from the radio compartment. The right waist gunner is the third man to leave the plane. He assists in launching the right dinghy. The flight engineer, fourth out, is shown sliding off of the fuselage onto the right wing. The ball turret gunner, seventh man out, is shown astride the fuselage receiving emergency equipment from the bombardier, who is last to leave the radio compartment. The pilot exits from the left window of the pilot's compartment and takes command of the left dinghy. The co-pilot leaves from the opposite window and takes command of the right dinghy.

### Abandoning the Plane

Dinghies should not be released until the plane has come to rest.

Crew members will exit in the order practiced. However, injured men will receive first consideration and assistance. Speed is important, but life raft equipment should be taken aboard.

If ditching occurs at night, landing lights may be turned on, providing the reflection does not confuse the pilot's vision. Bright lamps within the plane should be turned off. They may be snapped on after landing to guide nearby rescue parties.

Use care in launching the life raft to prevent punctures on ragged or sharp parts of the fuselage or wing surfaces.

If there are two dinghies, they should be tied together and should remain in the area until the plane has sunk. The outline of the aircraft will aid rescue planes.

Some of the equipment needed will be stored in the raft; the remainder must be hauled aboard and securely fastened. Check to insure the presence of all necessary items—ration kits, emergency radio and signalling equipment, etc. Parachutes will come in handy as cover from the sun and sails. Protection against hot tropical sun is important since it causes bad burns and dehydrates the body.

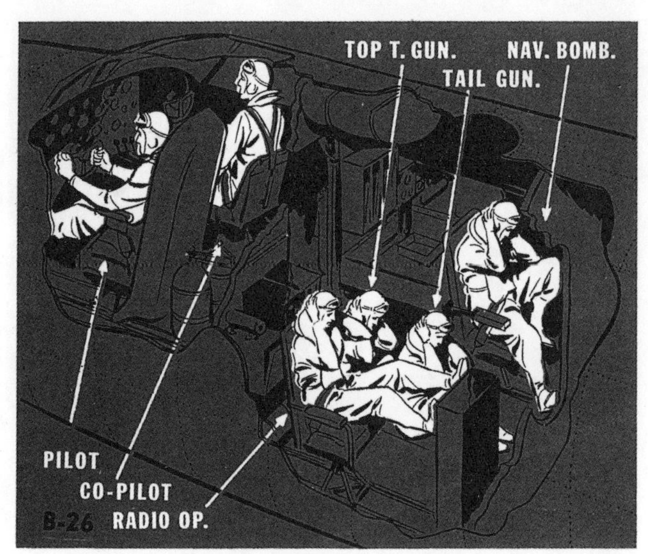


Ditching Positions in the B-26

**B-26.** Co-pilot pushes his seat to rear, allowing navigator-bombardier to exit from nose. Top turret gunner is braced against step leading to cockpit. His knees are drawn up, elbows close together to form a back support for tail gunner. All remain in position shown until plane comes to rest.

**B-17.** Navigator is braced against closed door, knees drawn up, hands on the back of bombardier's head. Bombardier braced against navigator's legs, holds head of man in front of him. Radio operator sits well back in seat, feet braced against radio equipment. Knees of all personnel in the supine position are slightly bent. All hold positions until plane comes to rest.

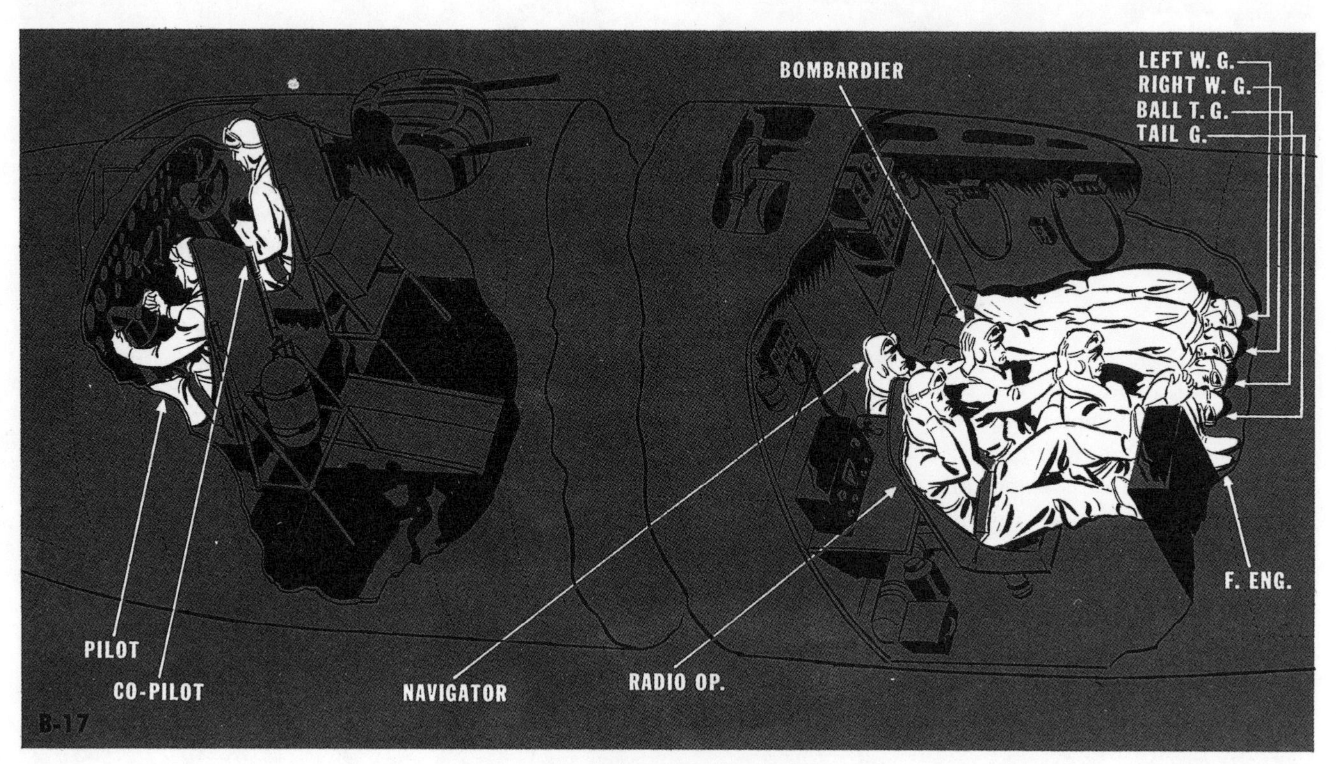


Preparing to Ditch the B-17

# DITCHING POSITIONS

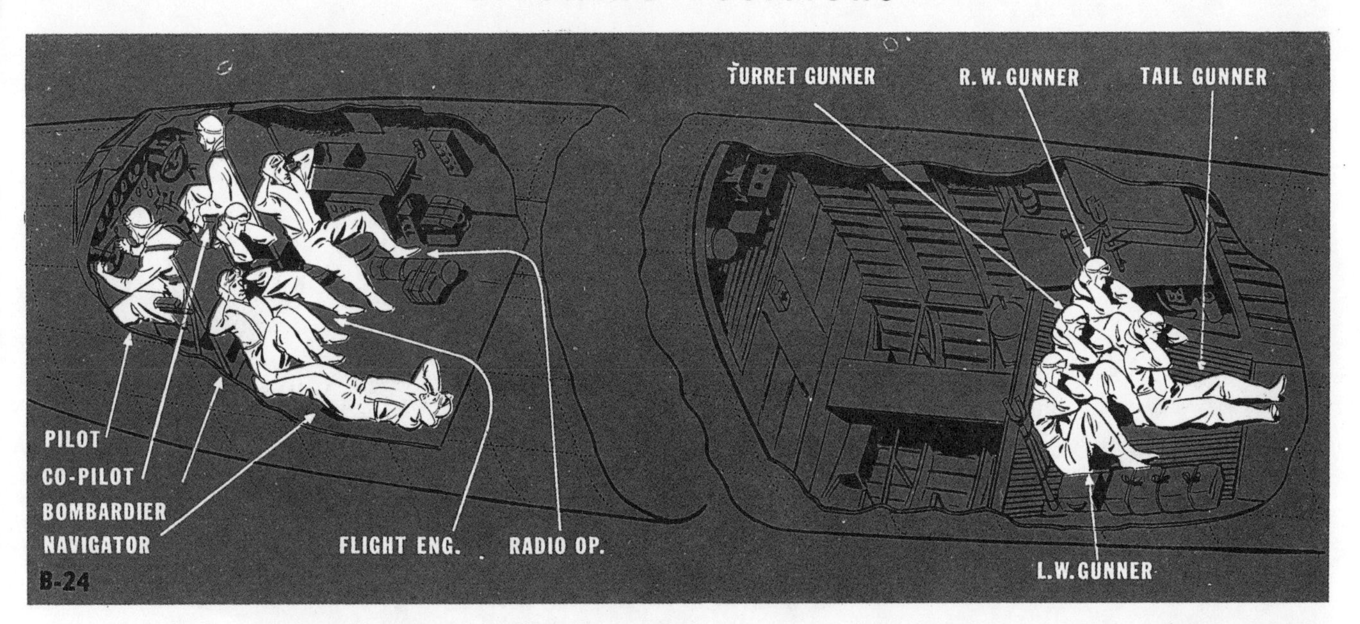

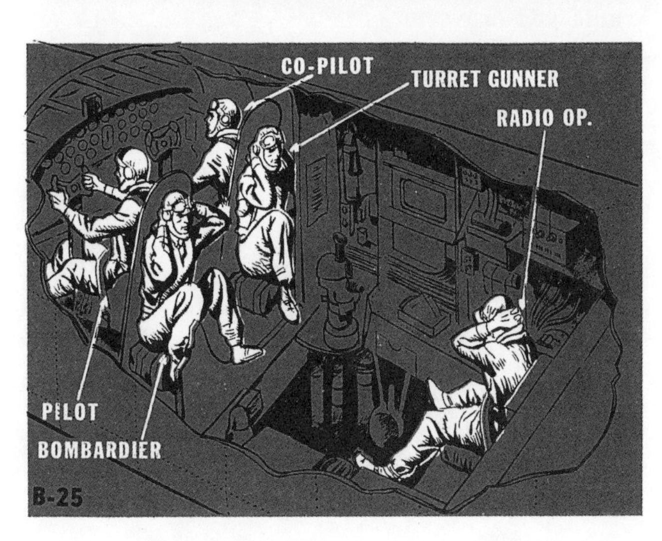


**B-24.** Ditching belt in the aft compartment is attached to gun mounts. Men brace against it as shown. Flight engineer, on cockpit step, also should use a ditching belt. Otherwise he should lie on floor, feet against step, knees bent slightly. Fasten down all loose equipment. Hold positions until plane stops. If help is near, crew should use one-man raft parachute and bail out.

**A-20.** Bombardier should wear one-man dinghy parachute pack and bail-out of bottom nose hatch if rescue appears possible. Other crew members take positions shown and maintain them until plane comes to rest.

**B-25.** If there is sufficient time and altitude, crew members in the rear should crawl through the bomb bay to ditching positions in forward part of plane. Otherwise they should remain in the after compartment. Positions shown should be maintained until plane comes to rest.

## INSPECT YOUR PARACHUTE

Before taking the parachute up, see that the rip cord pins are not bent and that the seal is not broken. **If the rip cord pins are bent or the seal broken, return the parachute to Supply.** See that the corners of the pack are neatly stowed so that none of the canopy is visible. See that the six or eight opening elastics are tight. Inspect each parachute you draw.

Inspect here. If pin is bent, return chute to supply.

## IN FLIGHT

If you find yourself in serious trouble, prepare to put your bail-out plan in operation. The plan will depend to some extent on whether you are alone or carrying a crew.

**If you are alone:**

1. Note your altimeter reading.
2. Estimate the altitude of the terrain below.
3. Decide on a minimum altitude or altimeter reading at which you can safely bail out, taking into consideration the flight characteristics of the plane and the kind of trouble you are having.
4. If you are still in trouble when you reach that minimum altitude—bail out.

**With a crew, consider these additional points:**

5. Remember that in general it is safer to jump your crew than to attempt a forced landing on hazardous terrain with a fully loaded plane.
6. Whenever there is serious question about a safe landing, warn your men that they may have to jump. Get them ready. If the situation improves, so much the better.
7. If the situation does not improve, pick the best available spot, slow the plane, and bail them out.
8. If necessary, bail out yourself. Whenever possible, set the automatic pilot to take the abandoned airplane away from inhabited areas in a prompt descent.

## THE BAIL-OUT

Every pilot should know the emergency exits provided for the airplane he commands and should see to it that his crew knows them and understands how and when to use them. Bail-out posters are supplied for most bombardment types and are now being prepared for other aircraft.

Practice making exits while wearing full equipment when the airplane is on the ground. It is the direct responsibility of the pilot to drill his crew in a standard bail-out procedure, including warning signals and exit signals.

> **DRILL IS ESSENTIAL**
> **Each crew member must know when, where, and how he is to leave the airplane.**

**Jumping from Single-Engine Trainers and Fighters**

1. Dump the cockpit canopy, or open the emergency exit, or pull the side window as low as possible.
2. Slow the airplane as much as possible.
3. Disconnect your radio headset, etc.
4. Release safety belt and shoulder harness.
5. Dive out and down. Often you can go out flat, onto the wing, and slide head first off the trailing edge.

**In certain types of fighters it is sometimes desirable to turn the airplane upside down:**

1. Dump the canopy.
2. Invert your plane.
3. Disconnect equipment.
4. Release safety belt and shoulder harness and dive straight down.

**Jumping from Twin-Engine Trainers, Bombers, and Transports**

You will normally use an escape hatch, the bomb bay, or a door, depending upon circumstances. Slide yourself to the edge of the opening and go out head first and straight down.

## CLEARING THE AIRPLANE

Probably the most important single act, in any parachute jump, is opening the parachute only **after you are clear of the plane. Wait** until you are well away from the airplane before you pull the ripcord. Keep your eyes open. Look around. If you have enough altitude, wait at least five to ten seconds before pulling the ripcord.

## PULLING THE RIPCORD

There is nothing complicated or difficult about getting your parachute safely open. Just:

1. Straighten your legs and put your feet together to reduce the opening shock, and to avoid tangling your harness.
2. Use both hands to grasp the ripcord pocket.
3. Grab the ripcord handle with the right hand, and yank! Keep your eyes open and look at the ripcord as you pull it.

## THE DESCENT

About two seconds after you have pulled the ripcord, you will feel a sharp, strong tug as the canopy opens and bites the air.

Look up to see that the chute is fully open. If a suspension line traverses the top, or the lines are twisted, manipulate the lines to remedy the fault.

Do not worry about oscillations. They will almost certainly occur on your way down, but are of minor consequence. Do not attempt to check them or to slip the parachute, as such maneuvers are useful only to experts, and are dangerous below 200 feet.

Make a quick estimate of your altitude by looking first at the ground below and then at the horizon.

You will descend approximately 1000 feet per minute.

Observe your drift by craning your neck forward and sighting the ground between your feet, keeping your feet parallel and using them as a driftmeter.

Face in the direction of your drift.

While you cannot steer your chute, you can turn your body in any desired direction. **The body turn is the most useful maneuver you can learn because with it you can make certain that you land facing in the direction of your drift. It is simple and easy. Note carefully exactly how it is done.**

STUDY THE PICTURES. Practice the body turn in a suspended harness if you get the chance. This description may sound backward to you. Note with special care how these turns are executed and simply say to yourself:

"To turn right, right hand behind my head."

"To turn left, left hand behind my head."

# HOW TO MAKE BODY TURNS

## TO TURN YOUR BODY TO THE RIGHT:

**1**

Reach up behind your head with your right hand and grasp the left risers.

**2**

Reach across in front of your head with your left hand and grasp the other risers. Your hands are now crossed, the right hand behind, and in each you have two risers.

**3**

Pull simultaneously with both hands; this will cross the risers above your head and turn your body to the right. You can readily turn 45°, 90°, or 180° by varying the pull.

**To turn to the left, reverse this procedure.**

In the descent, start your body turn high enough to allow you to master it. Once you have made the turn, you will find that you can control your direction of drift perfectly. Hold the turn, or slowly ease up if necessary, to bring you in facing downwind. Continue to hold the risers, whether you have had to twist them to make a body turn or not, and ride right on into the ground this way.

# THE LANDING

## NORMAL LANDINGS

Whether you have made a body turn or not, keep your hands above your head, grasping the risers.

Look at the ground at a 45-degree angle, not straight down.

**Set yourself for the landing by placing your feet together and slightly bending your knees,** so that you will land on the balls of your feet.

**Don't be limp; don't be rigid.**

Relax, and keep your feet firmly together with your knees slightly bent, and your hands grasping the risers above. Now hold everything and ride on into the ground, drifting face forward.

**At the moment of impact, fall forward or sideways in a tumbling roll to take up the shock.**

## ABNORMAL LANDINGS

### High Wind

If there is a strong wind blowing across the ground when you land, do two things.

First, make certain that you carry out the procedures described above for a normal landing, **including the body turn to face you exactly in your direction of drift.**

Second, once you are down, roll over on your abdomen and haul in hand over hand on the suspension lines nearest the ground. Keep right on pulling them in until you grab silk. Then, drag in the skirt of the canopy to spill the air and collapse the chute. If you can't manage this maneuver on your face, go over onto your back, but haul in the suspension lines until you have got the bottom edge of the canopy, then spill the chute.

### Trees

Tree landings are usually the easiest of all. If you see that you are going to come into a tree, drop the risers, cross your arms in front of your head, and bury your face in the crook of an elbow. You can see under your folded forearm. Keep your feet and knees together. If you get hung up high in a tree, consider first the possibility of immediate rescue before you try to climb down. Failing that, get out of the harness and cut the lines and risers to make a rope for climbing down.

Safe water landings are simple if you know what to do. The ability to swim is an advantage but not a prerequisite. Follow the procedure outlined here for all types of parachutes except the QAC and the single-point quick release, instructions for which are given below. Begin preparing for the water landing as soon as the parachute is open.

1. Throw away what you won't need.
2. Pull yourself well back in the sling by hooking your thumbs in the webbing and forcing the sling downward along your thighs.
3. Undo your chest strap by hooking a thumb beneath one of the vertical lift webs, pushing firmly across your chest to loosen the cross webbing so that you can undo the snap. **This must be done before you inflate the Mae West, as the chest strap cannot be released over an inflated life vest.**
4. If you are securely seated in the sling, and have time, free the leg straps by doubling up first one leg and then the other, unsnapping the fasteners. Keep your arms folded, or hang onto the risers, so that you will not fall out of the harness. You can remove the leg straps in the water either by unsnapping them or by working them down over your feet if you have been unable to free them in the air.
5. As soon as you are in the water, inflate your Mae West, one half at a time. Either half will support you. **Remember, never inflate your life vest until you have unfastened your chest strap.**
6. Get clear of the parachute promptly, and stay clear.

Carry an accessible, serviceable knife to cut harness and suspension lines if necessary.

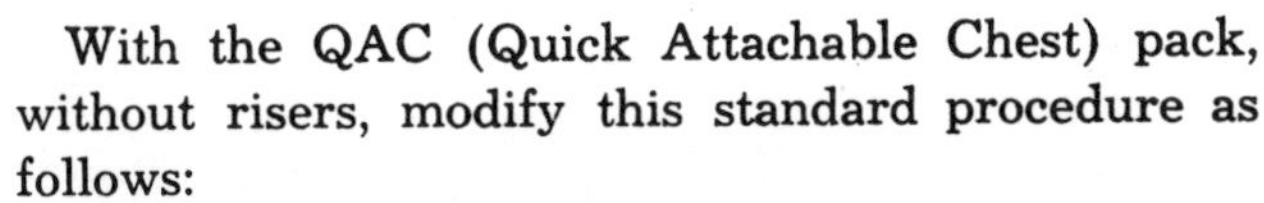

With the QAC (Quick Attachable Chest) pack, without risers, modify this standard procedure as follows:

1. Pull yourself well back in the sling and undo the leg straps.
2. As soon as you are in the water, release one side of the chest pack from the harness.
3. Then unfasten the chest strap.
4. Inflate the Mae West, one half at a time, but never until the chest strap is unfastened.
5. Get away from the parachute and shroud lines and stay away.

With the single point quick release harness, modify the standard procedure as follows:

1. Before reaching the water, turn the locking cap 90° to set the release mechanism for immediate operation.
2. As soon as you are in the water, but not before, pull the safety clip, and press hard on the cap to release the lock. The harness will then slide off.
3. Inflate the Mae West, one half at a time, but never until the harness has been released.
4. Stay clear of the parachute.

See Life Vest, PIF 8-10-2 and Life Rafts, PIF 8-11-3.

## NIGHT JUMPS

As soon as you are in the chute, prepare for a normal landing. Since you cannot see the ground on a dark night, you want to be ready to make contact at any moment. **Get your feet and knees together, your legs slightly bent. Hang onto the risers above your head and wait for contact.**

## HIGH ALTITUDE JUMPS

A successful jump has been made from above 40,000 feet using oxygen on the way down; and from as high as 35,000 feet without bail-out oxygen apparatus. Free falls of from 10,000 to 30,000 feet have been made before the ripcord was pulled, with no harmful effects whatever. This evidence means that intelligence and self-control, plus normally-functioning equipment, can get you down safely from any altitude to which you can fly.

**The chief hazards of high-altitude jumping are:**

1. Intense cold.
2. Lack of oxygen.
3. Excessive speed.

The last may, of course, be encountered in any jump, but is particularly likely to accompany emergencies at high altitude. All three of these hazards are greatly reduced by making a long free fall before opening the parachute. Also, exposure to gunfire, falling debris, etc., is practically eliminated.

A free fall from 40,000 feet takes just over three minutes. The same descent by parachute takes about 24 minutes. If the parachute is opened at approximately 5,000 feet above the terrain (when the earth "begins to look green," details appear, the horizon "spreads," and the ground appears to rush up), the descent takes about seven minutes.

Remember that in many emergency jumps you may leave the airplane at a speed likely to exceed the terminal or maximum velocity of a free fall. Hence, you should wait to slow down to terminal velocity before pulling the ripcord to avoid injuring yourself or damaging the parachute. In from five to 15 seconds, you will reach terminal velocity in any fall, and will actually slow down all the rest of the way, from 320 feet per second (218 mph) at 40,000 feet to 160 feet per second (109 mph) at sea level. With the parachute open, your rate of descent varies between 40 feet a second at 40,000 and 20 to 24 feet a second at sea level.

**If you have bail-out oxygen equipment:**

1. Take several deep breaths of oxygen from the plane's supply.
2. Turn on your bail-out oxygen supply.
3. Grip your pipestem between your teeth, or check your mask.
4. Dive clear of the airplane.
5. Wait as long as you safely can before pulling the ripcord.

**If you do not have bail-out oxygen equipment:**

1. Take several deep breaths of oxygen from the plane's supply.
2. Hold your breath and dive out.
3. Continue to hold your breath as long as you possibly can.
4. If you are above 30,000 feet, try to wait at least one minute before pulling the ripcord.

**Except in extreme emergency, do not attempt a bail-out without bail-out oxygen equipment above 30,000 feet.**

### FREE FALL AND OPEN PARACHUTE DESCENT FROM 40,000 FT

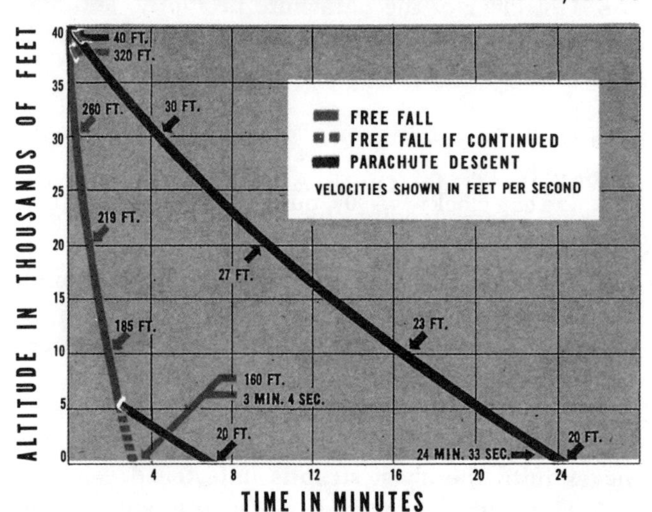


The graph compares parachute descent from 40,000 feet, and combined free-fall and parachute descent with delayed opening of the parachute at 5,000 feet.

REFERENCE: Technical Order 13-5-2.

# Parachute Types

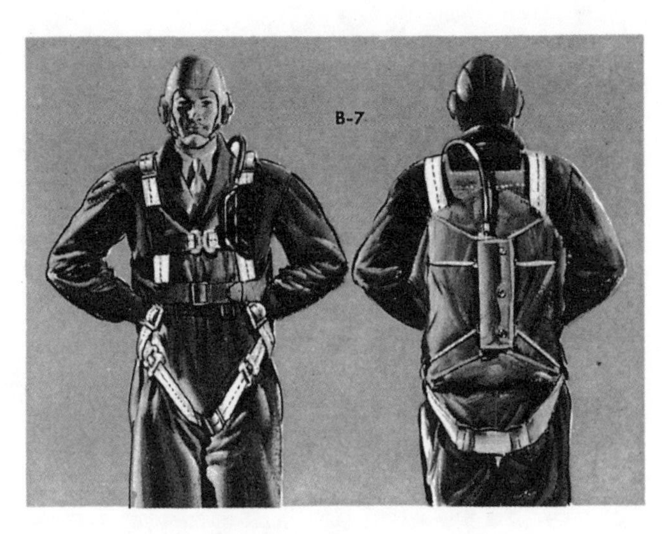


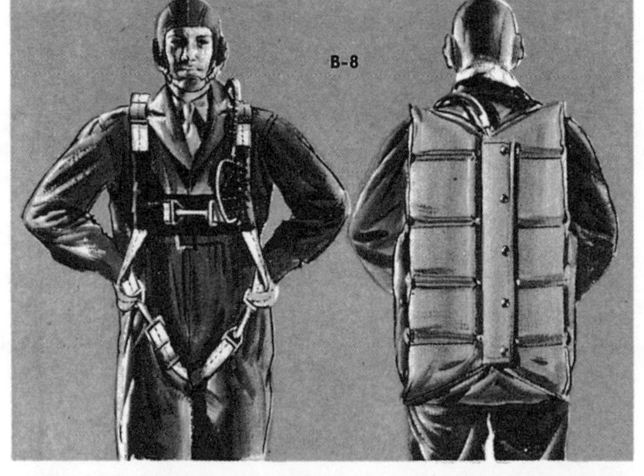


### BACK-TYPE PARACHUTES

**Type B-7** (AN6512). The chest straps and leg straps have bayonet type or snap fasteners. Note that parachute belt is worn outside harness to hold webbing snug.

**Type B-8.** Flexible back pack with bayonet type fasteners on chest and leg straps. Older type B-8 parachutes have snap fasteners.

**Type B-9.** Flexible back pack on single point Quick Release harness. To get out of Quick Release harness turn the cap clockwise 90°, pull safety clip, and strike the cap a sharp blow with the hand.

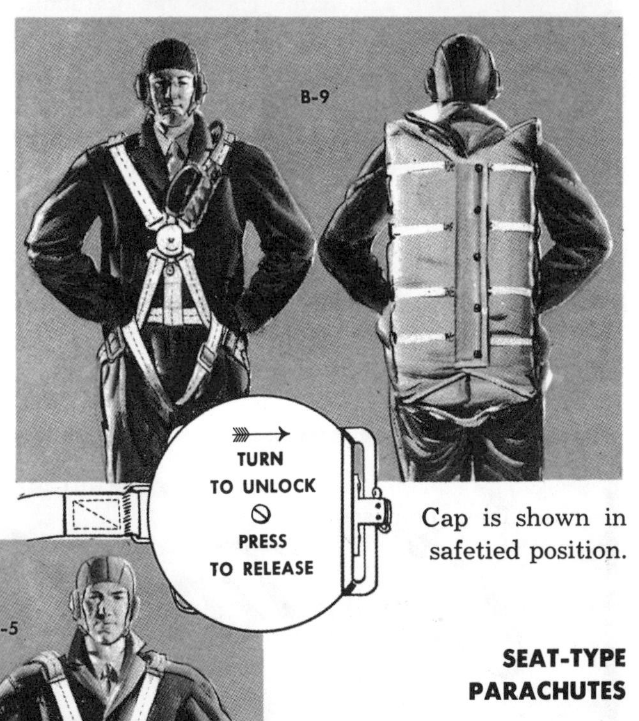


Cap is shown in safetied position.

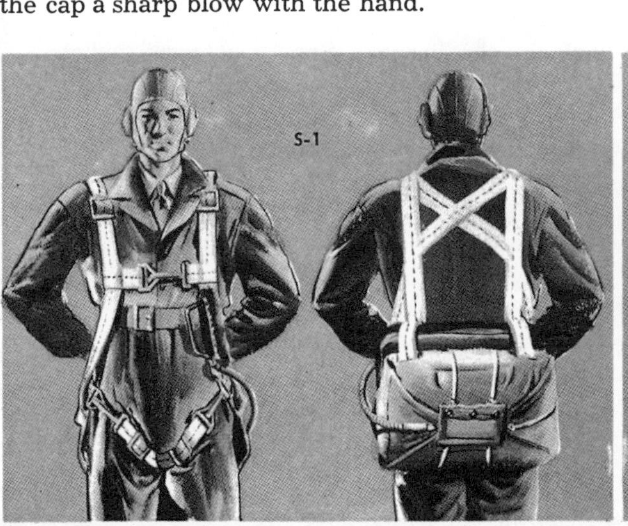


S-5

### SEAT-TYPE PARACHUTES

**Type S-1, S-2,** AN6510, and AN6511. Harness has back and seat pad. Chest and leg straps have snap or bayonet fasteners.

**Type S-5.** Same chute as S-1 with single point Quick Release harness.

## ATTACHABLE CHEST-TYPE PARACHUTES

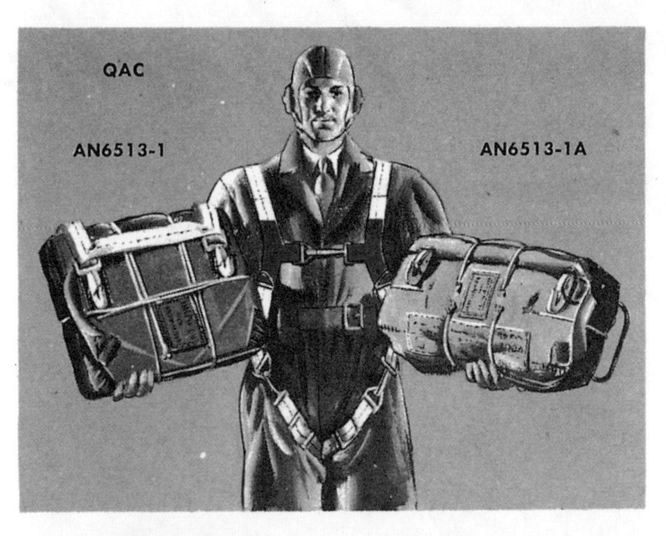


### Group 1 Assemblies

**Type QAC (AN6513-1).** Quick attachable chest-type parachute with square pack. Harness has snap fasteners on chest and leg straps. It has D-rings for attachment of pack.

**Type QAC (AN6513-1A).** Quick attachable chest-type parachute with barrel-type pack. Harness has snap fasteners on chest and leg straps. It has D-rings for attachment of pack.

**Note: On both AN6513-1 and AN6513-1A parachute assemblies the snaps are on the pack and the D-rings are on the harness.** Either of these packs can be used with the harness shown.

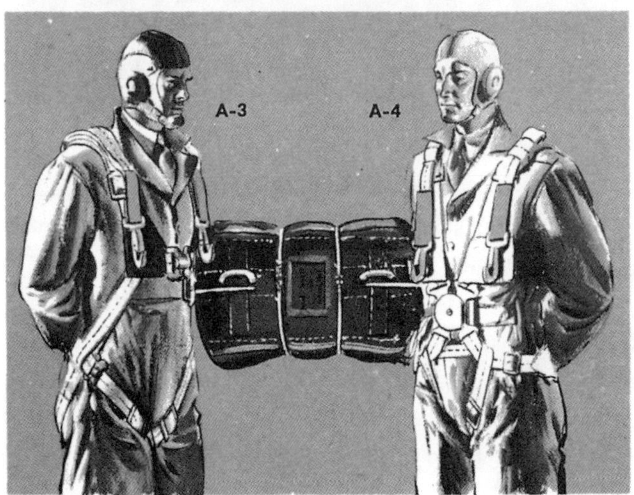


### Group 2 Assemblies

**Type A-3.** Quick attachable chest-type parachute with barrel type pack. Harness has bayonet type fasteners.

**Type A-4.** Quick attachable chest-type parachute with barrel-type pack and single point Quick Release harness.

**Note: On the A-3 and A-4 parachute assemblies the rings are on the pack and the snaps are on the harness.** This pack can be used with either of the harnesses shown.

**Parachutes of Group 1 are not interchangeable with parachutes of Group 2.**

The pilot is responsible for prevention of mismatching quick attachable chutes in his airplane.

Before the airplane moves for takeoff, inspect **all** attachable parachutes to see that the pack fits the harness. Snap each pack to its harness to make certain it matches.

If you find any pack which does not fit the harness, change either pack or harness to get the correct assembly.

Each group is to be identified by a color. The same color must be on both pack and harness.

Red identifies Group 1.
Yellow identifies Group 2.

**Be sure all packs and harnesses in your plane match.**

REFERENCE: Technical Order 13-5-39

# PYROTECHNIC PISTOLS

When radio communication is inadvisable or when radio equipment has failed, brief coded messages may be sent with pyrotechnic signals. Do not use pyrotechnic signals to control important operations unless no other means is available. The various colored signals which are available for use with M2 and AN-M8 pyrotechnic pistols are assigned different meanings under a code that will be changed at frequent intervals in each edition of Signal Operation Instructions. The M11, red star parachute signal, however, is always used as a distress signal to be fired from the ground or from a life raft.

## M2 Pistol

The M2 pyrotechnic pistol has a strong recoil. Use both hands to fire it if practicable. The signals themselves burn with an extremely hot flame; observe every reasonable precaution while handling or firing them.

1. **Fire signals only from airplane in flight** with the exception of the M11 distress signal.
2. Point the pistol in such a way as to prevent signals from striking any part of the airplane.
3. If a signal fails to ignite on the first attempt, try at least twice more. **If third or final try fails, keep the pistol pointed overboard and clear of all parts of the airplane for at least 30 seconds, then discard signal.**
4. **Discard a misfired signal, if possible, without handling the signal itself.** One method is to hold the pistol over an opening in the airplane and release the cartridge by pressing on the latch and allowing the signal to fall clear under the force of gravity. The force of the air blast prevents holding the pistol on the outside of most airplanes. **Be careful to prevent discarded signal from striking any part of the airplane.**
5. Do not discard misfired signals when flying over populated areas.
6. Fire the M11 distress signal as nearly straight up as is practicable.

## AN-M8 Pistol

The AN-M8 pyrotechnic pistol is replacing the M2 pistol. It is fired by inserting and locking the barrel in a type M-1 mount. This mount is really a little "door," fastened rigidly to the airplane, that permits the pistol barrel to extend through the airplane outer skin. The mount absorbs the recoil of the pistol. Observe these precautions in using this pistol:

1. Place cartridge in chamber after pistol is inserted in mount, and only when immediate use is anticipated.
2. Since the pistol is cocked at all times when the breech is closed, **never leave a live signal in the pistol when it is removed from the mount.**

# SMOKE GRENADES

Airplanes to be flown over sparsely settled regions on cross-country, patrol, or ferry missions will be equipped with either an M8 or an M3 smoke grenade. In the event of a forced landing, use the grenade as a marker to aid searching parties in locating the airplane which otherwise might be difficult to find.

Pilots observing smoke of the type produced by M8 or M3 smoke grenades will immediately attempt to locate the source.

The M8 smoke grenade burns about 3½ minutes, giving off a dense **gray** smoke, and is intended to be used primarily in heavily forested regions. It is easily distinguished from wood fires which give off a blue-gray or black smoke.

The M3 smoke grenade is designed to be used in snow-covered regions. It gives off a dense **red** smoke for 2 minutes which can be distinguished against a white snow background for about 4 miles by a person in an airplane.

### Method of Firing M8 Smoke Grenade

1. Grasp the grenade with lever held firmly against grenade body.
2. Withdraw safety pin, keeping a firm grip around the grenade and lever.
3. Either throw the grenade with a full swing of the arm, or place on the ground and release.
4. As the grenade is released from the hand, the lever drops away, allowing the striker to fire the primer.

### Method of Firing M3 Smoke Grenade

1. Pull the 3 vanes on the side of the grenade up and away from grenade body.
2. Place grenade in snow so that it is supported by the vanes in an upright position.
3. Keep lever held firmly against grenade and withdraw safety pin.
4. Release lever.

### Safety Precautions

**To avoid a fire, do not throw or place the grenade within 5 feet of dry grass or other readily inflammable material.**

After the grenade is ignited, **stay at least 5 feet away from the burning grenade,** as heavy smoke develops and there is a tendency to throw off hot particles of residue.

**Keep these smoke grenades dry.** If the chemical contents of a grenade become wet, it will ignite. Future procurement of these grenades for the Army Air Forces will be packed in individual waterproof containers.

All smoke grenades will be shipped and handled in accordance with Interstate Commerce regulations. These regulations prohibit the shipment of these smoke grenades in personal baggage.

REFERENCE: T. O. 01-1-38, dated April 14, 1942

# Panel Signals

Many of the emergency kits now supplied contain a large signal panel (roughly 10 ft. by 10 ft.). It is arc fluorescent yellow on one side and blue on the other. Immediately after you are forced down this panel should be spread out on the ground flat—yellow side up on dark backgrounds and blue side up on light backgrounds—the color will help rescue pilots to find you. Once a rescue pilot has located you, messages can be transmitted by folding the panel as indicated in the illustrations on these pages. If it is windy, hold the folds in place with rocks, sand, sticks, or improvised stakes if it is necessary. If several messages are to be transmitted don't change the folds too quickly—allow enough time for the pilot of the rescue plane to read each signal and indicate that he understands it (generally by dipping the nose of his plane several times). These same signals can be transmitted with the square yellow-and-blue sail now a part of the equipment supplied with the large inflatable rubber life raft.

The emergency signal panel also can be used as a tent since its blue side is coated with a waterproof compound. Also, the blue side can be used as an excellent camouflage cover for a life raft if enemy aircraft are sighted.

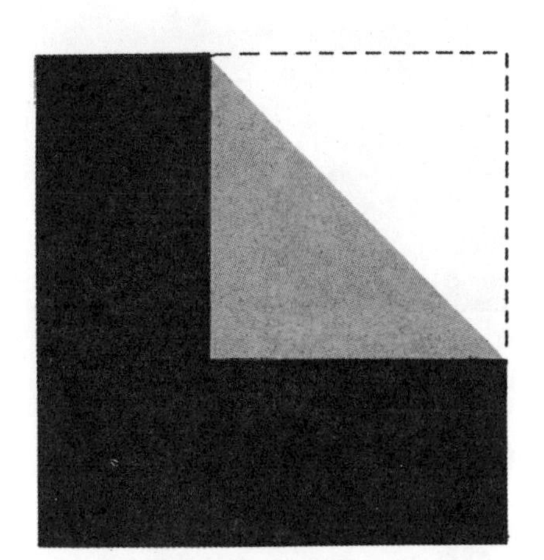

Need Gasoline and Oil, Plane is Flyable

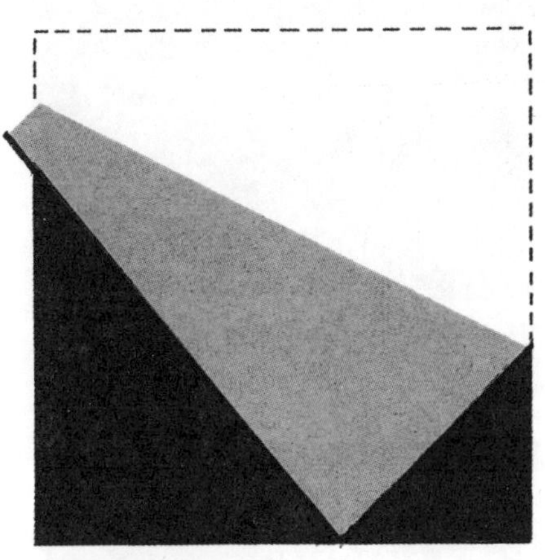

Need Tools, Plane is Flyable

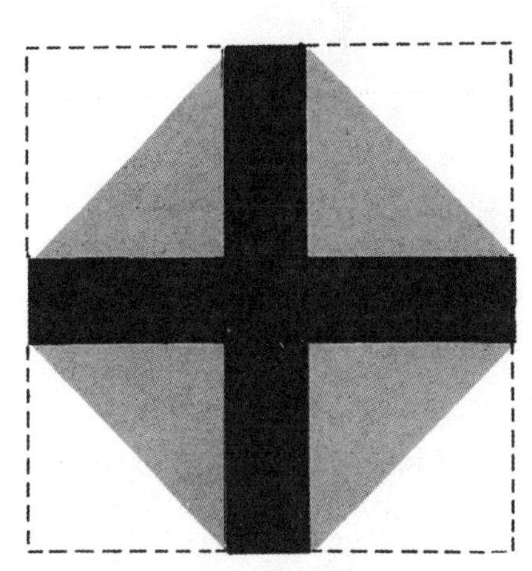

Need Medical Attention

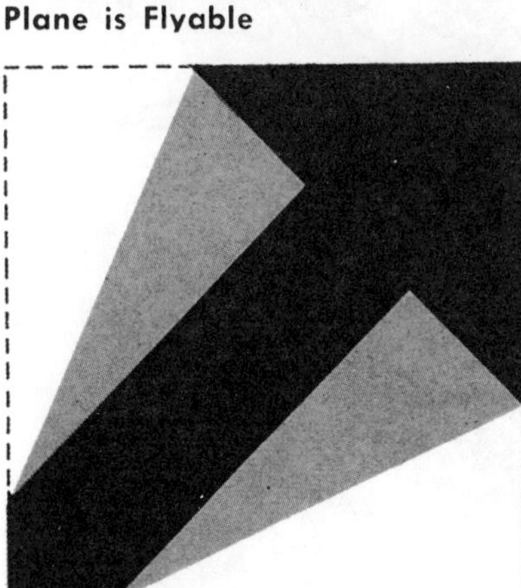

OK to Land, Arrow Shows Landing Direction

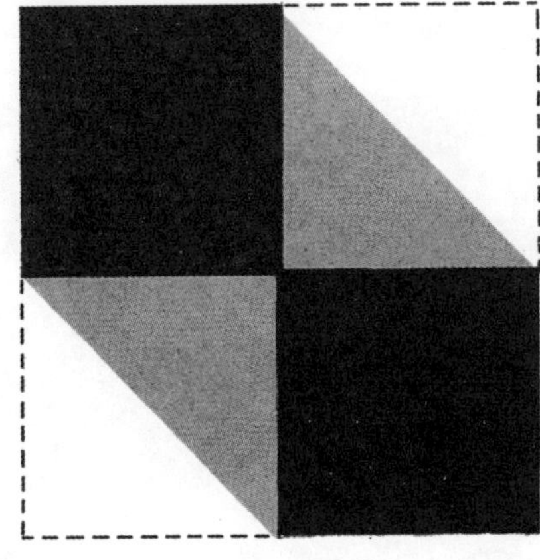

Do Not Attempt Landing

Indicate Direction of Nearest Civilization

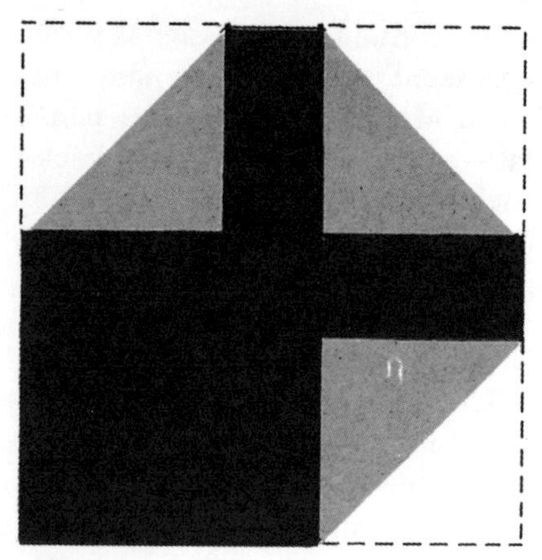

Need First-Aid Supplies

Need Quinine or Atabrine

Should We Wait For Rescue Plane?

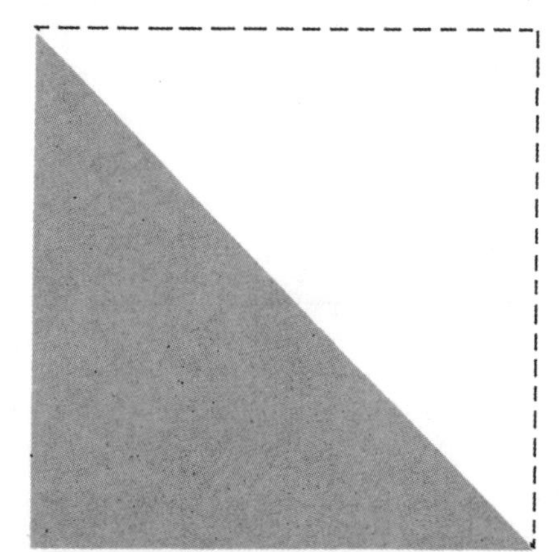

Need Food and Water

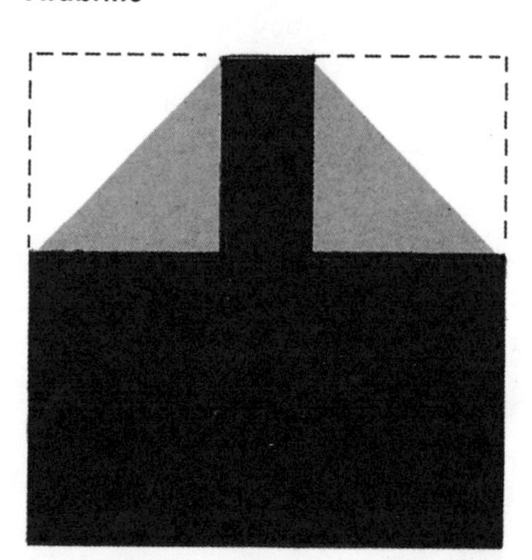

Need Warm Clothing

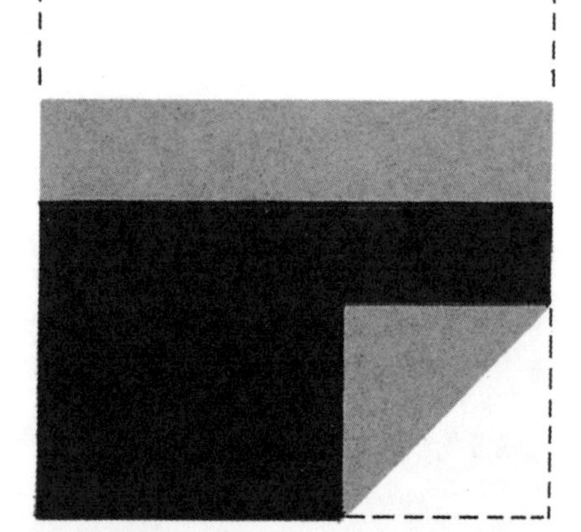

Have Abandoned Plane, Walking in This Direction ➡

# *Body Signals*

If a rescue plane flies low and circles your location and you are sure that you have attracted the pilot's attention, messages can be transmitted by the emergency body signals shown on this page. When performing the signals stand in the open, make sure that the background as it will be seen from the plane is not confusing, make the motions deliberately and slowly, and repeat each signal until the pilot indicates that he understands.

NEED MEDICAL ASSISTANCE—URGENT (Lie prone)

ALL O K
DO NOT WAIT

CAN PROCEED SHORTLY—
WAIT IF PRACTICABLE

NEED MECHANICAL HELP
OR PARTS—LONG DELAY

PICK US UP—
PLANE ABANDONED

DO NOT ATTEMPT
TO LAND HERE

LAND HERE (Point in
Direction of Landing)

OUR RECEIVER
IS OPERATING

USE DROP
MESSAGE

AFFIRMATIVE (Yes)

NEGATIVE (No)

## HOW PLANE ANSWERS

The pilot of the rescue plane will answer your messages either by dropping a note or by dipping the nose of his plane for the affirmative (yes) and fishtailing his plane for the negative (no).

AFFIRMATIVE (Yes) DIP NOSE OF PLANE

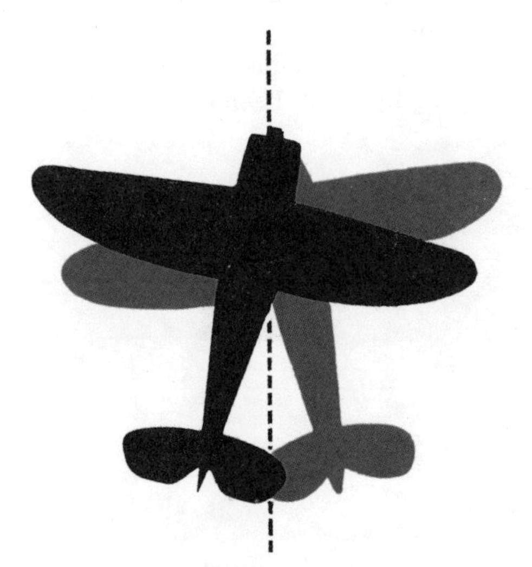

NEGATIVE (No) FISHTAIL PLANE

# EMERGENCY KITS

Emergency kits developed for the use of pilots, crew members, and ground forces fall into three basic groups—personal kits which are attachable to the pilots and crew members of an airplane, airplane kits which are to be carried in the airplane, and dropping kits to be carried by rescue airplanes. The development of these emergency kits was based on the assumption that all flying personnel have the basic back or seat pad type of kit. The rest of the kits contain additional equipment and rations to supplement that contained in the basic kit.

The following kits have been developed:

1. Parachute Emergency Kit (Type B-2)—This kit is the back pad type, intended for use when flying over jungle territory. Basis of issue is one per parachute in jungle and tropical areas outside Continental United States. The components of this kit include the following: Waterproof Match Container, Compass, Field Ration, Fishing Kit, Mosquito Lotion, Machete, Stone Sharpener, Signal Flare, Mechanics' Gloves, Mosquito Net, First-Aid Kit, Emergencies Manual.

2. Parachute Emergency Kit (Type B-4)—This kit is a seat or back pad type, and is intended for

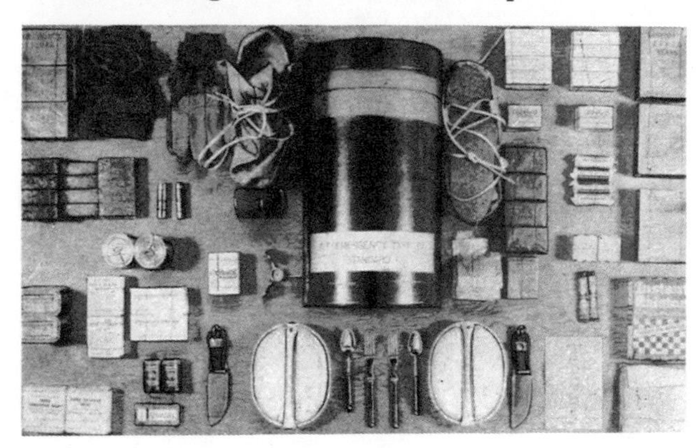

E-1 Emergency kit.

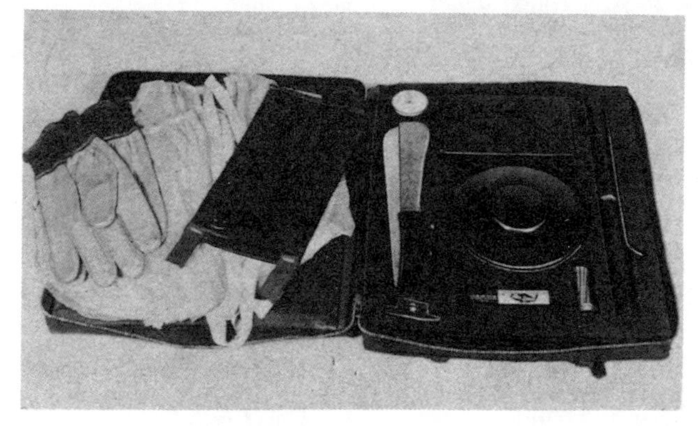

B-4 Emergency kit.

universal use. The components of this kit include the following items: Water Container Assembly, Signal Flares, Machetes, Signal Panel, Special Parachute Kit Ration Unit made up of Field Ration K components, Match Case with Matches, Cooking Pan Assembly, Compass, Pocket Knife, Fishing Kit, Can of Solid Fuel, First-Aid Kit, Mosquito Headnet, Goggles, and Gloves, Emergencies Manual.

3. Emergency Sustenance Kit (Rations) Type E-1—This kit is intended to supply the necessary rations for operations in northern climates. The basis of issue for this kit is one kit for every two men. The components of this kit include the following items: U. S. Army Mountain Rations, Drinking Water in Cans, Match Box with Matches, Hunting Knife, Mess Kit, Mosquito Headnet, Canvas Gloves, Fork and Spoon, Muckluce, Ice Creepers, Solid Fuel and Grill. Sewing Kit, War Department Manual FM31-15, Mosquito Repellent, Tobacco and Cigarette Paper.

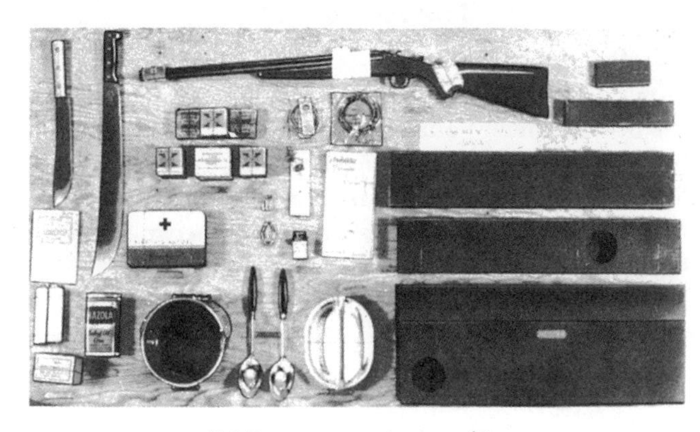

E-2 Emergency sustenance kit.

4. Emergency Sustenance Kit (Implements) Type E-2—This kit is intended to supply the necessary implements for use in northern climates. The basis of issue for this kit is one kit for each large airplane. The components of this kit include the following items: Combination .22-Caliber and .410-Gage Shot-Gun, Ammunition, Matches in Match Box, Camphor Cubes, Generator-Operated Flashlight, Candles, Fry Pan, Stew Pan, Large Spoon, Butcher Knife, Cooking Oil, Machete, Fishing Kit, and Signal Flares.

5. Emergency Sustenance Kit (Cooking Type E-4)—This kit is an emergency cooking unit employing a pressure type gasoline stove burning 100 octane gasoline. Two Stew Pans, a Fry Pan, Hot Pan Holder and Shield, and a Fabric Gasoline Bag are provided. The basis of issue of this is proposed as one kit per large airplane.

6. Emergency Sustenance Kit (Type E-5) (Over Water)—This kit is intended for carrying in large airplanes operating mainly over water. The basis of issue is one kit per large airplane. The components of this kit include the following items: Field Ration K, Drinking Water in Cans, Flashlight, Bailing Bucket, Compass, Matches, Knife, Hand Axe, Mirror, Candles, Fishing Kit, Flares, Rope, Paulin, Sea Markers, First-Aid Kit, Iodine Swabs, Boric Ointment, and Tomato Juice. This kit is used in addition to the kit supplied in the life raft.

E-6 Individual bail out rations kit.

7. Emergency Sustenance Kit (Type E-6) (Individual Bail Out Rations)—This kit is intended to supply additional rations and can be snapped on the parachute harness before bailing out. The contents of this kit include two units of Field Ration K.

8. Emergency Kit (Type E-7) (Individual Bail Out Water)—This kit is the same as the Type E-6;

except it contains two cans of drinking water in accordance with Specification AN-W-5a.

9. Emergency Sustenance Kit (Type E-8) (Desert and Tropic Implements)—This kit is intended to supply the necessary implements for use in desert and jungle. The components of this kit include the following items: Compass, Paulin, Combination .22-Caliber and .410-Gage Gun, Ammunition, Generator Flashlight, Machete and Sheath, Flares, Mirror, First-Aid Kit, Sewing Kit, Soap and Sunburn Ointment.

10. Emergency Sustenance Kit (Type E-9) (Desert and Ocean Rations)—This kit contains the necessary rations for desert, jungle, and ocean. The contents of this kit include Drinking Water in Cans, Field Ration K, Paulin, and Sun Hats.

11. Emergency Sustenance Kit (Type E-10) (Tropical Aerial Delivery)—This kit is intended for use in tropical areas for dropping by parachute the necessary rations and equipment to aircraft personnel stranded in the desert or jungle. The components of this kit include the following items: Field Ration K, Drinking Water in Cans, Tent, Sun Hat, Neckerchief, Paulin, Solid Fuel and Grill, Insect Repellent, Generator Flashlight, Mirror, Flares, Combination .22-Caliber and .410-Gage Gun, Ammunition, Scout Knife, Compass, Sewing Kit, Machete and Sheath, Atabrine Tablets, Benzedrine Sulphate Tablets, Soap, Tea Tablets, Matches, Sunburn Ointment, Aerial Delivery Container Assembly, Aerial Delivery Canopy Assembly, Vitamin Capsules B and C.

13. Emergency Sustenance Kit (Arctic Aerial Delivery)—This kit is now in process of development. This kit is intended primarily for use on the Greenland Ice Cap. The purpose of this kit is to supply stranded aircraft personnel on the Greenland Ice Cap and similar desolate territory with the necessary rations, clothing, and equipment to sustain life at the scene of a forced landing until a guide can be dropped by parachute to lead the party out on foot. The components of this kit include the following items: Parka and Trousers, Sleeping Bag, Tent, Gasoline Stove, Gasoline Container, Ice Saw, Ice Axe, Goggles, Mittens, Socks, Wristlets, Type E-1 Ration Kit and Type E-2 Implement Kit. The implements and rations are dropped in a parachute container, the clothing is dropped without a parachute container.

### Protect Yourself

**Before taking off on an extended flight over mountainous country, the Arctic, the jungle, the desert, or the ocean, check your emergency equipment. If your plane is not equipped, check with your supply officer. Also acquaint yourself with the Emergency Manuals on jungle, desert, Arctic, and ocean survival methods.**

E-7 Individual bail out water kit.

# LIFE PRESERVER VEST

Wear your life vest whenever you fly over water.

When the vest is issued to you, put it on, inflate it by the mouth tubes. Adjust the straps. **With the vest inflated the waist strap should be tight, the crotch strap snug.**

Deflate the vest by opening the valves at the base of the mouth tubes. Roll the vest up to deflate completely. Be sure to close the valves tightly to prevent leak on automatic inflation.

Wear the vest over the clothing and **under the parachute harness.** Tuck the vest under the collar of your flight jacket.

To inflate, pull one cord at a time so that if the mouth valves have been left open you will discover the error before you have discharged both $CO_2$ cartridges. One compartment will support you and will interfere less with swimming.

If the vest leaks, or fails to inflate completely from the $CO_2$ cartridge, fill by blowing into the mouth tubes. Open the valves while filling the vest by mouth, then reclose the valves tightly.

**Note: cutting off or bending the mouth tubes flush with the retaining loop will prevent possible injury to your eye at the time your parachute opens.**

Before each flight remove the cap from the inflator cylinder and inspect the $CO_2$ cartridge. If the seal at the tip is punctured replace the cartridge. With the lever which actuates the puncturing pin in the up position, parallel to the container, insert the new cartridge, seal end down. Always check the container cap to be sure it is screwed down tightly.

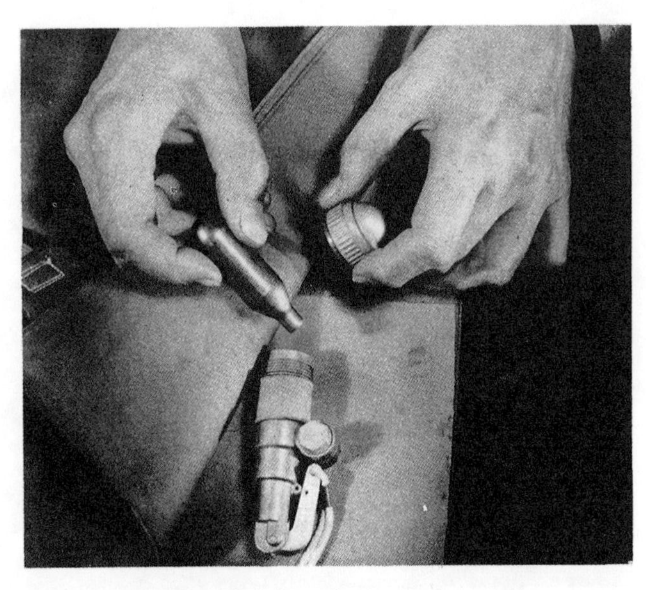

Inserting $CO_2$ inflator. Screw cap down tight.

**Sea Marker Packet**

A sea marker packet is cemented to the life vest. When friendly airplanes approach, release the packet by pulling down on the tab. The dye will form a large green area lasting three to four hours. This will help airplanes to find you.

**Caution**

Before takeoff be sure your life vest cartridge containers are loaded with live $CO_2$ cartridges, **and that the container caps are screwed down tightly.** (See illustration.)

Always make certain that the mouth inflator valves are tightly closed before pulling the inflating cords.

Turn in your life vest for inspection every six months.

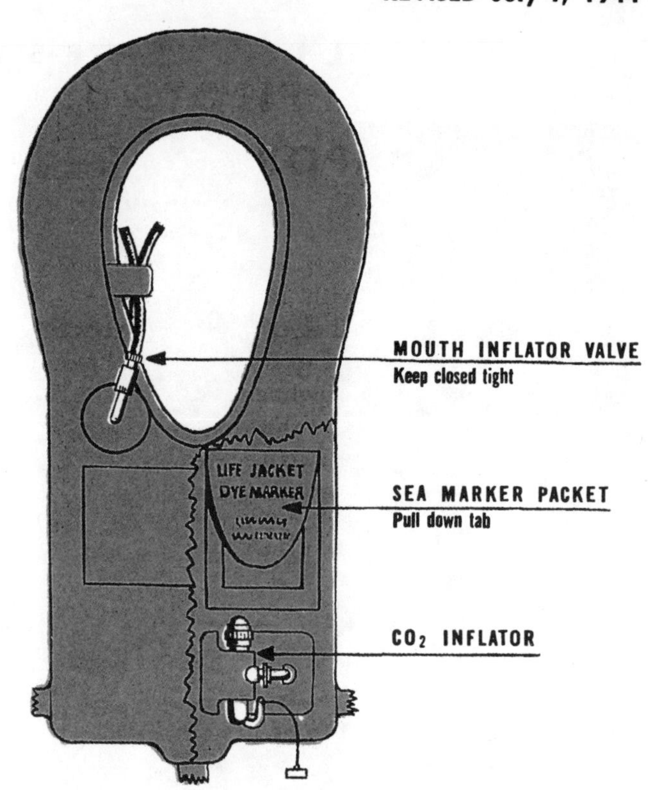


## WARNING: STAY AWAY FROM YOUR CHUTE IN THE WATER

After parachuting into water you will have a tendency to drift downwind into the fallen parachute as soon as you inflate your life vest. To avoid entanglement with harness and shroud lines, work upwind, away from the chute, and stay clear. If you have a raft, salvage your parachute for sail, cover, and extra lines. If not, get away from the chute and stay away.

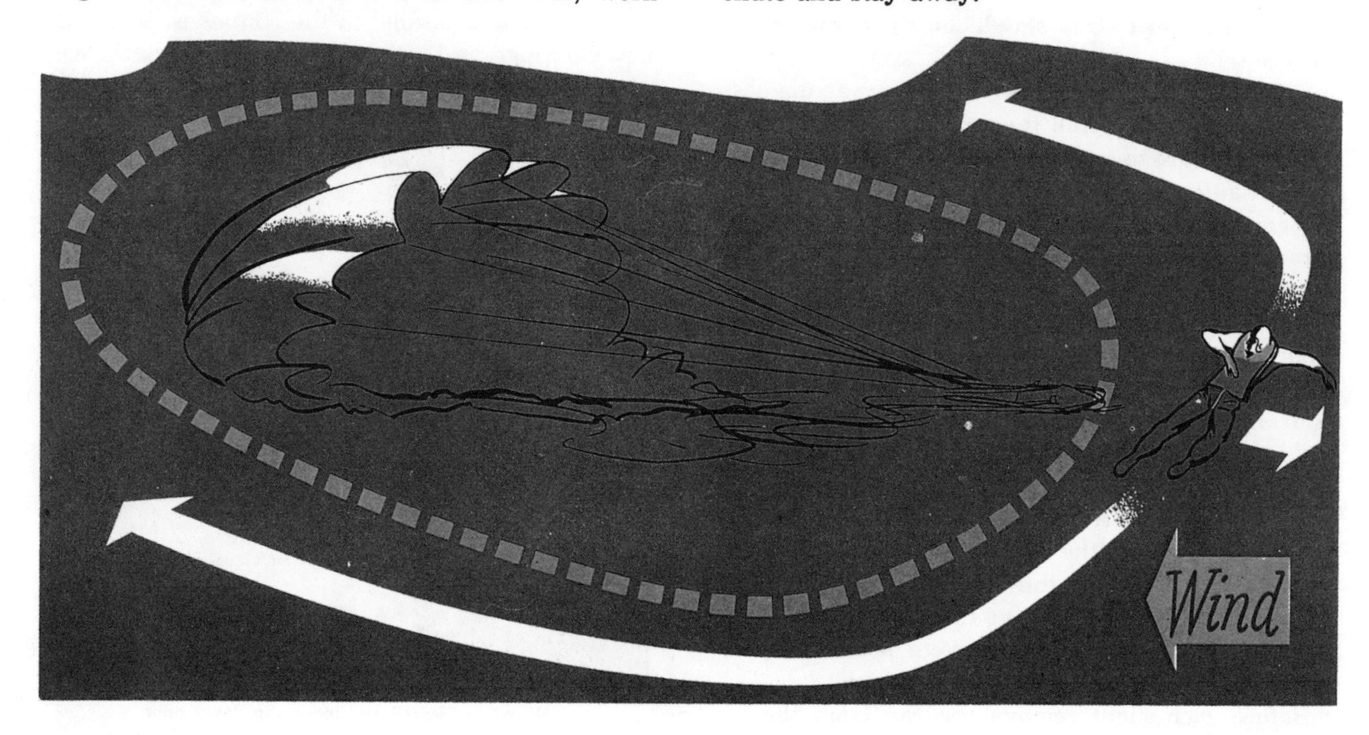

# SWIMMING THROUGH FIRE

When an airplane is ditched at sea there is always the possibility that a smashed wing tank and engine will spread flaming oil and gasoline on the water. By using the following procedure, however, you can swim to safety through such a fire, even when you wear a life vest.

1. Jump feet first upwind of your airplane. Cover your eyes, nose and mouth with both hands. Take a deep breath. Hold breath until you rise to the surface.

2. Just before you reach the surface, make a breathing hole in the flames. Swing your arms overhead to splash flames away from head, face, and arms.

3. Swim into the wind. Use the breast stroke. Before taking each stroke splash water ahead and to the sides. Keep mouth and nose close to the water. Duck your head every third or fourth stroke to keep it cool. If there are several men, swim single file. Let the strongest swimmer splash a path so the rest can follow safely in his wake.

## Swimming Under Water

If the heat is too intense or flames too high, swim underwater—out of the danger area. To do this:

1. Splash flames away from body.
2. Hold head near water level.
3. Deflate life vest by releasing valves.
4. Take a deep breath but do not inhale fumes.
5. Sink beneath the surface, feet first.
6. Swim upwind as far as possible.
7. Splash away the flames as you come to the surface. Take a deep breath and submerge again. Repeat procedure until you are beyond the fire.
8. Re-inflate life vest by mouth.

# LIFE RAFT KIT

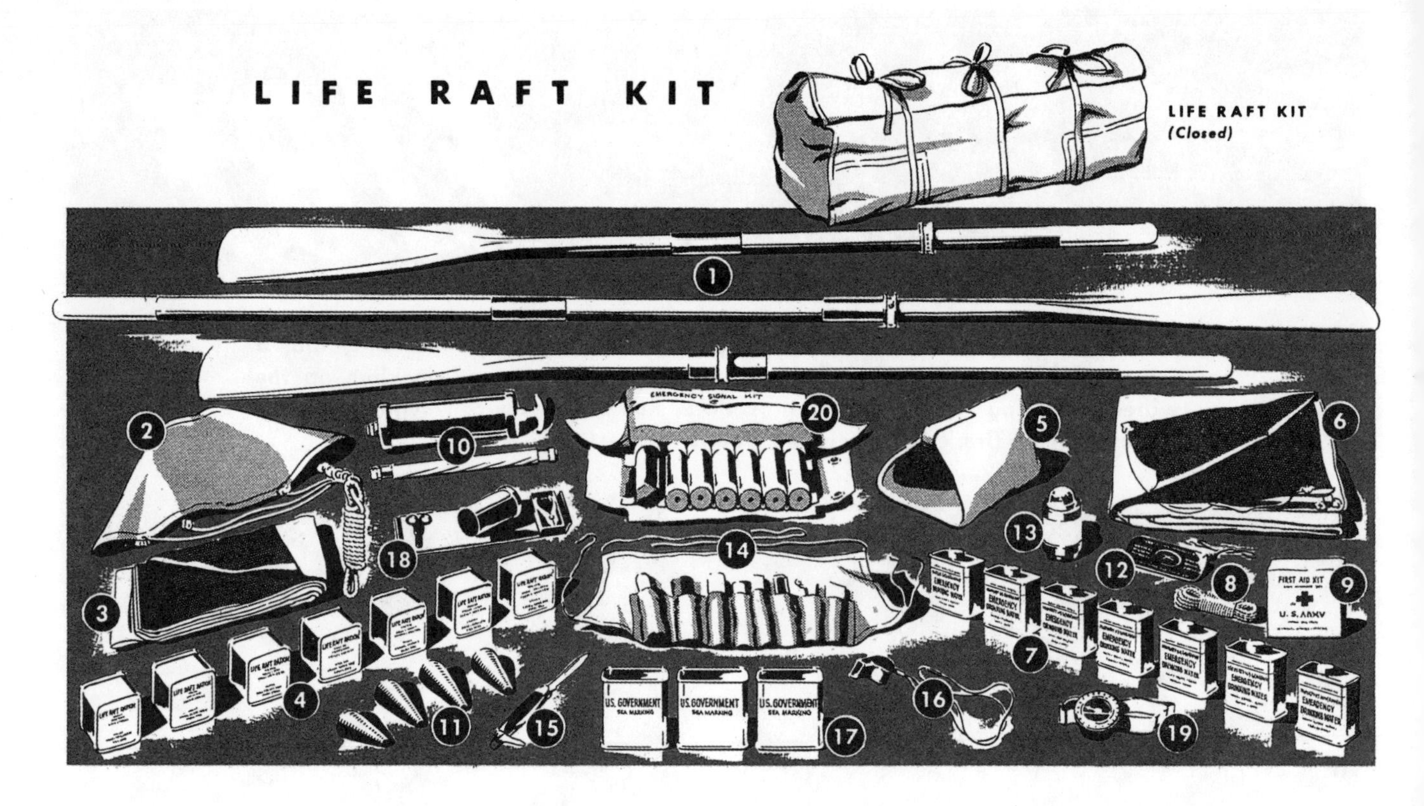


1. Oars. Use also as masts for sail.
2. Canvas Bucket Sea Anchor.
3. Sail. Use also for catching water if it rains.
4. Ration A.
5. Bailing Bucket. Use also for urination; don't try to stand in raft.
6. Shade and Camouflage Cloth. Yellow side for signal, blue for camouflage. Use it also for sun, wind, and spray protection and for catching rain water.
7. Emergency Drinking Water. Don't open or refill cans before flight. This water has been sterilized. Ordinary water will not keep. Save the cans for storage of rain water. Chemical sea water purification kits may replace some of the water cans.
8. Line. Tie down all loose equipment.
9. First-Aid Kit. Use as directed.
10. Inflating Pump. To replace with air the $CO_2$ lost by leakage and to inflate seats.
11. Puncture Plugs. Screw in plugs to seal leaks.
12. Signal Mirror. Best bet for signalling plane in daytime. Instructions for use on back of mirror.
13. Flashlight. Floating water-proof type. Tie it to the raft with the cord provided. Let it float as a night signal when an airplane is sighted. The battery will last 24 hours.
14. Fishing Tackle. Instructions included. Don't let hooks puncture raft. Dry lines and hooks.
15. Knife.
16. Whistle. Use to attract attention if ship or plane is near. Don't bother shouting.
17. Sea Marker. When you see a plane, pour the marker on the water. It will form a large colored slick which will last one to two hours. Get it overboard quickly when a plane is sighted as it takes a while for the color to spread. Stir the chemical with an oar to hasten spread.
18. Repair Kit.
19. Wrist Compass.
20. Signal Kit. Contains flare pistol and five or six flares. Fire proper flare (day or night) when plane is sighted. Aim high and ahead of plane.

**Warning**

When deflating life raft before storing it away in the airplane, use the deflating pump for that purpose. **Be sure the life raft is completely deflated.** If any air or $CO_2$ is left in the raft it will expand at altitude and become a dangerous hazard.

# Life Raft Discipline

### Equipment

Familiarize yourself now with the use of the equipment provided with the various life rafts. Ask your Personal Equipment Officer for demonstrations and instruction in its use.

If there are two or more rafts, connect them with the line provided to keep from becoming separated. Remain in the vicinity of the plane if it stays afloat, but not so close that the raft might be damaged by tossing against a sharp projection. Securely fasten the kit and all loose gear to the raft, with tight but easily untied knots.

Get the emergency radio into operation as soon as weather permits. Instructions are on the set. Keep all signalling equipment where you can get at it quickly. Keep flares and Very Pistol and cartridges as dry as possible. Use the flares only when a ship or plane is near. Fire the pistol almost vertically for maximum height, ahead of the plane so that the shot will be within the visibility range of the pilot.

Use the tarpaulin yellow side up for a signal, blue side up for camouflage from enemy.

Keep the sea anchor out. It will head you into the wind or check your drift.

### Water and Food

Take no food and water for 24 hours. Then ration it carefully. The pilot is in charge of rationing. In general don't eat food if you have no water.

Have all members of the crew drink all the water they can before any over-water flight. Don't take chances on starting your raft expedition thirsty.

You can collect rain water in the tarpaulin or sail. Drink as much as you can and store the remainder in empty water cans and other containers.

**Never drink sea water or urine.**

Take good care of your fishing kit.

### Protection

**In the tropics protection from the sun is vital.** Rig the oars and tarpaulin as a canopy and stay in the shade. Keep arms, legs, and head covered. Wet yourself, clothes and all, with bucket, sponge, or by immersion, but be careful to keep salt water out of your mouth.

Don't overexert. Perspiration will result and you will require more water.

**Continued exposure to cold sea water plus loss of circulation** may bring about a condition known as "Immersion Foot." To guard against it, keep your feet as dry as possible. Move your feet around and wriggle your toes to encourage circulation. If feet become swollen and sore, don't rub them. Rubbing will make the condition worse. Sprinkle open sores with sulfanilamide powder.

Large salt water burns or boils should be covered with sulfanilamide ointment and a light bandage. Don't prick or squeeze boils.

Don't worry about the absence of bowel movement or urination. It is a natural situation. Never take a salt water enema or a laxative.

If there is more than one man aboard, establish a watch routine. Keep a man on alert at all times. **Tie the man to the raft with at least ten feet of slack.**

# One man LIFE RAFT

Raft lanyard goes *under* harness and clips to life vest ring.

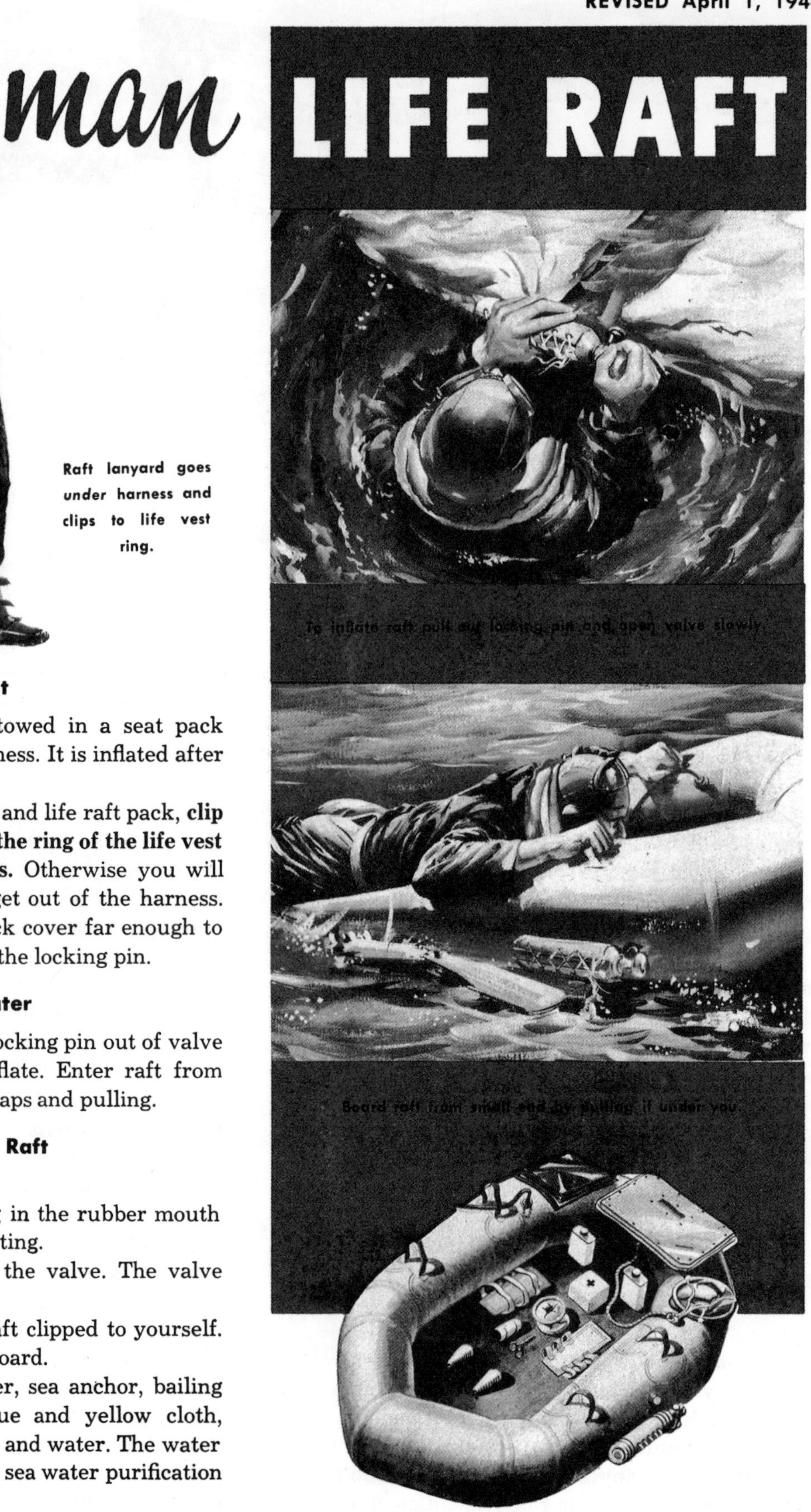

To inflate raft pull out locking pin and open valve slowly.

Board raft from small end by pulling it under you.

**Pre-Flight**

The one-man life raft is stowed in a seat pack attached to the parachute harness. It is inflated after the jumper strikes the water.

When you put on parachute and life raft pack, **clip the lead strap from the raft to the ring of the life vest waist strap under the harness.** Otherwise you will lose the raft pack when you get out of the harness.

Before flight unsnap the pack cover far enough to expose the $CO_2$ cylinder. Test the locking pin.

**In the Water**

Pull open pack cover. Pull locking pin out of valve handle and open valve to inflate. Enter raft from small end by grasping hand straps and pulling.

**Aboard the Raft**

Keep your life vest on.

Top off inflation by blowing in the rubber mouth tube. Tighten valve after inflating.

Keep the $CO_2$ cylinder on the valve. The valve might leak if exposed.

Keep the lead strap from raft clipped to yourself.

Fasten down everything aboard.

The raft contains sea marker, sea anchor, bailing bucket, bullet-hole plugs, blue and yellow cloth, first-aid kit, repair kit, paddles, and water. The water may be replaced by a chemical sea water purification kit in some rafts.

# FIRE FIGHTING

## ALL FIRES ARE THE SAME SIZE WHEN THEY START— CATCH THEM EARLY

### Fires While Starting Engines

Both the pilot and the ground crew will be **fire conscious** and prepared to take the necessary action in event of fire while engines are being started.

If possible, a ground crew, or a member of the airplane crew where ground crew is not available, equipped with adequate fire extinguishing equipment will stand by outside the airplane while engines are being started.

When neither a ground crew nor a member of the airplane crew is standing by at starting of engines, **have the fire extinguishing equipment at hand ready for instant use if a fire occurs.**

Backfiring while starting sometimes causes fires in the induction system due to the excess of fuel present from priming.

By allowing the engine to continue running, such fires often are sucked through the engine and thereby extinguished. However, if the fire still exists, use built-in carbon dioxide extinguisher while engine is running. Shut off fuel supply to engine.

If a fire starts in the induction system and engine does not continue to run or cannot be kept turning over by the starter, the ground crew will use portable fire extinguishers. Pilot will shut off fuel supply to engine. If necessary, pilot will use built-in carbon dioxide extinguishing equipment.

### Fires During Flight

In case of fire during flight:

1. **Pilot will warn all crew members to have parachutes attached** in readiness for possible emergency use, and to stand by for further instructions.
2. **Use all applicable fire extinguishers.**
3. If flying low, **gain as much altitude as possible.**
4. Pilot or airplane commander will **determine whether landing will be attempted or the airplane abandoned.**

### Engine, Fuel Tank, and Amphibian Hull Fires

1. If the airplane is equipped with a built-in fire extinguishing system, **set distributor valve to proper position and pull release handle sharply.**
2. In case of engine fire, altitude and other conditions permitting, first **shut off supply of gasoline to engine and fully open throttle.**
3. **Open emergency exits.**
4. **Land as soon as possible,** if fire is extinguished, to determine and correct the cause before continuing the flight.
5. If the airplane is not equipped with a built-in fire extinguishing system, follow the procedure outlined in (2), (3), and (4) above.

### Cabin Fires

1. **Close all windows and ventilators. Use hand fire extinguisher on the fire.** When using carbon tetrachloride extinguisher in confined spaces, **stand as far from fire as possible, and open windows and ventilators immediately after using.** The effective range of the carbon tetrachloride extinguisher is approximately 20 feet.

2. If an electrical fire, **turn main switches OFF.**
3. If a leaking fuel or oil line, **shut off valves.**
4. If gasoline, oil or other similar combustible liquids are involved, **use hand carbon dioxide extinguisher** if available.

### Wing Fires

1. **Turn all switches controlling landing or navigation lights "OFF."**
2. **Open emergency exits.**
3. **Attempt to extinguish fire by side-slipping** the airplane away from fire where possible. If the fire continues to burn, the pilot or airplane commander will **determine whether a landing will be attempted or the airplane abandoned.**

### Flare Fires

1. In case of ignition of the flares in the rack, immediately **release the flares.**

### General

All personnel participating in aerial flights will familiarize themselves with the location and proper use of fire extinguishing equipment installed on aircraft. In case of the smaller tactical airplanes, the fire extinguishers provided are primarily for use on the ground. These extinguishers are of the hand type using carbon tetrachloride as the extinguishing medium. Hand-type fire extinguishers are unsuitable for combating a fire outside of a fuselage in flight.

In some of the larger airplanes, hand-type carbon dioxide fire extinguishers are provided in addition to the carbon tetrachloride extinguisher for combating fires within the cabin. The carbon dioxide extinguishers are particularly effective in combating gasoline and oil fires. **If fabric, wood, etc., are involved, the carbon tetrachloride extinguishers should be used in conjunction with the carbon dioxide extinguisher.**

Some types of airplanes are also provided with built-in carbon dioxide fire extinguishing systems, so that the engine compartments, hulls of amphibians, or gasoline tank compartments can be flooded with the carbon dioxide gas in case of fire. A distributing valve mounted within easy reach of the pilot is set to direct the gas to the desired location and a pull handle is operated to release the gas. The operating controls of this equipment are clearly marked to indicate their method of use.

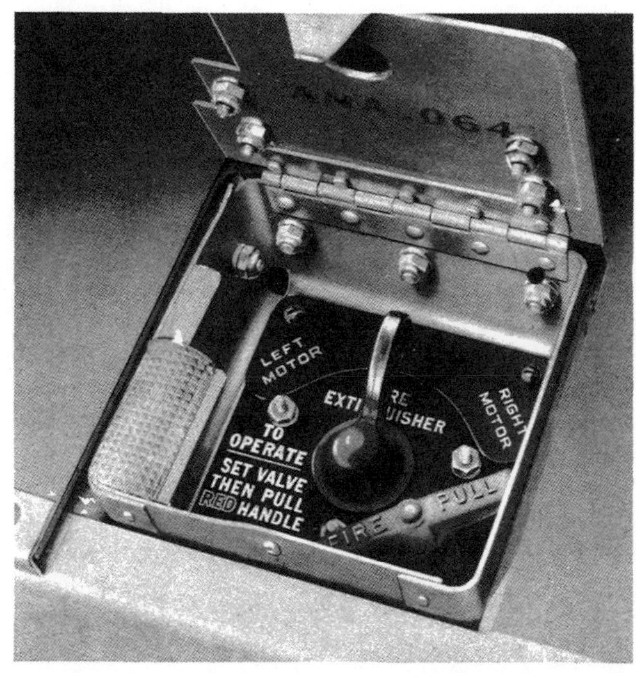


Fire extinguishing equipment carried in aircraft is adequate for combating fires in their earliest stages only. **Use the equipment as soon as the fire is detected.**

### Precautions

Carbon tetrachloride is a volatile fluid, the gases of which, when inhaled in large amounts, act as an anesthetic, causing drowsiness, dizziness, headache, excitement, anesthesia and sleep. If small doses of the fumes should be breathed in over a period of time, the first probable effect would be drowsiness, inability to keep the eyes open, followed by sleep or perhaps headache and nausea.

If the odor of carbon tetrachloride is detected while flying, investigate to determine its source. **If a fire extinguisher is leaking, correct it at once.** If unable to stop the leak, place the extinguisher where it will not leak into the cockpit or cabin, if possible. As a last resort, empty the extinguisher or dispose of it.

Carbon tetrachloride is poisonous if taken internally. Even one-quarter of a teaspoonful may prove fatal. The symptoms of poisoning do not appear for several days after the fluid is taken into the stomach, thus giving a false sense of security. Any one who ingests some of the fluid will report to the Surgeon immediately for advice and necessary treatment.

When carbon tetrachloride is sprayed on a fire, a poisonous gas, phosgene, is produced. Inhalation of even a small amount of phosgene may produce harmful effects, and may be fatal if a sufficient quantity is taken into the lungs. Therefore, **stand back as far as possible when using the carbon tetrachloride extinguisher, and avoid breathing the fumes produced. Open windows and ventilators as soon as the fire is extinguished.**

Carbon dioxide is a non-poisonous gas and breathing it will not adversely affect a human being either at the time it is inhaled or afterwards. If the concentration of carbon dioxide gas is high enough, it will have a smothering effect, due to the exclusion of oxygen, but the quantity of carbon dioxide gas contained in a hand fire extinguisher installed in aircraft is not sufficient to raise the concentration in an airplane cabin to this point.

When using the portable type carbon dioxide extinguisher, **avoid contact with the horn or the chemical itself,** as the carbon dioxide is at an extremely low temperature and may cause severe burns.

*Be sure to know the location of fire extinguishers!*

# KIT, FIRST-AID, AERONAUTIC

## Installed in Military Aircraft

**Medical Supply Catalog No. 97765**

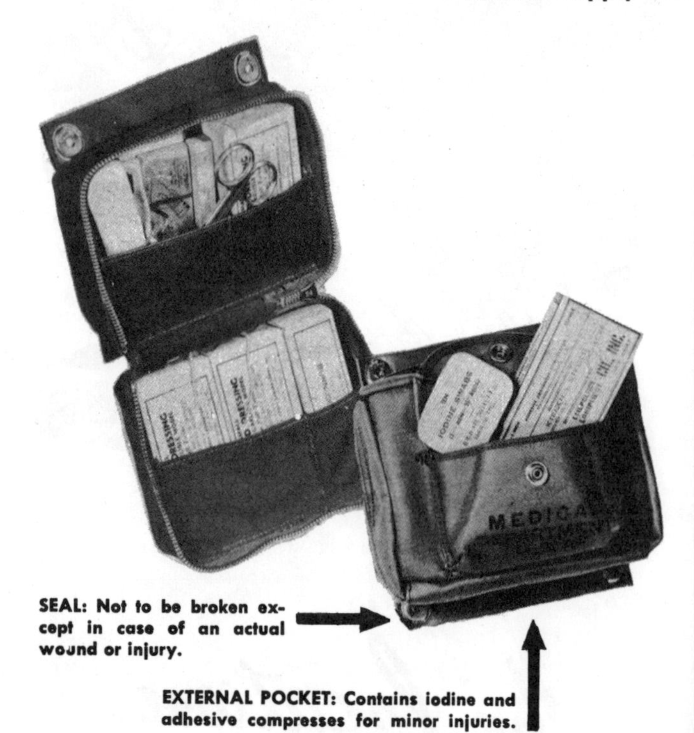


1. In the case of a wound, first stop the flow of blood. The clothing should be cut away and a compress or wound dressing applied after the sulfanilamide powder has been sprinkled into the wound. If a firmly applied dressing will not cause the bleeding to stop, or if there is actual spurting of blood from an artery, the tourniquet should be applied. A tourniquet must be released every twenty minutes and removed as soon as hemorrhage stops.
2. To relieve severe pain open the small cardboard container and follow directions given there in the use of the hypodermic syrettes of morphine. Do not hesitate to use the hypodermic to relieve suffering.
3. In case of head injury have the man lie quietly with head slightly elevated.
4. In the event of marked blood loss with shock and/or unconsciousness have the man lie horizontally or lie with the head down, if possible.
5. An adequate supply of oxygen is doubly important in case of serious injury. Use it generously.

### CONTENTS

1. Tourniquet, (1)
2. Morphine syrette (2)
3. Wound dressing, small, (3)
4. Scissors, (1 pair)
5. Sulfanilamide crystals, envelope, (1)
6. Sulfadiazine tablets, (1 box of 12 tablets)
7. Burn ointment, (1 tube) (Boric or 5% Sulfadiazine)
8. Eye dressing set
9. Halazone tablets
10. 1" Adhesive compresses (1 box) (Contents of small outer pocket)
11. Iodine swabs (10) (Contents of small outer pocket)

# NOTICE: Drugs Contained in This Kit are Potent and Must Be Used Correctly. Follow Directions!

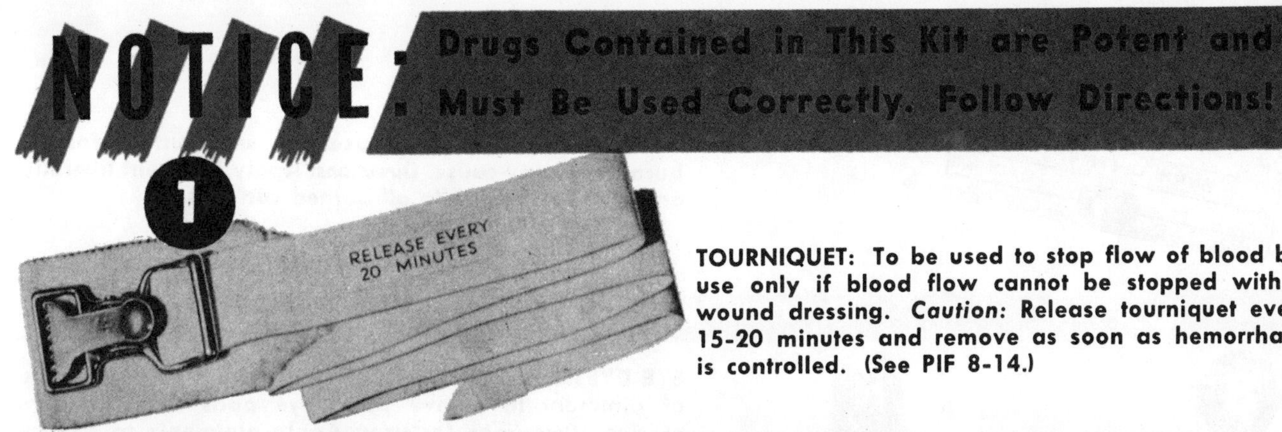


**TOURNIQUET:** To be used to stop flow of blood but use only if blood flow cannot be stopped with a wound dressing. *Caution:* Release tourniquet every 15-20 minutes and remove as soon as hemorrhage is controlled. (See PIF 8-14.)

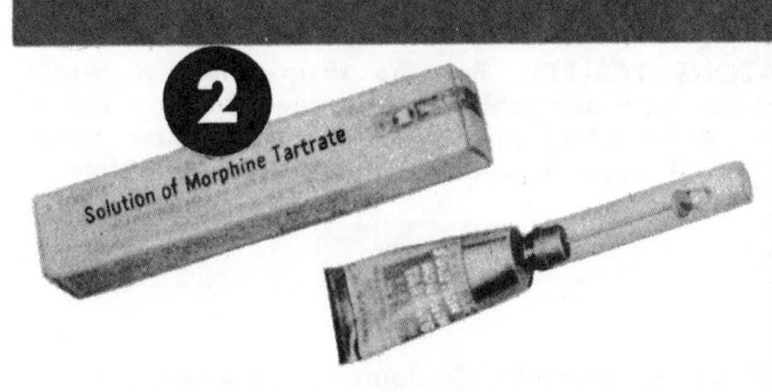


**MORPHINE SYRETTE:** To be used to relieve pain and should be employed without hesitation to prevent suffering. *Directions for use:* Remove transparent hood, grasp wire loop and push wire in to pierce inner seal, turning if necessary. Pull out and discard wire, thrust needle through skin at least half its length and inject solution by slowly squeezing the syrette from the sealed end. In extreme cold, warm syrette by holding under clothing next to skin.

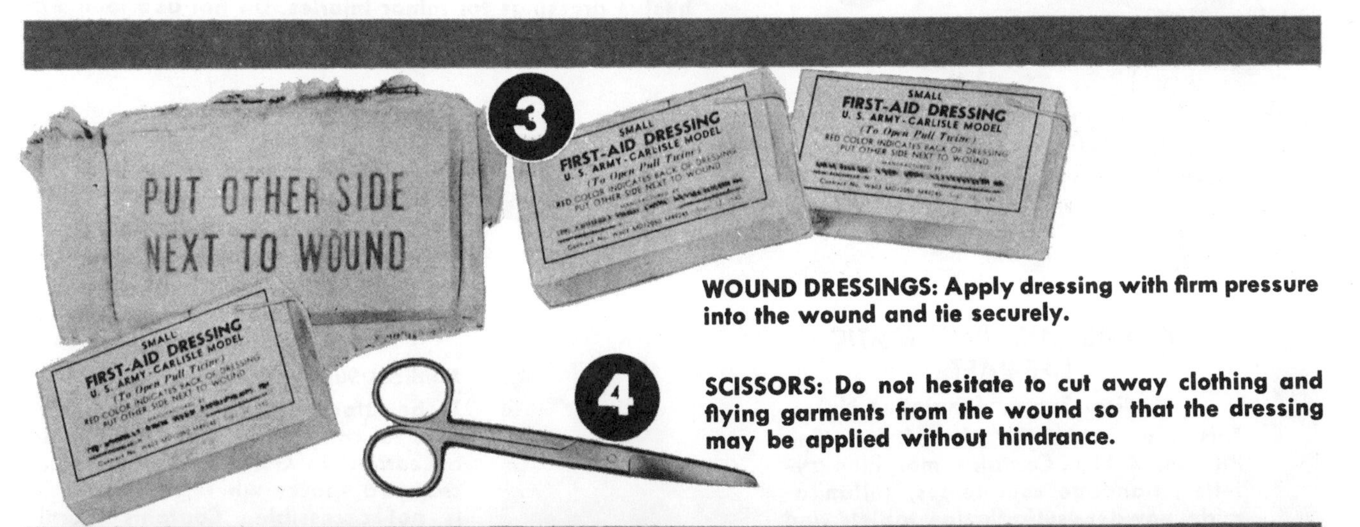


**WOUND DRESSINGS:** Apply dressing with firm pressure into the wound and tie securely.

**SCISSORS:** Do not hesitate to cut away clothing and flying garments from the wound so that the dressing may be applied without hindrance.

**SULFANILAMIDE POWDER:** Sprinkle on the wound to prevent infection.

**SULFADIAZINE TABLETS:** To be taken internally if wounded. *Directions for use:* Take two tablets with water every five minutes until all twelve tablets are taken. Swallow whole without chewing.

**BURN OINTMENT:** To be used on skin surface for all burns from any cause. *Directions:* Apply ointment liberally and without friction to all burned surfaces.

**EYE DRESSING SET:** Contains tube of burn ointment, tube of ointment to relieve pain, eye pads and adhesive plaster. *Directions for use:* Apply ointments to surface of eyeball and inner surfaces of lids; then cover eye with pad and secure in place with adhesive plaster.

**HALAZONE TABLETS:** For the disinfection of water. *Directions:* Add one tablet to a canteen full or a pint of water. After tablet dissolves wait 30 minutes before drinking. If water is greatly polluted use two tablets.

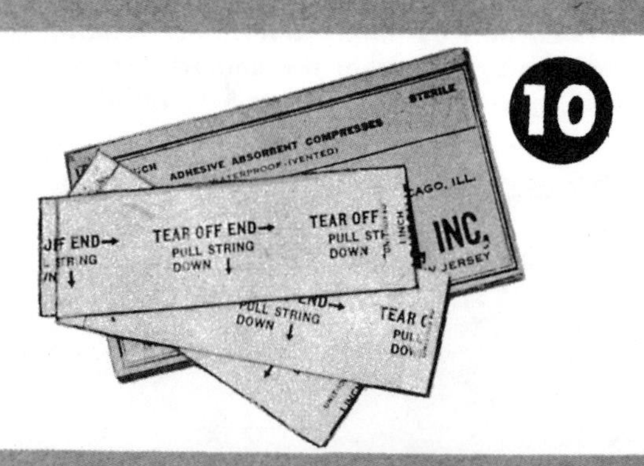


**EXTERNAL KIT POCKET:** Contains iodine swabs and adhesive dressings for minor injuries. Do not use iodine in serious wounds. (Use sulfanilamide powder.)

## KIT, FIRST-AID, PNEUMATIC LIFE RAFT

**Medical Supply Catalogue No.**

**This is a part of the life raft kit. (See PIF No. 8-11.) Contains morphine syrettes, bandage compresses, sulfanilamide powder, sulfadiazine tablets and burn ointment as illustrated in photos Nos. 2, 3, 5, 6, and 7.**

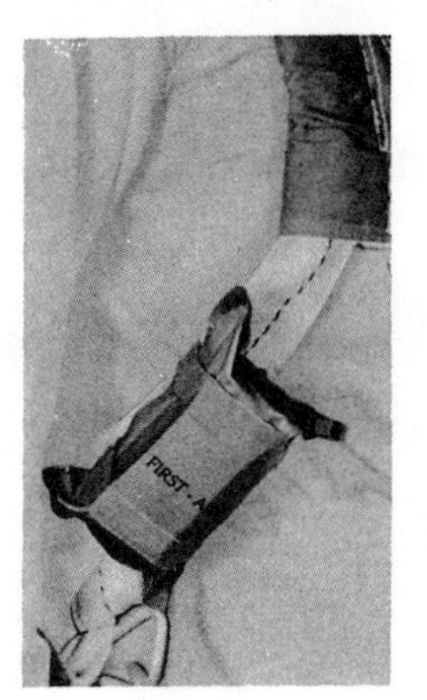

## KIT, FIRST-AID, PARACHUTE

**Medical Supply Catalogue No. 97785**

**To be attached to the parachute harness for constant availability. Should be carried in Gun Turrets and other cramped spaces where the larger kit is not accessible. Contains tourniquet, morphine and wound dressing as illustrated in photos Nos. 1, 2, and 3.**

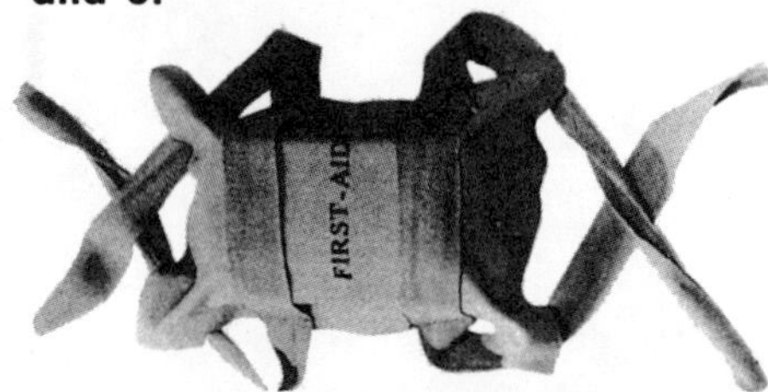

Your airplane is a good first-aid station. You have the Kit, First-Aid, Aeronautic, and the Packet, First-Aid, Parachute. Oxygen is frequently available. Splints, or splint materials, are at hand. Hot drinks are often carried in thermos jugs. In certain bombers you will be provided with blood plasma. Familiarize yourself thoroughly with the first-aid supplies which you carry, and get clearly in mind just what you can do with them.

### Wounds and Injuries

Wounds and injuries involve one or more of these problems: **pain, cuts, bleeding, broken bones, burns, frostbite, shock, and unconsciousness. Generally you** will have to deal with combinations of these, such as cuts which are bleeding, burns that cause pain, broken bones associated with cuts or burns, and so on. Shock usually comes on after a good deal of blood has been lost either inside the body (where you may not be able to see it), or on the outside. Shock also accompanies deep or extensive burns. Unconsciousness may be produced by a head injury, may follow shock, or may occur as a result of failure to get enough oxygen.

In giving first-aid, try to size up the general situation accurately. Then attend to the most serious problems first. Above all, use common sense.

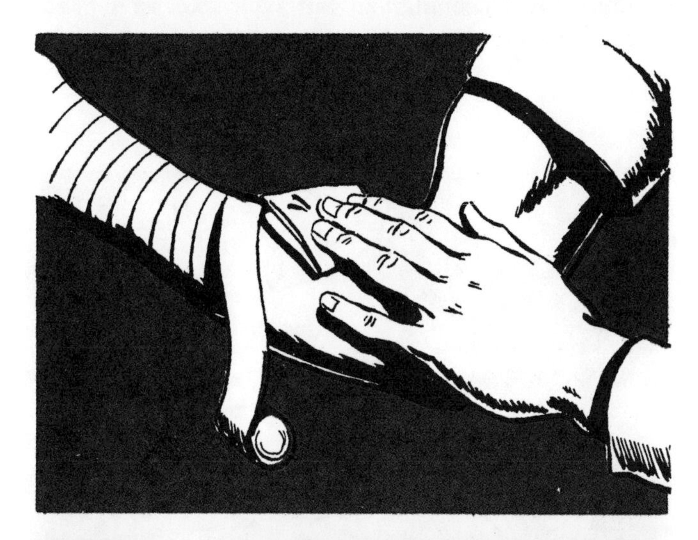

### Cuts and Bleeding

1. Expose wound by cutting nearby clothing with scissors.
2. Cover cuts with sterile dressings and apply firm pressure.
3. If this does not stop the bleeding, elevate the bleeding part.
4. If these measures fail to stop bleeding in arms or legs, apply a tourniquet in the middle of the upper arm or middle of the thigh. The tourniquet must be released every 15 minutes for at least a few seconds, depending upon the amount of bleeding.

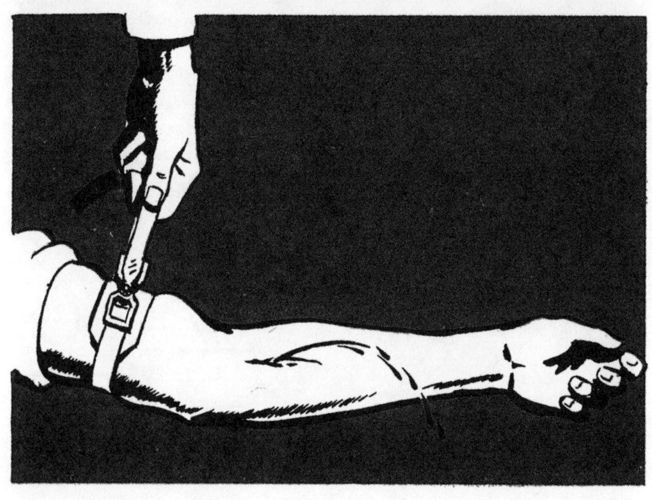

#### Tourniquet (Warning)

A tourniquet must be removed, or temporarily released, every 15 minutes. **Failure to release the tourniquet often enough or long enough to provide an adequate circulation to the blocked portion of the arm or leg may necessitate amputation later.**

### Pain

Use morphine at once for severe pain. This makes it possible for the patient to lie quietly, preventing aggravation of the injuries. Do not use more than one tube (½ grain) of morphine at any one time.

When giving morphine, mark down the time and dose on the patient's forehead or clothing with a pencil. Remember that an excess of morphine can be fatal. Do not give morphine to a person who is unconscious, who has a head injury, or who is breathing less than 12 times per minute.

### To Give Morphine

1. Paint any small area of skin with iodine.
2. Remove the transparent cover from the morphine syrette.
3. **Push in the wire loop to puncture the inner seal;** then pull the wire out.
4. Thrust the needle through the skin, using care not to press morphine out of the tube while doing so.
5. Squeeze the tube slowly to inject the morphine.

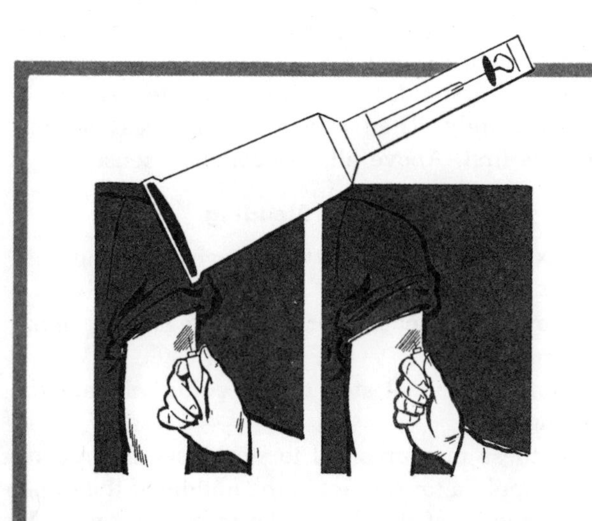

**Give Morphine:**

1. To stop pain.
2. To decrease shock.
3. To facilitate moving the patient.

**Don't Give Morphine:**

1. To an unconscious person.
2. To a person with a head injury.
3. To a person who is breathing less than 12 times per minute.

### Shock

You can tell when a patient is in shock by the total picture he presents rather than by any single sign. Usually he will have:

1. Lost considerable blood, or
2. Suffered severe burns, or
3. Been subjected to intense pain, or
4. Received a head injury.

**His skin is pale,** cold, clammy, or moist.

**His breathing is shallow,** and may be irregular.

**His pulse is weak,** rapid, thready, and often difficult to find.

Sometimes there is nausea and vomiting.

**Treat shock by doing the following things as promptly as possible:**

1. Stop any obvious bleeding.
2. Give pure oxygen to breathe. (Automix "OFF.")
3. Give morphine. (Exception: Head injury.)
4. Keep the patient warm with blankets, extra clothing, or a sleeping bag, but avoid excessive heat.
5. Loosen any tight clothing.
6. Place the patient with his head slightly lower

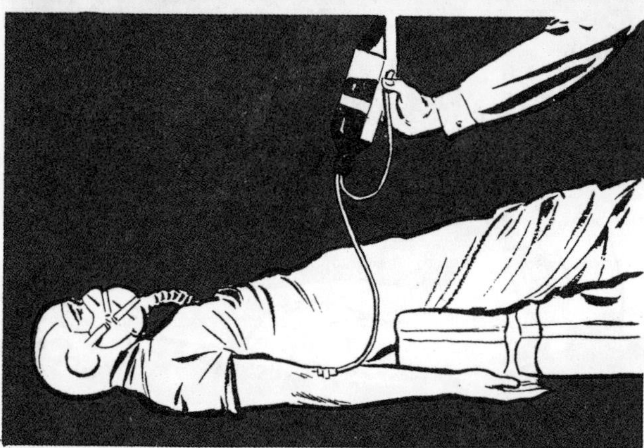

than his feet, to promote better circulation to the brain.

7. Inject plasma, when it is available, in accordance with the directions on the plasma package.

### Fractures

1. If a broken bone is associated with a cut, sprinkle with sulfa powder and cover with a sterile dressing. If the dressing is firmly bound in place it will almost always stop the bleeding.

2. Give morphine.

3. Apply a temporary splint to the part, using wood, strips of metal, heavy cardboard, or any convenient pieces of equipment such as a machine-gun barrel or fire ax.

4. **Do not attempt to set the bone. Manipulation causes shock.**

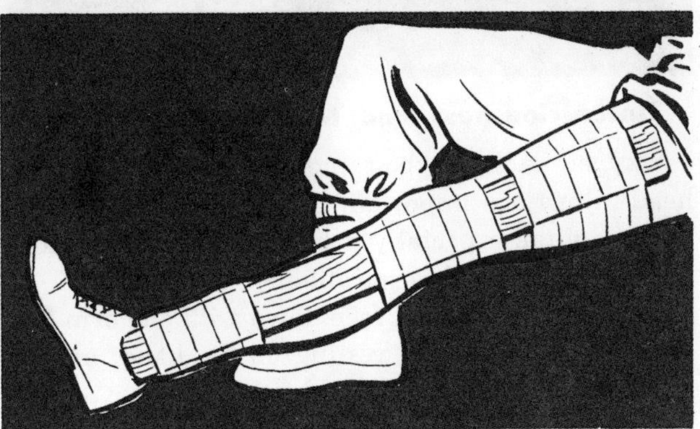

### Burns

**For minor burns:**

Squeeze burn ointment onto a sterile dressing. Then cover the burn gently with the dressing.

**For severe burns:**

1. Give morphine.

2. Treat shock. (Oxygen; plasma, if available.)

3. Apply burn ointment on sterile dressings, **and bind the dressings gently but firmly in place.**

4. Never open blisters resulting from burns.

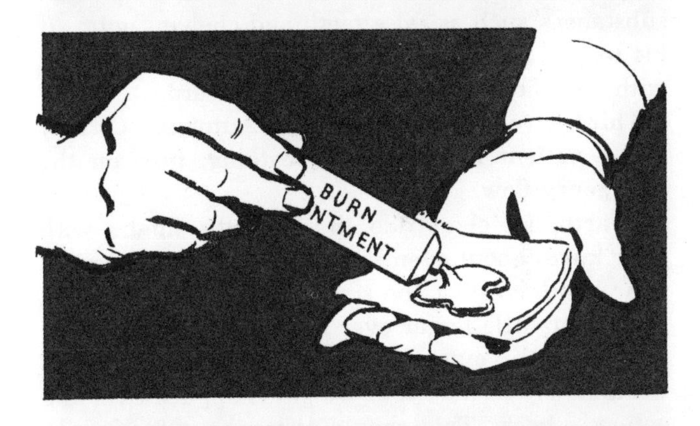


### For eye burns

Apply Metaphen ophthalmic ointment directly to the eyeball. Then apply the boric acid ointment to the inner surface of the eyelid. Cover the eye with a dressing and secure in place with adhesive strips, provided the skin around the eye is not burned. Do not touch the eye with your fingers, and do not rub it—either before or after the ointment has been applied.

### Transportation of Wounded

If it becomes necessary to move an injured crew member improvise a litter with 2 poles and a pair of flying jackets. Turn the sleeves inside out and insert the poles through them. Then close the jacket over the outside of the poles. Additional support can be obtained by using boards or cardboard splints inside the jackets. Litters can also be improvised with poles and blankets. Take great care to be as gentle as possible in moving an injured person onto a litter. Keep his body as flat as possible at all times. Have 3 or more persons move and support him by placing their arms under his legs, buttocks, back, shoulders, and head.

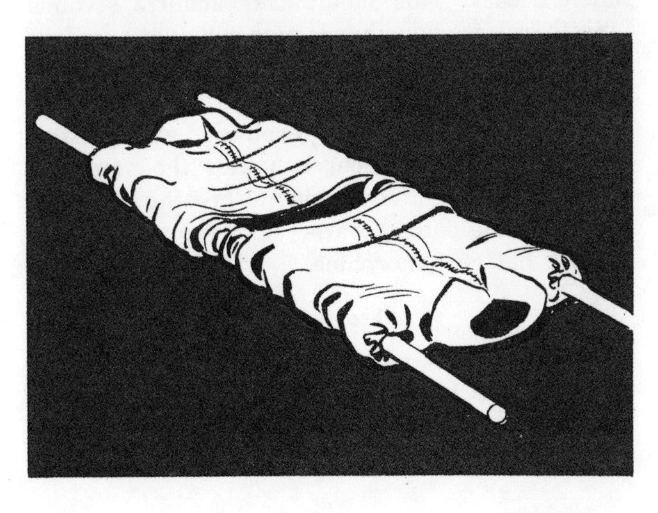

### Unconsciousness and Near-Unconsciousness

Oxygen lack, carbon monoxide poisoning, and head injury are important causes. Immediate treatment is vital, especially if breathing has stopped.

1. Give artificial respiration:

**First,** lay the patient face down with one arm bent at the elbow, his face resting on his hand, and his other arm extended beyond his head.

**Second,** open his mouth and remove all foreign substances such as false teeth and chewing gum. If his tongue has fallen back into his mouth, grasp it with your fingers and pull it well forward.

**Third,** give him pure oxygen. (Automix "OFF.") **If the patient has stopped breathing, turn on the emergency flow.**

**Fourth,** kneel astride the patient's thighs with your knees about even with his. Place the palms of your hands against the small of the patient's back, with your little finger over the lowest rib.

**Fifth,** with your arms stiff, swing your body forward slowly so that your weight is applied over the patient's back. This should take about 3 seconds.

**Sixth,** release your hands with a sudden snap and swing backward to remove all pressure from the patient. After about 2 seconds repeat the cycle.

**Continue giving artificial respiration without stopping for 2 hours or longer, unless the person to whom it is being given begins to breathe normally.**

2. Keep the patient warm.
3. Do not give morphine.

### Frostbite

1. Fingers, toes, ears, cheeks, chin, and nose are the parts most frequently affected.
2. Numbness, stiffness, and whitish discoloration are the first symptoms.
3. Wrinkle your face to find out if it is numb; watch for blanched faces of your crew mates.
4. If frostbite occurs, warm the affected part gradually. Never rub or attempt to thaw it rapidly.
5. If blisters develop, do not open them. (See HEAT AND COLD, PIF 4-7-3.)

### Failure of Oxygen Supply

If a crew member's oxygen supply fails above 10,000 feet, make every effort to replace his equipment or give him an emergency supply. If this is not practicable, descend to 10,000 feet as fast as safe operation permits. Loss of oxygen above 20,000 feet is critical, but there is no need for panic. **Get oxygen, or get down.**

### Wound Disinfectants

1. Sprinkle Sulfa powder in open wounds.
2. Use iodine only for small cuts and scratches, which should not be covered by a dressing.
3. Never put iodine on or into large or deep wounds.

# *Pilots' Information File*

# INDEX